# Microcontroller Projects *for Amateur Radio*

BY JACK PURDUM, W8TEE, AND ALBERT PETER, AC8GY

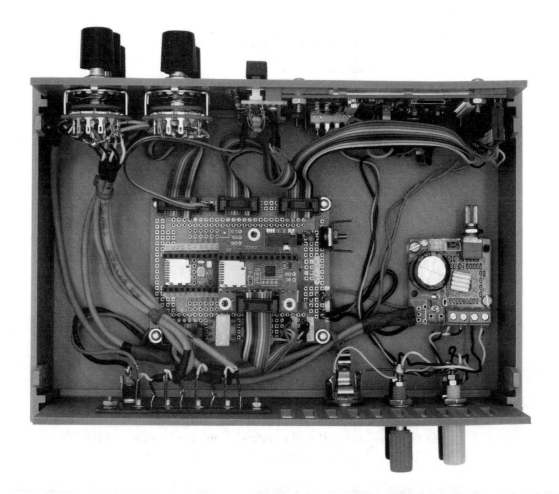

**Production**
Michelle Bloom, WB1ENT
David F. Pingree, N1NAS
Maty Weinberg, KB1EIB
Jodi Morin, KA1JPA

Copyright © 2020 by
The American Radio Relay League, Inc.

*Copyright secured under the Pan-American Convention*

All rights reserved. No part of this work may be reproduced in any form except by written permission of the publisher. All rights of translation are reserved.

Printed in the USA

*Quedan reservados todos los derechos*

ISBN: 978-1-62595-128-1

First Edition

## Online Support

Additional information about the projects in this book, the latest software and other support files, and general discussion of microcontroller projects may be found on the authors' website, **groups.io/g/SoftwareControlledHamRadio**. If you think you've found an error, please visit the group's site and post your discovery. If you're already into a new project and have difficulty, group members may be able to lend a hand.

# Contents

About the ARRL

About the Authors and Acknowledgements

About this Book

1  Getting Started

2  An Overview of C

3  A Gentle Introduction to C++ and Object Oriented Programming

4  Displays

5  Projects Power Supply

6  Mini Dummy Load

7  Morse Code Tutor

8  Programmable Bench Power Supply

9  100 W Antenna Tuner with Graphical SWR Analyzer

10  A CW Messenger

11  DSP Post Processor

12  DSP Audio Mic-Processor

13  Signal Generator

14  Double-Double Magnetic Loop: A Luggable Portable HF Antenna

15  Controller for Double-Double Mag Loop

16  Finishing Your Projects

Appendix A — Products and Component Sources

## About ARRL

We're the American Radio Relay League, Inc. — better known as ARRL. We're the largest membership association for the amateur radio hobby and service in the US. For over 100 years, we have been the primary source of information about amateur radio, offering a variety of benefits and services to our members, as well as the larger amateur radio community. We publish books on amateur radio, as well as four magazines covering a variety of radio communication interests. In addition, we provide technical advice and assistance to amateur radio enthusiasts, support several education programs, and sponsor a variety of operating events.

One of the primary benefits we offer to the ham radio community is in representing the interests of amateur radio operators before federal regulatory bodies advocating for meaningful access to the radio spectrum. ARRL also serves as the international secretariat of the International Amateur Radio Union, which performs a similar role internationally, advocating for amateur radio interests before the International Telecommunication Union and the World Administrative Radio Conferences.

Today, we proudly serve nearly 160,000 members, both in the US and internationally, through our national headquarters and flagship amateur radio station, W1AW, in Newington, Connecticut. Every year we welcome thousands of new licensees to our membership, and we hope you will join us. Let us be a part of your amateur radio journey. Visit www.arrl.org/join for more information.

225 Main Street
Newington, CT 06111-1400 USA
Tel: 860-594-0200
FAX: 860-594-0259
Email: membership@arrl.org

**www.arrl.org**

# About the Authors

## Jack Purdum, W8TEE

Dr. Purdum is a retired professor from Purdue University's College of Technology. He sincerely enjoys writing...this is his 19th textbook, augmented with dozens of articles and presentations. He has been continually licensed since 1954 and a Life Member of the ARRL since 1972. Like many of us, work and family tempered his mid-life ham radio activity, but he discovered QRP when he retired and has been a champion for low-power operation ever since.

## Dedication

*To my grandkids: Hailey, Spencer, Liam, and Luke*

## Acknowledgments

No one writes a book without the help and support of others. First and foremost, I would like to thank my co-author, colleague, and dear friend, Al Peter (AC8GY) for his vision and talents in bringing this book together. Some of you have heard me say that when we are together, I stand next to Al and try to look smart — there is more truth in that statement than I care to admit!

I would also like to thank Dave Vest, Dr. John Weiner, Konrad Kristensen, John Strack, and Bev and Joe Kack for their encouragement and support during this project. I'd also like to thank Mark Wilson (K1RO) and the ARRL staff for their work that improved this book immeasurably. Finally, thanks to you, the readers who actually buy books. Without your support, this entire activity collapses...a most sincere thank you!

# Albert F. Peter, AC8GY

Al Peter's educational background is in Physics and Engineering, with degrees from University of Cincinnati and University of Michigan. Al has been an electronics builder-experimenter since high school, and he earned Amateur Extra Class ham license in 2010. He was founder and CEO of SDRC, a major consulting firm specializing in Mechanical CAD and related engineering and manufacturing software. Al also has extensive mainframe and microcontroller software experience, and enjoys the meshing of his varied interests of electronics design/building, short wave listening, computer (PC) construction, and investigating new technologies such as software defined radio.

His technical experiences have merged into the design and construction of MCU-related projects for ham radio. His understanding of both analog and digital electronics design come together with MCU programming to create unique and very functional projects for the ham radio enthusiasts. Current projects include those in this book, as well as an add-on board for the µBITX transceiver, a scalar network analyzer, a graphical real-time antenna analyzer, several antennas for VHF and HF, an automated MCU-controlled dust collector system for his woodworking shop, plus other varied economical high-function tools for the ham shack.

Al resides in Cincinnati, Ohio. His other interests include still photography, videography, amateur astronomy imaging, woodworking, and fly fishing. His biggest challenge is finding time to fit it all in. He is also a co-founder of the Greater Cincinnati Builders Group. Al has several published articles in various magazines on technical/amateur radio subjects, including on a dual-band compact 2-meter/ 440-MHz, Slim-Jim antenna (*QST*, December 2011), a QRP BCI filter, a 150 W dummy load/watt meter project published in *QST* (with co-author W8TEE), and an article on the Double-Double Mag Loop antenna (*Rad Com*, February 2020) described in this book.

## Dedication

*To my grandkids: Jacob, Josh, Leah, Nolan*

## Acknowledgments

To my dear wife Jeanie, who supported my efforts completely while I was absorbed in bringing this book to light.

# Introduction

# About This Book

In just the five years since the *Arduino Projects for Amateur Radio*[1] book was published, microcontroller technology has continued to march forward at a blistering pace. Indeed, the use of microcontrollers in amateur radio equipment has increased exponentially. Likewise, the variety of microcontrollers from which to choose has also increased significantly. That increased variety has benefited all of us through competition: Prices have generally fallen, the processing power has risen, or both. For that reason, we are not limiting our sphere of experimentation in this book to just the Arduino family of microcontrollers. We now include the Teensy (PJRC.com), STM32F series, and ESP32 family, too.

Truth be told, we could probably have just used the Teensy controllers for all of the projects in this book. However, the Teensy 3.6, while extremely powerful, is also a little pricey at $30. As this book is being written, a Teensy 4.0 has just been introduced with a $20 price point. Other than pin count, it is a much more powerful microcontroller than the Teensy 3.6. In some projects, using either Teensy would be an H-bomb-to-kill-an-ant approach to the project. A $3 Arduino Nano works just fine in many projects. However, we didn't want to rule out the STM32F103 (aka "Blue Pill") either, as it has features similar to the Nano, yet can be purchased for less than $2. The STM32 also has more memory and higher clock speed than the Nano. The ESP32 also has more memory and faster clock speed, but also throws in Wi-Fi and Bluetooth capability. There are other microcontrollers available, but we are sticking to these because they can all be programmed in the Arduino programming environment (the Integrated Development Environment, or IDE) and most have the same libraries available, too. (Chapter 2 provides more details on these microcontrollers so you can see the feature set of each.)

Yet, despite the explosive growth of microcontrollers in ham radio projects, we still hear hams complain that they don't try such projects because "I don't know how to program."

Yeah...so what?

When you started to study for your amateur radio license, did you know what a Colpitts oscillator was? Did you know anything about the various classes of power amplifiers? Probably not, but you learned enough to get your license. Likewise, we think you'll pick up enough while reading through this book to make small changes in the code. Even something as simple as changing the startup, or splash, screen is fun to do and adds a personal touch to your projects. Just swap your name and call sign for ours, recompile and upload, and it's now "your" project. Such simple changes can be done easily in a few minutes...at most!

We also get the "old-dog-new-tricks" excuse for not learning about and using microcontrollers. Really? Taken together, we (the authors: Jack, W8TEE, and Al, AC8GY) have completed more than 152 laps around the sun, so don't tell us about being "too old." Training and background? Al's the electrical engineering (EE) brains behind the projects and has an extensive background in physics. Jack is an economist by training who just happened to fall in love with programming more than 40 years ago. Together, Al does all the EE work while Jack stands in the background and tries to look smart. Jack then takes Al's EE designs, mixes in a little software, and presents both to the user. The end result of this mixing of backgrounds are projects that we think you will enjoy building and using.

## Why You Should Buy This Book?

First, we think there is a sufficient variety of projects in this book that at least several of them should appeal to you. The projects result in equipment that is both useful around the shack and inexpensive to build when compared to commercial counterparts. Not only that, but we are pretty sure that many of you will have an "ah-ha" moment where you will think of extensions of these projects, or perhaps even new projects along the way. If so, we hope you'll share your ideas with our group (SoftwareControlledHamRadio, **groups.io/g/SoftwareControlledHamRadio**). We want these projects to serve as a base from which others can improve and expand each one.

Second, there is an inner satisfaction that comes from building and using your own equipment. Most of you have probably built some piece of equipment that's currently sitting in your shack, so you know what it feels like when you complete a build of your own. To those of you who haven't had that experience, you're in for a treat. It's sorta like the feeling you had the night before Christmas when you were a kid. Building your own equipment isn't that hard, either, and there are likely dozens of hams in your club or otherwise nearby who can help you if you need it.

Third, this book is a little different than most: it is designed to teach and well as show you how to construct the project. There are a lot of project books out there, but most are of the "cookbook" variety — projects are presented in a way that is similar to baking a cake. That is, you gather a bunch of parts (ingredients), arrange them on a perf board (combine the ingredients), solder them together (bake in an oven at 350 degrees), and use the piece of equipment (eat the cake). That's fine as far as it goes, but what did you learn in the process to make the next project easier or more enjoyable? Having a better understanding of the project also makes it easier for you to maintain and/or extend the project.

Al takes the lead on explaining the "what's" and "how's" of the project's hardware design and Jack walks you through the software that complements the hardware. The explanations not only help you understand the project under construction, they make it easier to understand the next project. That is, this book begins with a very simple project and gradually increases in complexity as you progress through the book. Yet, we realize that not all of you want to learn enough about programming to make changes in the project's software. To that end, the book is written so you can skip over the software section of each chapter if you choose to do so. While we think you're missing out on a major benefit of the book, you can complete and use the new piece of hardware without needing to delve into the how's and the why's if you don't want to.

We know from feedback from the *Arduino Projects* book mentioned earlier that many of you want more details on the software end of things, but in a "digestible" way. To that end, we've added a chapter on Object Oriented Programming (OOP). While this may seem strange in a projects book, an understanding of why C++ is used for almost every library should help you use those libraries. Also, even a small understanding makes using any library easier.

Fourth, we think the projects presented in this book will augment your arsenal of tools to repair and maintain your equipment. Probably all of you have a power supply in your shack that plugs into your transceiver. That said, it's nice to have a power supply capable of providing a variety of voltages for testing purposes, too. Even better: How about a programmable bench supply? These can cost hundreds of dollars, but you can build your own for a small fraction of that.

Finally, when you finish reading this book, we feel confident that you will have a better understanding of what microcontrollers are all about and how easy it is to write the software that augments their power. Indeed, some of you are going to find that this area of experimentation adds a new dimension of pleasure to the hobby we all enjoy.

## What Do I Need to Use This Book?

We have a more complete answer to this question in Chapter 1. The short answer, however, is that you need a microcontroller board(s) and the software necessary to program the various microcontrollers. Appendix A is a reference list of suppliers that we have used for buying various components and parts for building the projects in this book. If you are just getting started, we'd suggest buying several Arduino Nanos and several STM32F103s (often called the "Blue Pill", or BP) and their prototype I/O extension board, as seen in **Figure A**. (By the way, note the USB connector on the left edge of the STM32 in Figure A. The Nano Pro Mini, while cheaper, does *not* have a USB con-

**Figure A — STM32 "Blue Pill" and I/O extension board.**

nector, making it much less convenient to program. We would suggest you do *not* buy the Nano Pro Mini.)

If you look closely at Figure A, you can see the STM32 sitting in a socket surrounded by a forest of pins. Each one of those pins is tied to one of the Blue Pill's Input and Output (I/O) pins. Therefore, an extension board makes it easy to wire a microprocessor into a project for experimenting and testing without actually soldering it in place. Al has created extension boards for the STM32 and ESP32. They are available from QRP Guys (**qrpguys.com**) by the time you read this.

Special jumper wires, called DuPont wire cables (see **Figure B**), have plastic tips that can be plugged onto the pins of the I/O extension board, making it very easy to build and modify circuits. Note that the bottom set of cables in Figure B have female connectors on the right end of the cable. Those female connectors are easily attached to the I/O pins you see in Figure A, so make sure the DuPont cables you buy have at least one end with a female connector. Costs: about $2 each for the STM32, $12 for the I/O extension board, and $1 for 40 DuPont cables. (Note that these cables are best used for prototyping and testing a circuit. Once you are satisfied with the design, the DuPont cables should be replaced with permanently-soldered connections.) In many cases, when hooking up components like displays that have many connections, it is best to go ahead and use soldered connections for these fixed components. We use #30 AWG solid insulated wire for most of our digital connections.

The software needed to program the microcontroller is free. (Details on how to download the software is provided in Chapter 1.) The Arduino family of microcontrollers are part of the Open Source movement, which means the

**Figure B — DuPont cable connectors.**

software is available at no cost. When you download the Arduino software, you are asked to make a donation to support further development of the software. We hope you'll agree that it is a worthy effort and will donate what you can.

In addition, you also need the usual electronic construction components — soldering iron, resistors, capacitors, wall cube power supplies, wire cutters, and so on. Most hams have a junk box that provides many of the components needed to complete a project. In those rare instances where we don't think we can expect you to have a particular item, we provide a potential source for it. Some pieces of test equipment (such as an oscilloscope) are nice to have, but expensive and not required. If you don't have such equipment, most high school or college physics labs will and are very good about letting you use it. Also, don't forget members of your ham club. Chances are good that one of them has the piece of equipment you need and would be happy to let you use it. Appendix A has a list of many parts suppliers we have used as well as some of the equipment we used to test our projects.

## Building Circuits

At the present time, we plan to have printed circuit boards (PCBs) made for most of the projects in this book. Very simple circuits, such as the Mini Dummy Load project, don't warrant a PCB. Other projects are sufficiently complex that using the PCB is a lot simpler than point-to-point wiring. QRP Guys (**qrpguys.com**) will be selling the PCBs. However, you can also make the projects using perf board without the need for a PCB. Indeed, some of you may want to make your own PCB if you are so inclined.

Our PCBs may use thru-hole or SMD components and we have not finalized that decision. However, Al and I are starting to see the light about using surface mounted devices (SMDs). Most support chips (such as Si5351 or AD9851) are SMDs. Also, SMD resistors and capacitors are dirt cheap, and require fewer nano-acres of PCB space. A lot of hams think that SMDs are only for the young with steady hands. Not so! Al and I are two years younger than dirt, but we both work with SMDs all the time.[2] (True, we prefer the larger 1206 component size, but can work with smaller sizes, too.) More and more we are leaning towards an SMD PCB for the projects.

So, if you're going to roll your own PCB, why not try one using an SMD layout instead of thru-hole? It's probably going to be a lot easier than you think and less expensive, too. Also...no holes to drill!

## How Should I Read This Book?

Actually, that's a better question than it may appear at first blush. If you are an experienced engineer who works daily on microcontroller projects, you probably don't even need this book. Some of you do have an electronics background, but when it comes to software...not so much. Likewise, some of you have considerable software experience, but are a little thin on the hardware. All of the rest of us fall somewhere in between. We feel that everyone in between these two extremes will find something informative and enjoyable in the book. So, *how* should you read this book?

If you have considerable hardware and software experience and still bought this book, feel free to skip to the project that piqued your interest the most. For everyone else, we hope you read the book from cover-to-cover in the order in which the chapters are presented, even though you may not really be interested in building that chapter's project. Why? First, the book is "layered" in that earlier projects (for example, displays) are often incorporated into later projects. Understanding one simple project makes it easier to understand more complex projects. Second, each project has its own little nuggets of information that just might be useful to you down the road. For example, Chapter 2 has a lot to say about programming and perhaps you feel that's not your thing. We've heard that many times before, yet, once they got into it, they really enjoyed the challenge and reward that came from expanding their software horizon. Probably the same experience can be said about hardware, too.

Sometimes you have to read topics that may not be your primary interest. For example, Chapter 3 has a lot to say about Object Oriented Programming (OOP) and we can actually sense some readers' eyes glazing over. Yet, when they start to expand their experimentation, they find that almost all of the libraries that make programming so easy are based on OOP concepts. Only after that penny has dropped do they go back and read the OOP material in Chapter 3. We believe people tend to be Effort-Reward oriented. We think you will find a little effort up front reading what you think may not be of interest will actually yield noticeable rewards later on.

For all these reasons, we hope you will read the book from start to finish. Yet, at the same time, we assume there is no urgency on your part in reading this book. Take your time and enjoy the trip.

## Errata and Help

First, the authors, Beta testers, and editors at the ARRL have scoured this book from cover-to-cover in every attempt to make this book perfect. Alas, despite the best efforts by all, there are bound to be some hiccups along the way. Also, Jack doesn't profess to be the world's authority on software development nor does Al presume he's cornered the market on brilliant hardware design. As hiccups show up, we will post the solutions on our group site (**groups.io/g/SoftwareControlledHamRadio**). Also, rather than type in the code from the book, you should download it from our group site. That way, you know you have the latest version of the software. Likewise, if you think you've found an error, please visit the group's site and post your discovery. If you're already into a new project and have hit a brick wall, perhaps *all* of us who are active on the website can lend a hand. That's the real purpose of our group: to augment, enhance, and create projects and perhaps help with any problems that come up during your journey.

## Conventions Used in This Book

We have tried to write the book in a fairly consistent manner. The first part of a chapter discusses the project under consideration. The second part concentrates on the electronic elements of the project. The final section of the chapter discusses the software used to drive the project.

We use italics to refer to C syntax items, such as function or variable names. We also use italics on C keywords to distinguish them from the narrative, such as the C programming keywords *for* or *while*. Usually, variable names begin with a lower case letter using camel notation. That is, we try to use descriptive variable names, capitalizing sub-name words, such as *ledPin*, *presentHeading*, *myTransmitFreq*. Function names that *we* develop begin with a capital letter and also use camel notation. When functions are referred to in the book's narrative, function names are followed by parentheses, as in *DisplayFrequency()*. Capitalizing the first letter of a function name and adding parentheses at the end of the name instantly distinguishes it from a variable name. When array variables are referenced in the text, they are followed with array brackets, as in *myMessageData[ ]*. We believe these conventions make it easier to read the narrative associated with the software. These conventions also help you when testing and debugging the software.

Note that many microcontroller libraries follow the convention of having C++ method names start with a lower-case letter and then camel notation (for example, *Serial.readBytesUntil()*). This is one way to distinguish C functions from C++ methods within a class. When used with a library, all class member and method names follow the "dot" operator, so it's pretty easy to know something is part of a class. (Note the dot operator between *Serial* and *readBytesUntil()* in *Serial.readBytesUntil()*) If this sounds like Greek to you now, reading Chapter 3 should result in a warm, cozy, feeling when you're done.

Code listings and fragments are presented in `a different font` than text (this is called `Courier`). This helps to make the narrative stand apart from the code when being discussed in a chapter. In some chapters, especially the later ones which have more complex software, code listings are not complete. The listings are simply too long and would represent an unnecessary denuding of the nation's forest to have a complete listing. Instead, partial listings (code fragments) are presented when they represent something different or crucial that's being done in the software. In some cases, the listing is shown because it may be different for different types of equipment. It is helpful to download the project code from our group site before you start reading the chapter.

Even if you don't think you want to know anything about software, we hope that you will take the time to read Chapters 2 and 3. It should help you better understand how each project works and we think it will add to your experience with the book. You paid for those pages, so you may as well read them.

Ok...now the fun begins.

## Notes

[1] *Arduino Projects for Amateur Radio*, Jack Purdum with Dennis Kidder, McGraw-Hill, Nov. 2014.

[2] You can find hundreds of videos on using SMD components, but we think this one is very good: **www.youtube.com/watch?v=RODp8HSlFPA**.

# CHAPTER 1

# Getting Started

Our goal in this chapter is to discuss the environment in which the majority of your programming efforts take place. From now on, rather than spelling out the word microcontroller, we use μC as an abbreviation instead.

## Which μC to Buy?

As we pointed out in the Introduction, there are many controllers from which to choose. The first *Projects* book only used μCs from the Arduino family.[1] This book, however, expands to include other μCs. Projects in this book use one of the following, as shown in **Figure 1.1**:

- the Arduino Nano and Mega 2560 Pro Mini;
- PJRC's Teensy 4.0;
- the STM32F103 (aka the "Blue Pill" [BP]); and
- the ESP32.

All of the μCs can be programmed using the Arduino IDE (Integrated Development Environment) software. Once you understand how to use the Arduino IDE, there is only a very small learning curve to programming the other controllers in the same IDE. (See the Arduino Software section later in this chapter for more on the IDE.)

We continue to use the Nano simply because it is inexpensive, has a robust body of Open Source code already written for it, and a large number of peripherals can be attached to it either directly or via a supported interface (such as I2C or SPI). However, many projects in this book require more resources than the Nano provides. **Table 1.1** shows a comparison of the major resources for each μC used in this book.

Table 1.1 has a lot of information in it, but what are the most important resource factors you need to consider? Well, to answer that, we need to know what the resources mean. We only consider the primary factors here, but expand the discussion on the other resources as needed throughout the book.

### Processor Bits

Inside each microcontroller are a number of digital registers, which are electronic elements where binary data can be stored. The Nano uses an 8-bit architecture, which means each register is designed to hold 8 binary digits (a 1 or a 0). Each binary digit is referred to as a bit. If you group 4 bits together it's called a *nibble*. (Seriously, although I admit it doesn't come up in conversation

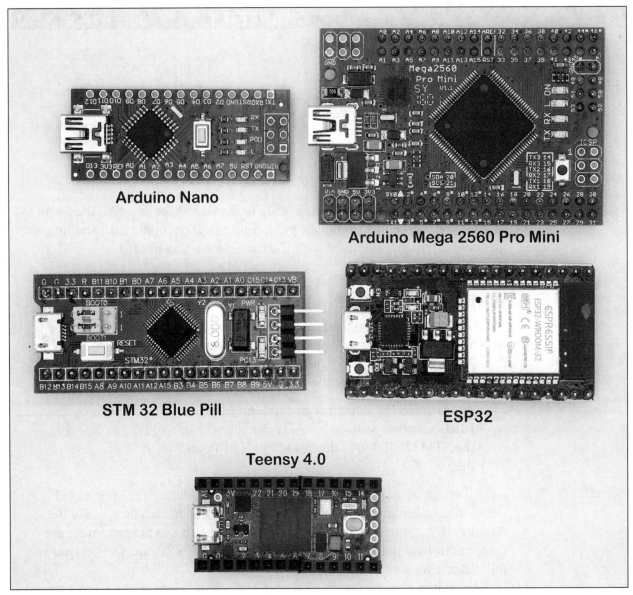

Figure 1.1 — µCs used in this book.

much.) Collectively, 8 bits make up a basic unit of computer storage called a *byte*. Therefore, a *byte* can represent 256 distinct values ($2^8$, or 0 – 255). You can probably figure out what kilobyte, megabyte, gigabyte, and terabyte mean. (I've used the term *mega-munch* of memory to denote a computer with a very large, but unspecified, amount of memory.) Note that the Teensy, Blue Pill, and ESP32 µCs use 32-bit architectures.

Why is the difference important?

Suppose you want to add two *long* data type variables together. In C, each *long* data type variable requires 4 bytes (32 bits) of storage. Because each byte is comprised of 8 bits, we need 32 bits of register storage inside the microcontroller to process a *long* data type variable. For the Nano, the task of adding the two numbers is conceptually spread out over 8 registers because we need four 8-bit registers for each *long* data type variable. However, the other two

**Table 1.1**
**Resources by Microcontroller**

| Resources | Arduino Nano | Arduino Mega 2560 Pro Mini | Teensy 4.0[5] | Blue Pill (STM32F103) | ESP32 NodeMCU WROOM 32 |
|---|---|---|---|---|---|
| Processor bits | 8 | 8 | 32 | 32 | 32 |
| Flash[1] | 32 KB | 256 KB | 2048 KB | 64 KB - 128 KB | 1.3 MB[4] |
| SRAM | 2 KB | 8 KB | 1024 KB | 20 KB | 380 KB[4] |
| EEPROM | 1 KB | 4 KB | 64 KB[2,3] | ?[3] | ?[3] |
| Processor Clock Speed | 16 MHz | 16 MHz | 600 MHz | 72 MHz | 240 MHz |
| I/O pins | 14 | 54 | 40 | 26 | 25 |
| Interrupts | All 14 mappable, 2 external | 6 | All digital pins | All 26 mappable | All 25 mappable |
| Timers | 3 (2 8-bit, 1 16-bit) | 6 | 16 | 14 | 4 |
| SPI | 1 | 1 | 3 | 1 | 2 |
| I2C | 1 | 1 | 3 | 1 | 2 |
| DAC resolution (bits) | 10 | 10 | 12 | 12 | 12 |
| Analog pins | 8 | 16 | 14 | 5 | 15 |
| Price | $3 | $8 | $20 | $3 | $10 |

**Notes**
1) Figures are for total flash memory. Some portion is taken by the bootloader and therefore is not available for program development.
2) The Teensy doesn't actually have EEPROM, but rather emulates it by setting a block of flash memory aside that can be accessed as EEPROM memory.
3) Teensy, BP, and ESP32 can emulate EEPROM. See main discussion
4) Uses both internal and external memory. These are expected minimums.
5) The Teensy 4.0 (T4) just came out and is a real powerhouse at a lower price of $20. We have modified those projects" that used the Teensy 3.6 to use the T4.

microcontrollers only need 2 registers since each 32-bit register can hold a *long* data type variable. Manipulating 8 registers inside the microcontroller takes more time than manipulating only 2 registers. As a result, *ceteris paribus* (other things being equal), the 32-bit processors are going to do most tasks faster than an 8-bit processor. Also, $2^{32}$ can represent a much larger range of unsigned values (0 – 4,294,967,295).

## Flash Memory

This is where the instructions derived from your program source code live within the µC. Your program is responsible for providing the instructions that tell the µC what to do. A program called a *compiler* converts your English-like program source code written in C into the binary (1s and 0s) instructions that the µC understands. These binary instructions vary from one µC to the next... they are *not* binary compatible. The Arduino IDE (explained below) takes those compiled binary instructions and sends them via the USB cable for storage in the µC's flash memory.

Flash memory has perfect memory recall. That is, when you turn off the power to the microcontroller, flash memory retains whatever was in that memory at the time the power was removed. When you reapply power, the content of flash memory is the same as it was when you removed the power. It's the same type of memory you've used with thumb (flash) drives that you plugged into your USB port. This is why you don't have to reload the program each time you power up the μC. On the down side, flash memory has a finite write life. That is, you can reliably rewrite the flash memory about 10,000 times. After that, it may become unreliable.

Obviously, the more flash memory the microcontroller has, the larger and more complex can be the program that the microcontroller is tasked to perform. When it comes to μCs, bigger (flash memory) is better.

## SRAM Memory

SRAM stands for Static Random Access Memory. Unlike flash memory, SRAM goes stupid when power is removed. When power is reapplied, SRAM will contain some random bit pattern that is most likely not in a state to do anything useful. Yet, SRAM is at least as important as flash memory when it comes to performing a program task. Indeed, the amount of SRAM often is more of a limiting factor on program development than the amount of flash memory.

The reason for the importance of SRAM memory comes from the fact that all of a program's data must be stored in SRAM. (The Arduino people recognize this fact and created the *PROGMEM* directive to help lessen the Arduino's SRAM restrictions. It's a clumsy way to address the limited SRAM the Arduinos have, but it does work.) Think about it. In its simplest terms, all a computer does is take data into it in one form, crunch on it for a while, and then output new data based upon the data that it was fed. Having a place to store this data while it is being processed is crucial.

Some types of data that you store in SRAM hang around for the entire duration of the program, while other types of SRAM data can come to life when needed, be discarded when their task is completed, and then reused again later in the program. Knowing how to code for "hang around" data versus "temporary" data often is the difference between having enough memory for a program to run or not. Because the amount of SRAM memory available is a real bottleneck in many programs, we discuss programming techniques in this book to help you overcome the SRAM limitation.

## EEPROM Memory

Electrically Erasable Programmable Random Access Memory (EEPROM) is very similar to flash memory, but is a generally a little slower to access and store (read/write) data than flash memory. Neither the Teensy, the Blue Pill (BP), nor the ESP32 have "real" EEPROM, although they can emulate it. EEPROM is often used to store configuration data. That is, data that is not likely to change that often or does not need to be fetched for some processor-intensive calculation where access time might be important. For ham radio applications, EEPROM is often used to store configuration data, such as your favorite CW sidetone frequency, keyer speed, filter settings, favorite frequencies and operating mode.

As you can see in Table 1.1, the Teensy sets aside 64 KB of flash memory for use as EEPROM. Given the designed purpose of EEPROM, 4 KB is usually more than enough for most μC applications. The BP is more interesting. Instead of dedicating a chunk of flash memory for (perhaps unneeded) EEPROM use, the BP lets you decide how many 1 KB pages of flash memory you wish to use. This allows you to adjust the EEPROM footprint a little to better suit your needs. Nice! The ESP32 with 1 MB of flash by default allocates 1 sector for EEPROM (for example, 4096 bytes). Given what EEPROM is best suited for, that's probably enough. None of the projects in this book use anywhere near any of those EEPROM limits.

One more thing: recall that flash memory can get Alzheimer's after about 10,000 writes. Because the non-Arduino µCs use flash memory to emulate EEPROM, you probably shouldn't use EEPROM for "common" data storage that's going to be updated often. If you have such a need, you could add an EEPROM card or an SD card reader to the project. Many TFT displays have a card reader built into them.

## Processor Clock Speed

The rate at which your microcontroller can plow through and transform a program's data depends in large measure upon the processor's clock speed and the native size of the µC's registers. Think of the processor's clock speed as the rate at which a stop light changes to control the speed of the data traffic. Instead of making your drive home miserable, these traffic lights control how fast your data can move through the system. Moving data between registers and across a data bus takes time. That time is measured in processor clock cycles. Perhaps the Blue Pill needs the same number of clock cycles as the Teensy to perform some task. However, because the Teensy's clock is more than 8 times faster than the Blue Pill, we should expect the Teensy to boot-scoot the data around the system much faster and, hence, accomplish a given task in less time. The ESP32's clock is 3 times faster than the Blue Pill! Alas, fast boot-scooting isn't free. There's a tradeoff between speed and cost.

For reasons we don't understand, the Arduino family of (Atmel) processors has not kept up with either the clock speeds or register sizes of its competitors. Perhaps their recent buyout by a competitor has something to do with it. Atmel does have some faster processors (such as the Leonardo and Due), but they are disproportionately expensive and the resource boost is less than dazzling. Does that mean the Arduinos are going the way of the dinosaur? No, if for no other reason than there is probably a bazillion lines of free source code out there available for you to use in your own projects. Also, some programming tasks benefit from a super-fast traffic light; others don't. A processor that waits a lot for humans to type something in or respond to a button press looks like continental drift speeds to a µC. The good news is that at least you have a choice at prices that make sense.

## Additional Information

**Table 1.2** presents additional information about the microcontrollers that may impact your microcontroller decisions. Most microcontrollers have built-in voltage regulators. Column 2 in the table shows the allowable voltage range that the regulator can handle. As a general rule, most common voltage regulators work when the input voltage is a couple of volts higher than the desired regu-

**Table 1.2**
**Microcontroller Voltage and Current Information**

| Microcontroller | Vin Volts | µC Voltage Volts | GPIO Voltage Volts | GPIO 5 V Tolerant? | GPIO current Per Pin | Max 3.3 V Regulator Current |
|---|---|---|---|---|---|---|
| Nano | 7 - 9 | 5 | 5 | Yes | 40 mA | n/a |
| Mega 2560 | 5 - 12 | 5 | 5 | Yes | 40 mA | n/a |
| STM32 | 5 | 3.3 | 3.3 | No 3.3 V Max | 10 mA | 1000 mA |
| ESP32 | 5 | 3.3 | 3.3 | No 3.3 V Max | 40 mA | 1000 mA |
| Teensy 4.0 | 5 | 3.3 | 3.3 | No 3.3 V Max | 10 mA | 300 mA |

lated voltage (for example, the LM7805 likes 7 V in to get a regulated 5 V out).

Column 3 in Table 1.2 is the voltage used by the microcontroller. Notice that only the Arduino family swim in the 5 V pool. The same is true for the voltage appearing on the I/O pins (Column 4). Column 5 says that putting 5 V on the STM32, ESP32, or Teensy runs the risk of damaging the microcontroller. The last two columns are concerned with the current on the I/O pins (Column 6) and the built-in regulator (Column 7).

## Interfacing Digital Modules

In this book we present to you a number of combinations of microprocessors, displays and other support modules. Not all of these digital wonders use the same voltages to talk to each other. This can be a significant problem, because the wrong (too high) voltage at a GPIO pin can damage the device. One solution is to use only devices that have common voltage levels. That severely limits your choices, however. A better way is to interface these devices using little modules called "level shifters," which translate the digital signals on one end to the proper value at the other end. Most of the available level shifters are bi-directional, meaning that they don't care which is the input or output, they just transmit the signal to the other end with the proper voltage level. For more information see **core-electronics.com.au/tutorials/how-to-use-logic-level-shifters-converters.html**.

Figure 1.2 — Logic level shifter.

The units we use are available for about $1 each for a 4-channel unit. **Figure 1.2** shows a typical example by HiLetgo. The physical construction of level shifters varies significantly among manufacturers, so use their instructions for adding it to a circuit. All that is needed is +3.3 V and +5 V supply connections and the various GPIO pins on each side. Problem solved!

## Which Processor Should I Buy?

You already know the answer: It depends. If you're trying to develop a device that needs to sample something a five million times a second, you need some serious clock speed to do that. If you're involved with a project that only needs a sample once a minute, but there are hundreds of sensors, perhaps the number of I/O pins is more important than clock speed. The nature of the problem you're trying to solve is probably going to be the determining factor.

Still, you need an answer. First, if you purchased this book because of some specific project, read that chapter (after reading the first three chapters!) and purchase the recommended µC. If you are just starting the learning process, the Nano is a good choice. It's inexpensive, has enough resources to serve as a good learning platform, and has a boatload of free software. The BP and ESP32 do not have the Open Source software depth the Nano has, but they are gaining ground. Also, the BP costs about the same as the Nano, can be programmed in the Arduino IDE, and has a growing body of available libraries. Both are cheap enough I'd consider purchasing several of each.

The Teensy 4.0 (T4) is another issue. As you can see in Table 1.1, the Teensy has a deep resource pool and a clock that is more than 8 times faster than the BP (and over twice as fast as the ESP32). The problem with the Teensy is its cost. The Teensy 4.0 ($20) costs less than the Teensy 3.6 ($30), but gives up some I/O pins and a few other things in the process. If your project needs to do DSP or audio processing, or uses FFTs or other processor-intensive actions, the Teensy 4.0 is a great choice. However, if all you want to do is blink an LED or two, it's an expensive choice. We have designed the DSP projects to use the new and less expensive T4.

The ESP32 has some benefits, too. It has built-in Bluetooth and WiFi connectivity, a growing body of source code, a very fast clock controlling a rich resource base, and costs under $10. While it is overkill for many applications, if we had to pick just one µC, it would probably be either the STM32 or ESP32, especially if we are controlling a reasonable number of I/O pins. Also note that the different flavors of the ESP32 affect the resources available to you and can lead to some confusion when using it. Some ESP32s have 64 – 120 KB of flash, others have 4 MB. Some come in 30-pin DIP packages while others have 38 pins and bring more of those pins out for you to use. Another source of confusion is that some I/O pins cannot be used for output. (Pins 34, 35, 36, and 39 are input-only.) Other pins do weird things when their state changes (for example, they do a reboot). At first, we preferred the 38-pin variety with a 1 MB flash. However, further inspection shows that most of those additional pins cannot be used for external I/O port uses. For that reason, we think the NodeMCU is the preferred ESP32.

Another reason we are a little lukewarm on the ESP32 is that its pin map varies according to who makes the board. Take an Arduino Nano from any manufacturer and its pins are compatible with everyone's pin layout. Not so with the ESP32. Therefore, make sure you have access to a pin map for any ESP32 board you buy.

## Some eBay Caveats

Virtually all of the Nanos for sale on eBay are clones of the original Arduino Nano. Because the developers made it an Open Source device, anyone is free to clone the Nano. We buy Nanos 10 at a time and usually pay less than $3 each. Check fellow hams and club members to see if you can combine an order to take advantage of a quantity discount. We do *not* buy the Nano Pro Mini. It's cheaper, but does not have the USB connector, which makes it less convenient to program.

## Arduino Nano Clone Problems

The Arduino Nano and Mega 2560 Pro Mini clones may also have one or both of the following problems:

1) Typically, clones are produced with the least expensive hardware possible, which often means the USB-to-serial converter needs the CH340 device driver. (A "real" Arduino Nano uses the FTDI device driver.) If you find that your Nano locks up when it tries to upload the program code from your PC to the Nano, it could be you have the wrong device driver. Just do an internet search on "CH340 device driver download" to download and install the device driver (**www.wch.cn/downloads/CH341SER_ZIP.html**).

2) Another potential problem area is that, as the Arduino compiler continues to improve, some of the Nano clone manufacturers have not updated the software they put in the Nano. In particular, there's a program called the *bootloader* that must be embedded in the Nano's flash memory for it to work properly within the Arduino IDE. It's the bootloader's responsibility to take the data coming from your PC via the USB cable during an upload and tuck it away in the proper places within the Nano's flash memory. If your Nano isn't uploading your program properly, look at the top menu line of the Arduino's IDE and select: Tools → Processor:"ATMega328P" → ATMega328P (Old Bootloader). (See Figure 1.5 later in this chapter.) This choice tells the compiler that the Nano has an old bootloader program, so play by the old rules. A clone still works just fine; it just handles the incoming data from your PC a little differently. Fortunately, you don't need to know about those internal differences.

The instructions above make more sense after you've read the next section. However, we didn't want you to think something was wrong simply because the clone might not load the program correctly on the first try. If that's the case, give the two adjustments discussed above a try. Chances are one of them will fix your problem.

## Arduino Software

A µC without software is about as useful as a bicycle without wheels. Like any other computer, a µC needs program instructions for it to do something useful. The Arduino programming environment has provided all the software tools within their (free) Integrated Development Environment (IDE) that you need to write program code. All of the programming tools you need (library manager, text editor, compiler, assembler, and linker) are integrated into the IDE. The programming language used by the Arduino IDE is a robust subset of the C programming language. While all of the features of C are not supported (such as the *double* data type), it is probably 99 percent complete. The remainder of this chapter discusses downloading, installing, and testing the software you need for project programming.

Start your internet browser and go to **arduino.cc/en/Main/Software**. There you will find the Arduino software download choices for Windows, macOS, and Linux. Click on the link that applies to your development environment. The latest Arduino IDE available at the time this is being written is Release 1.8.10. You are given the opportunity to make a contribution towards further development of the IDE. We hope you contribute what you can afford.

You are asked to select the directory where you wish to extract the files. I named my directory *Arduino1.8.10* and placed it off the root of the D drive (*D:\Arduino1.8.10*). Many of you may not have a D drive, which I use almost exclusively for programming purposes. In your case, use your main drive, which is likely drive C, and create the directory named *C:\Arduino1.8.10*.

This directory is different from the default Arduino installation path. You can use the default installation if you wish and the compiler will perform as expected. We prefer our installation path because we can install upgraded versions of the IDE, but keep the older versions unchanged should we need to "back up."

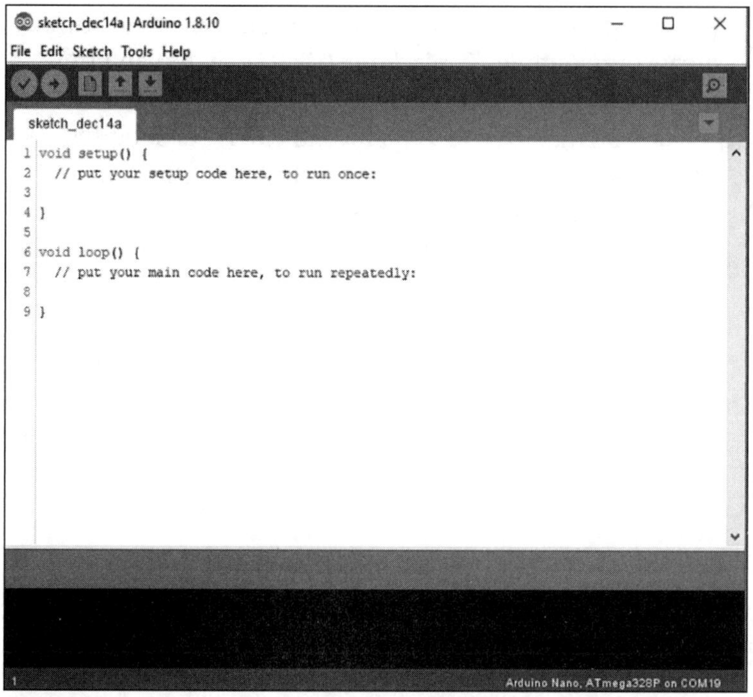

Now download the Arduino installation file and run the installer. The program tells you where the default installation is placed. If you use our directory path, use the installer's Browse button to browse to the *C:\Arduino1.8.10* directory. When the download completes, there will be an executable (exe) file in your *Arduino1.8.10* directory.

Inside the Arduino directory you just created, double-click on the arduino.exe file that you just saved. This should cause the IDE installer to start. After the IDE loads into memory, your screen should look similar to **Figure 1.3**. You should have your Arduino board connected to your PC via the appropriate USB cable for your board.

**Figure 1.3 — The startup screen for the Arduino IDE.**

All that remains is to select the proper Arduino board and port. If you look carefully at the bottom of **Figure 1.4**, you can see that I was last programming a Maple Mini Blue Pill. Figure 1.4 shows how to set the IDE to recognize an Arduino Nano board. You can activate the board selection by using the menu sequence *Tools → Board → Arduino Nano*. (We use this notation throughout this book when using a menu selection sequence.)

As you can see, the one IDE supports a large number of the Arduino family boards. Click on the board that you are using. The sequence shown in Figure 1.4 means that I am switching from a Maple Mini

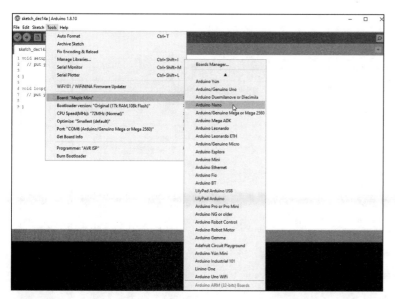

**Figure 1.4 — Replacing the Maple Mini board with the Arduino Nano.**

**Getting Started   1-9**

(STM32) board to an Arduino Nano board. (Compare the shaded text areas.)

After you make your board selection, you must now tell the IDE which port you will be using to move the compiled code via the USB cable from your PC to the μC board you just selected. In other words, the port you are about to select is used for communication between your PC and the μC. Use the *Tools → Port* menu options and select the appropriate port. This is shown in **Figure 1.5**.

Sometimes the Windows environment does not find the proper port. This is usually because the device driver for the port isn't found. If this happens, go to your installation directory and into the *drivers* subdirectory (for example, *C:\Arduino1.8.10\drivers*) and run the driver installation program that's appropriate for your system. (**Figure 1.6** shows the *drivers* subdirectory contents. You may need to run one of the dpinst-??.exe programs, depending on the system you are using. Try one and, if the port still isn't available, try the other program. It doesn't hurt anything if you run the wrong one.) If you are using a Nano clone, you will need to install the CH340 driver before the PC can establish a communications link to your Nano board.

Once the driver is installed, the program should now find your Arduino COM port.

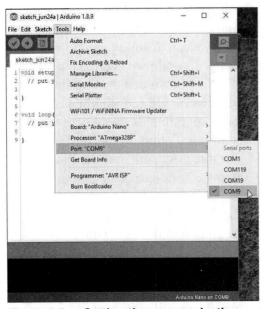

**Figure 1.5 — Setting the communications port for the Arduino Nano.**

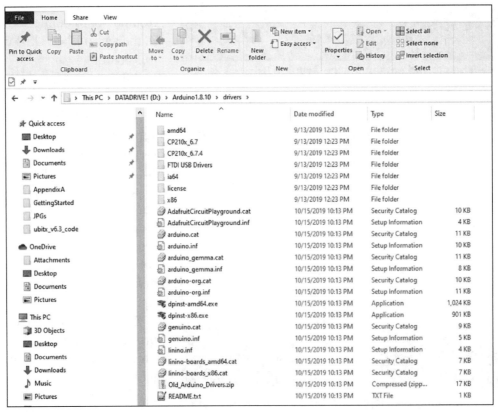

**Figure 1.6 — The drivers subdirectory.**

## The Integrated Development Environment (IDE)

The easiest way to test that everything installed correctly is to run a program. Your IDE has a sample program called Blink. You can find it using the menu sequence: *File → Examples → 01. Basics → Blink*. You can see this sequence in **Figure 1.7**. Once you click on Blink, the IDE loads it into the Source Code window. Once the Blink program is loaded, the IDE looks similar to **Figure 1.8**. We've marked some of the more important elements of the IDE in the figure.

### Source Code Window

The large white space in the IDE, the Source Code Window, is where you type in your program source code. Program source code consists of English-like instructions using the C programming language that tell the compiler what code to generate. Because you already loaded the Blink program, you can see the Blink source code in Figure 1.8.

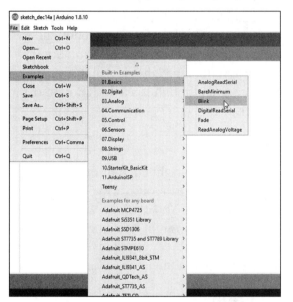

**Figure 1.7 — Loading the Blink program.**

### Compile Icon

The compile icon is exactly that: It compiles your program. It does not, however, necessarily generate an executable program. As you learn in the next chapter, there are all sorts of tasks that a compiler must consider before it can produce an executable program. A huge percentage of these tasks are already done for you and are just waiting to be used in your program. A sub-program that is part of the Arduino IDE is called the linker. It's the linker's responsibility to gather all of these sub-tasks together and arrange them in a form of a program that can be

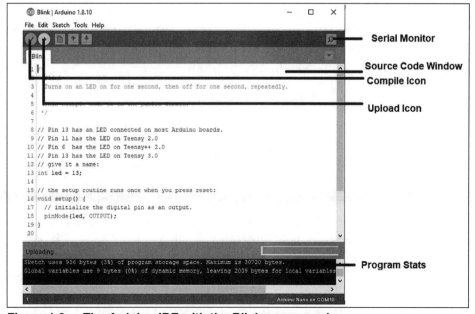

**Figure 1.8 — The Arduino IDE with the Blink source code.**

Getting Started    1-11

executed. The compile icon does not invoke the linker by itself. However, using the Compile Icon is a quick way to check your code for syntax errors. (More on these and related topics in Chapter 2.)

## Upload Icon

The upload icon has several tasks. First, it does call the linker into play to gather up all the pieces-parts of your program and arrange them into an executable form. The linker then takes that binary data and transmits it via the USB cable (using the port you selected) to the flash memory of the μC you are using. The bootloader in the μC's flash memory does a few housekeeping chores and then immediately starts to execute the program. If you skip the Compile icon but click the Upload icon, the Upload icon automatically both compiles *and* uploads your program code to the Arduino via the USB cable.

Note what all this means: Your PC is used to write, test, compile, and debug your program, but the code actually runs on the μC. The μC uses its bootloader to manage all of the communications between your PC and the μC. You don't need to worry about it. (If you compile a program on a Nano, it will tell you that you have about 30 KB of program space even though it's a 32 KB memory bank. The "missing" 2 KB is the bootloader.)

## Serial Monitor

The Serial Monitor icon shown toward the upper-right edge in Figure 1.8 invokes the *Serial* object. The *Serial* object lets you print out various data elements that are in your program on your PC's display. You click on this icon to see the output from any *Serial.print()* statements you might be using in the program via the *Serial* object. The output of the *Serial* object is displayed on your PC. We'll show an example of such statements in a moment.

## Program Stats

The Program Stats window tells you how much flash and SRAM memory your program is using. The window is also used to display error messages if the IDE detects something wrong with your program. (If you are using a μC other than an Arduino, the Stats window may show some additional supporting information about that controller.)

## Your First Program

Let's look at the Blink program and make a couple of simple changes to it. The code is reproduced in **Listing 1.1**, with all of the comments stripped away. Look at the element called *setup()*. The *setup()* program element is actually called a function. A *function* is a group of one or more program statements designed to perform a specific task. The purpose of the *setup()* function is to define the environment in which the program is to operate. Without getting too deep right now, *pinMode()* is also a function which has been predefined for you to establish how a given Arduino I/O pin is to operate in the program. In this case, we are saying that we want to use the *led* pin (*led* is a variable defined at the top of the program as pin 13) to output data.

However, every Arduino board happens to have a small LED available for

**Listing 1.1**
**Blink Source Code**

```
/*
  Blink
  Turns on an LED on for one second, then off for one second, repeatedly.

  This example code is in the public domain.
 */

// Pin 13 has an LED connected on most Arduino boards.
// Pin 11 has the LED on Teensy 2.0
// Pin 6  has the LED on Teensy++ 2.0
// Pin 13 has the LED on an Arduino
// give it a name:

int led = 13;

// the setup routine runs once when you press reset:
void setup() {
  // initialize the digital pin as an output.
  pinMode(led, OUTPUT);
}

// the loop routine runs over and over again forever:
void loop() {
  digitalWrite(led, HIGH); // turn the LED on (HIGH is the voltage level)
  delay(1000L);            // wait for a second
  digitalWrite(led, LOW);  // turn the LED off by making the voltage LOW
  delay(1000L);            // wait for a second
}
```

use on pin 13. The other boards also have a similar LED, but it may be mapped to a different pin number. A pinout image makes it easy to find the various pins. For example, in the lower-right corner of Figure 1.10 later in this chapter, you will see that the pin labeled as PC13 is the LED pin. Each manufacturer is free to use whatever pin they wish for this programmable LED. Consult your pinout image to find out which pin to use for your controller board. In the program in Listing 1.1, however, we are going to use pin 13 to output data. Therefore, the output data consists of blinking the Arduino's on-board LED.

One thing that is unique about *setup()* is that it is only called once. That is, when you first apply power to the Arduino or you press its Reset button, your code automatically calls (executes) the code in the *setup()* function. Once *setup()* has defined the program's operating environment, its task is complete and it is not called again unless you restart the program.

If you look at the code in *loop()*, the first statement is a call to the *digitalWrite()* function. This function is also pre-written for you. The *digitalWrite()* function has two pieces of information (called function arguments) passed to it: *led* (pin 13) and HIGH (+5 V for the Nano, +3.3 V for the other μCs). These two pieces of information (the pin number and its state) are needed by the *digitalWrite()* function to perform its task. The function's task in this case is to set the state of pin 13 (*led*) to HIGH. Setting the pin high has the effect of turning the LED on. The next program statement, the *delay()* function call, is another pre-written piece of code that has a number between its parentheses. This number (1000L) is also a *function argument* which is information that is passed to the *delay()* function code which it needs to complete its task. In this case, we are telling the program to delay executing the next program statement for 1000 milliseconds, or a total of one second. Note that we end the numeric constant 1000 with the letter 'L'. This is called a data type suffix and it is use to tell the compiler in no uncertain terms that the 1000 is a *long* data type. We tend to use data type suffixes more than most programmers because it helps document the program. The default numeric type is usually an *int* (integer) data type.

After one second, another *digitalWrite()* function call is made, only this time it sets the state of pin 13 to LOW (0 V). This turns off the LED tied to pin 13. Once the LED is turned off, the program again calls *delay()* with the same 1000 millisecond delay. So, collectively, the four program statements in *loop()* are designed to turn the LED on and off with a one second interval. That almost sounds like blinking the LED, right?

Now here's the cool part. After the last *delay()* is finished, the program loops back up to the top of the *loop()* function statements and re-executes the first program statement, *digitalWrite(13, HIGH)*, again. (The semicolon at the end of the line marks the end of a program statement.) Once the LED is turned on, *delay()* keeps it one for one second and the third program statement is again executed.

Because we're done reading the Blink program code, press the Compile-Upload icon and after a few seconds you will see a message saying the upload is done. If you look at your Arduino, it's now sitting there blinking its LED for you.

Your program repeats this looping sequence until: 1) you turn off the power; 2) you press the reset button (which simply restarts the program); 3) there is a component failure; or 4) the cows come home. If your program is performing as described here, you have successfully installed the IDE and compiled your first program!

## A Simple Modification

Let's add two lines to make this "your" program. Click in the Source Code Window, which moves the arrow cursor into the Source Code window. (See Figure 1.8.) The cursor changes from an arrow to an I-bar cursor. Now type in the shaded new lines shown in **Listing 1.2**.

Obviously, you should type in your name and call sign. The first line says we want to use the IDE's *Serial* object to talk to your PC using a 9600 baud rate. The *Serial* object has a function embedded within itself named *begin()*, which expects you to supply the correct baud rate as its function argument. That object method

**Listing 1.2**
**Blink Modifications**

```
void setup() {
  Serial.begin(9600);
  Serial.print("This is Jack, W8TEE, and Al, AC8GY");
  pinMode(13, OUTPUT);
}
```

sets things up so your PC can communicate with your program. The second line simply sends a message to your PC at 9600 baud over the *Serial* communication link (your USB cable). If you click on the Serial Monitor icon (see Figure 1.8), the Serial Monitor dialog box opens. At the bottom of the Serial dialog box are other baud rates that are supported by the IDE. The *begin()* baud rate and the rate shown at the bottom of the Serial monitor's dialog box must match. If they don't, your PC will blow up! Naw...just kidding. Actually, you may not see anything at all or you may get characters that look a lot like Mandarin.

If, after making sure you have the program's baud rate the same as the Serial monitor's baud rate, no output appears in the dialog box, it could be that your processor blew right by the *Serial.print()* statement before the *Serial* object was properly initialized. If that appears to be the case, add this line immediately after the *Serial.begin(9600)* statement:

```
while (!Serial)
  ;
```

This is an empty *while* loop that spins around until the *Serial* object is fully instantiated. (Instantiation is explained in Chapter 3. For now, just think of instantiation as the process of getting a variable to be fully functional in a program.)

If the compiled program shows the new message on your display screen, you know that you've installed the IDE correctly, the compiler works, the USB-PC connection is working, and the μC is working. Great!

## Installing the Various Microcontroller Patches

Patches? What is a microcontroller patch? Each microcontroller has its own set of binary instructions that cause it to perform some small task. For example, if you used an editor that could display information using the hexadecimal (base 16) numbering system, the flash memory of one of the μCs might have an entry that looks something like:

```
C5 50 00
```

While these hexadecimal numbers may make sense to the μC, humans need a little more help. We might examine the same couple of memory bytes with a programming tool called an assembler and those same bytes now look like:

```
JMP 5000
```

which forms a binary instruction that directs the μC to transfer program control to memory address 5000 and start executing the program instructions found at that address. An *assembler* is simply a programming tool that makes the hex codes easier to read by using abbreviations, or *mnemonics*, for the processor's instruction set.

The problem is that C5 might be the binary instruction for a memory JMP on one μC, the same C5 instruction might well be a LDA (Load Register A) instruction for a different μC. Each μC has its own set of binary instructions that cause that μC to perform some specific action. Those instruction sets are completely different for all of the μCs we use in this book.

So, how do we get the Arduino IDE to use the proper set of binary instructions for the different μCs? In a very real sense, that's exactly what you were doing when you made the Board selection depicted in Figure 1.4. While we are simplifying things a bit, a *"patch"* allows a given μC to add its binary instruction set as a choice for the Arduino IDE compiler. So, for example, when you compile the Blink program for the Arduino Nano, because you selected the Nano Board (Figure 1.4), the compiler uses the Nano's binary instruction set. So the compiler takes the C programming language instructions for the Blink program and converts them into the binary instructions used by the Nano.

If you take the C version of the Blink program, but have selected an ESP32 Board instead of the Nano, the compiler is smart enough to grab the binary instruction set for the ESP32 μC and generate the program code. This is what enables you to use the same IDE for different μCs.

The installation of a "patch" for each μC provides the means by which the Arduino IDE can find and use the proper binary instruction sets necessary to generate a program that can run on all of the different μCs. You might think of the IDE as having the ability to switch the *core* of the code generator that is buried within the compiler to use the binary instruction set for the Board you are using.

The remainder of this chapter provides information on how to install the patch for each of the μCs we use in this book. (μCs in the Arduino family don't need a patch since their instruction sets are native to the Arduino IDE.)

## Teensy

As you can see in Table 1.1, the Teensy has a lot going for it in terms of memory resources and clock speed. It also has some very powerful and useful libraries that you can use. The biggest downside is the cost. Members of the Teensy family cost about $20 to $30. The Teensy is not Open Source so the Teensy family is not second-sourced. Also, discounts on volume purchases are very scant. Still, for digital signal processing (DSP) tasks, there are a lot of things going for the Teensy, especially the 4.0 model.

Installing the patch for the Teensy family is well-explained at **www.pjrc.com/teensy/td_download.html**. You can find instructions for Windows, macOS, and Linux platforms at the site above. Repeating those instructions here is unnecessary as they are easily followed as presented. PJRC also supports an active Forum for questions about the Teensy. It can be found at **forum.pjrc.com/**.

The pinout for the Teensy 4.0 is shown in **Figure 1.9**. Note that some of the

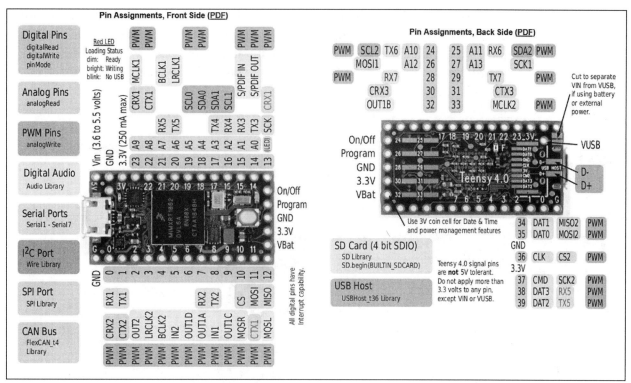

Figure 1.9 — Pinout for the Teensy 4.0.

pins are accessed from the bottom of the board. If your project has high I/O pin density, you may need to use the Teensy 3.6. The PJRC website has images for the Teensy 3.6, too.

## STM32

The STM32F103 is also called the "Blue Pill" (BP from now on) because of its color and small size. It's inexpensive but packs a more powerful punch than the Arduino Mega, Uno, or Nano. In some cases, information applies to other members of the STM32 family but is of interest to the BP users, too. In those more-generic instances, we refer to them as "STM32" rather than the more-specific BP.

An excellent discussion of how to use the IDE's Board Manager to install the core libraries for the STM32 family can be found at **github.com/stm32duino/wiki/wiki/Getting-Started**. The core can also be installed manually by downloading the library directly and installing it yourself. The library can be downloaded from **github.com/stm32duino/Arduino_Core_STM32**. The patch can be downloaded from **github.com/rogerclarkmelbourne/Arduino_STM32**. Roger Clark has forgotten more about the STM32 than we'll ever know. It appears that the differences between the two downloads is real, but that they function together pretty well. The sites above also have information about other places to find support for the STM32 running in the Arduino IDE. One that we have found very useful is **www.stm32duino.com** which is more directly aimed at using the BP in the Arduino IDE.

A pinout of the BP is shown in **Figure 1.10**. Note that different manufacturers of the STM32 boards may or may not bring out all of the pins shown in Figure 1.10. Also, some sell the STM32 without a bootloader, which makes it less convenient to program. Ask your STM32 vendor if the bootloader is already installed before you buy.

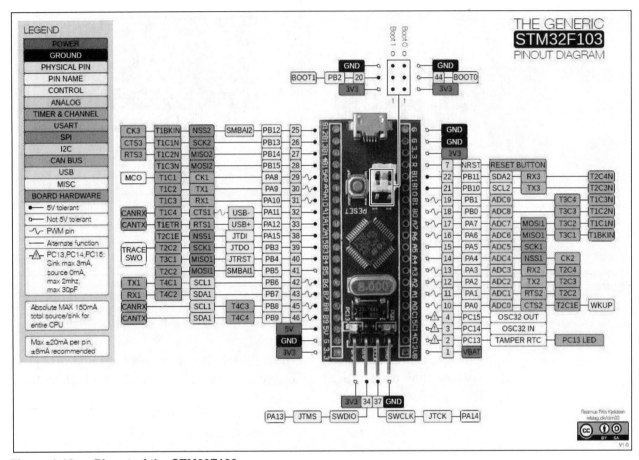

**Figure 1.10 — Pinout of the STM32F103.**

## STM32 Device Firmware Update

On some machines, when the Arduino compiler finishes compiling an STM32 program, you'll see a line in the Stats Window that says:

```
Searching for DFU device [1EAF:0003]...
```

DFU stands for Device Firmware Update to account for the fact that the STM32F supports other (for example, CAN, I2C, SPI) bootloaders than the standard USB bootloader that the Arduino family uses. When you see this message appear at the bottom of the IDE in orange print, you have about 5 seconds to press the Reset button, which you can see next to the yellow pin jumpers in Figure 1.10. Very quickly, you should see a new message that says:

```
Found it!
```

and the compiler proceeds to upload the file to the STM32. Usually, all subsequent compiles/uploads will run to completion without pressing Reset again. If you close and re-open the Arduino IDE, you will have to press the Reset button once again.

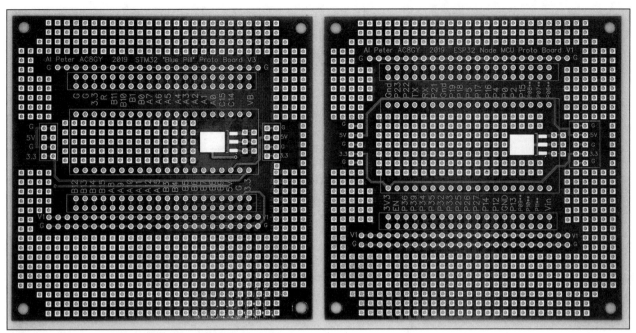

Figure 1.11 — Al Peter's expansion boards.

If you buy an STM32F103 without a bootloader, you can add one using the ST-Link programmer module. While there are literally hundreds of websites with instructions on using the ST-Link module, one of the better one is **idyl.io/arduino/how-to/program-stm32-blue-pill-stm32f103c8t6**.

Al has created expansion boards for the BP and ESP32 that make it much easier to breadboard circuits. These boards can be seen in **Figure 1.11** and are distributed by **QRPGuys.com**. Al's expansion boards make it very easy to use Dupont wires between the BP and the circuit being built. There are power (V1, V2) and ground (G) rails on the board and a small prototyping area, too. If you look closely at Figure 1.11, you'll see that the pin numbers are also silk screened onto both sides of the board. This saves you time by not having to constantly flip the board over to see which pin is which. We found the expansion board very useful while prototyping various BP circuits.

## ESP32

Unlike the other microcontrollers, the ESP32 boards bring out pins that have very specific functions or cannot be used for other common I/O functions. Some of the unusual aspects of the ESP32 are summarized in **Table 1.3**. For example, pins 34, 35, 36 and 39 can only function as input pins. General Purpose Input/Output (GPIO) pins 0, 5, 14, and 15 output a PWM signal when a boot is performed. **Figure 1.12** shows the pinouts for the ESP32 that we used during development, which is the HiLetgo ESP-WROOM-32. If you looked closely at Figure 1.12 (and you did, didn't you?), you'll notice that there are 19 pins on each side of the board — 38 total. If you read Table 1.3 closely, you noticed that pin 39 can be used as a general I/O pin. Wait a minute...how can a board with 38 pins use pin 39?

It can't.

Actually, most ESP32 boards use the ESP32-D0WDQ6 chip, which has 48 pins, but not all of those pins are brought out on the board. Another popular board is the ESP32 Development board, but it only provides access to 30 pins. While the ESP-WROOM-32 can be purchased for around $10, the ESP32 Development board is about half that price. Originally, we thought the additional cost of the ESP-WROOM-32 was worth the added expense. Now we're not so

**Table 1.3**
**General Purpose Input Output (GPIO) Pins for ESP32**

| GPIO Pin | Input | Output | Description |
|---|---|---|---|
| 0 | pulled up | OK | outputs PWM signal at boot |
| 1 | TX pin | OK | debug output at boot |
| 2 | OK | OK | connected to on-board LED |
| 3 | OK | RX pin | HIGH at boot |
| 4 | OK | OK | |
| 5 | OK | OK | outputs PWM signal at boot |
| 6 thru 11 | x | x | connected to the integrated SPI flash |
| 12 | OK | OK | boot fail if pulled high |
| 13 | OK | OK | |
| 14 | OK | OK | outputs PWM signal at boot |
| 15 | OK | OK | outputs PWM signal at boot |
| 16 thru 33 | OK | OK | |
| 34 | OK | | input only |
| 35 | OK | | input only |
| 36 | OK | | input only |
| 39 | OK | | input only |

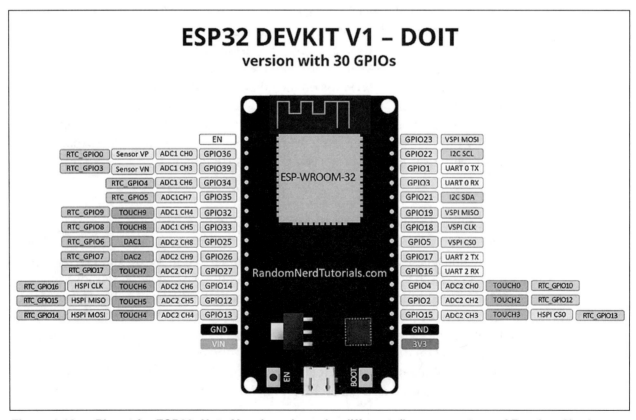

Figure 1.12 — Pinout for ESP32. *Note*: Your board may be different. (Image courtesy of Random Nerd Tutorials.)

sure, because most of those additional 8 pins are dedicated and cannot be used in a general sense. For that reason, we're now concentrating on the less expensive NodeMCU version of the ESP32.

Jack did a flat forehead mistake in that he used pins 3 – 5 while writing the menu demo for Chapter 4. He knew something was wrong because each time the IDE tried to run the program, the Serial monitor would show a reboot and a "Guru" message with a bunch of register information. Fascinated by the Guru messages, it took Jack way too long to remember that he should not be using those pins for general I/O use. We now think that maybe the 30-pin versions might not be too bad. In any case, pay attention to Table 1.3 when you are using an ESP32 board.

We couldn't decide on the best way to refer to the ESP32 pins. The reason for the indecision results from several factors. First, there are a number of boards that use the same basic ESP32 chip, but the boards vary on the pins that are brought out for use on the board. Some boards in the ESP32 family have 30 pins, some 36, and some 38. Further, the silkscreens on the boards also vary. Some use plain numbers (13), other prefix the pin number with a letter (D13).

However, all of the boards have an image file that uses the GPIOxx format to identify the pins. For example, in Figure 1.12, the second pin down from the upper left (physical pin 2) is reference as GPIO36 (General Purpose Input/Output Pin 36). This means in your code you could say:

```
#define SWITCHPIN    36
```

which creates a symbolic constant that ties your switch to physical pin 2, but is used in your code as SWITCHPIN, which is tied to I/O port 36. Therefore, to figure out which I/O pin your code should reference, use Figure 1.12 and simply strip away the "GPIO" which leaves the I/O pin number. That I/O pin number becomes your reference for your use of that pin in your code.

Unlike the other controllers, Espressif evidently has not required board makers using the ESP32 to use a standard pin configuration. As a result, the ESP32 you buy may not be able to use the pinout numbers shown in Figure 1.12. It also means you must pay attention to the pin assignments that exist for your particular board. The last set of boards we bought (from HiLetgo) includes a sheet with the mapping of each of the pins. Truthfully, the ESP32 boards are harder to tame, but the resource depth and performance are worth the effort it takes to learn how to use it, especially if you require the built-in Bluetooth or WiFi functions.

There are dozens of links that tell you how to download and install the patch for the ESP32, but the one at **circuitdigest.com/microcontroller-projects/programming-esp32-with-arduino-ide** is easy to understand and follow. It even gives an ESP32 Blink program that you can use to test whether the installation was installed properly. The method used is a little different than you've seen earlier because it uses the Board Manager to install the patch. It works fine. If you would rather use Github like we've done before, you can download the patch from **github.com/espressif/arduino-esp32**. Either way works fine.

Once you have installed the patch, connect your ESP32 to the USB cable and try to compile/upload the Blink sketch. You may have to hold the reset button the first time. If you've installed the 10 µF cap mentioned in the sidebar, you

> ## Odd...
>
> During development of our first ESP32 project, I would always have to press the Reset pin on the ESP32 board immediately after a compile or it would not upload the code. (It timed out after about 10 seconds of trying to upload the compiled code.) Not a big deal, you just had to make sure you hit the Reset button before it timed out. It was just one of those things you learn to live with.
>
> I took my code and breadboard over to Al's for some tests using his "big boy" test equipment. I made a small change to the code, pressed the compile/upload button and was ready to press Reset, but the code was uploaded without me pressing Reset. Odd, since Al and I are both using the same versions of the IDE and Windows 10. There is a difference in that Al's machine runs a fast Intel processor while mine's using an AMD Ryzen 7 processor. I mentioned this on our internet group and one reader suggested that I connect a 10 µF cap between the EN (Reset) pin (see Figure 1.12, upper-left) and ground. I did and, voila, no more pressing the Reset button to perform the upload! I'm not smart enough to figure out why it works, but I figure there's something different in the BIOS that required the reset. If you're having a similar problem, try this fix...worked for me.

should not have to press the Reset button again during this session. (You will have to press it the next time you load the IDE.) Select the *ESP32 Dev* module for the board and select the COM port for uploading. Now load and compile/upload the Blink program to make sure everything's working properly.

## Conclusion

This chapter should help you decide which µC you should buy for subsequent projects. If you're just starting out, buy a Nano, Mega 2560 Pro Mini, or a Blue Pill. (The Nano is the easiest to get started with.) They're cheap, yet can do almost everything this book can throw at it. The nice thing about a Nano is that, if you brick it (burn it up, thus releasing the Magic White Smoke and turn it into a silicon brick), not much is lost. Also, these things are like cockroaches: It takes a lot to kill them, so don't be afraid to experiment.

On the other hand, if you have a specific project in mind, at least read the first three chapters before reading your project of interest. Having done that, experiment a bit with that µC before you start building the project. It's like having a first date: You expect everything to go well, but sometimes it doesn't. Also, keep in mind that you could probably switch processors in many of the projects. Such a switch will involve some coding changes, sometimes extensive, but it can be done. That said, if you value your time at more than a penny an hour, it's cheaper to buy the processor that's used in the project.

You might also investigate your µC vendor's price list to see if they offer quantity discounts. Sometimes it's significantly less expensive to buy multiples and have the spares on hand if you need them. Keep in mind: This is a hobby! Enjoy it!

## Notes

[1] Jack Purdum and Dennis Kidder, *Arduino Projects for Amateur Radio*, McGraw-Hill, Nov. 2014.

## CHAPTER 2

# An Overview of C

Some of you were tempted to skip this chapter, weren't you? We're glad you didn't. Knowing something about what software does and how it works just gives you another tool to hang on your tool belt. However, be forewarned that, like any other shiny new tool, you're going to want to use it at some point. The purpose of this chapter is to provide you with some understanding of how all the software pieces-parts fit together. Just as a carpenter wouldn't refer to a Phillips screwdriver as a "Plus screwdriver" (because it looks like a + sign), there are some terms that we introduce so that you can more clearly read and discuss software ideas and issues on some of the popular Arduino and other software forums. We close the chapter with some conventions we use throughout this book plus some guidelines on writing code. To us, this is a fun journey. We hope you agree.

If you do have programming experience, and terms such as lvalues, rvalues, and symbol tables are familiar to you, you could skip this chapter. However, we have a different way of looking at programs and software in general. We hope you'll take the time to read this chapter.

## Programs

In the broadest of terms, a *computer program* (aka, computer *software*) is designed to perform a specific task. Virtually all software is designed to take a set of (verified) data input (from a sensor, a keyboard, a disk drive, a database, a radio signal, or other source) in one form, manipulate that data in a specific way, and output new data that was derived from the original input data. The process that dictates specifically *how* that data is changed from its input state to its output state is called an *algorithm*. This sequence is depicted in **Figure 2.1**.

A *software engineer* is often responsible for the design of the algorithm, while a *computer programmer* converts the English-like statements of the algorithm into some computer language (for example, C, C++, Basic, Java, Fortran, and so on) that can be converted (*compiled*) into the binary instructions the computer can process. The program can then be run by a user who collects the program input, processes it, and outputs the results.

The language we use in this book is called C. In computer terms, C is an old language that's been around since the mid-1970s. While I've tinkered with (and taught) about a dozen languages over the years, C remains my favorite. Why?

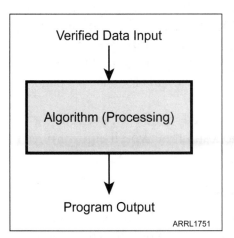

**Figure 2.1 — Basic steps of a computer program.**

## Good Things About C

First, C is a simple, yet robust, programming language. C contains only 32 keywords — words/terms the compiler understands. Other programming languages can have hundreds of keywords (Visual Basic, for example). Despite the relatively few keywords it uses, C is often the first programming language chip manufacturers initiate on a new computer because C is capable of performing most computer tasks efficiently without eating up a lot of the computer's resources. Also, because C does have a long history, there are shell compilers for C that make it easy to move the C compiler from one processor chip to a new one. For the most part, porting the compiler to a new processor involves little more than changing the code generator to reflect the changes in the instruction set of the new processor. The other parts of the compiler require relatively minor changes.

Second, the developers of C (primarily Dennis Ritchie and Ken Thompson) ingeniously left the input and output (I/O) aspects of the language out of its formal specification. Moving the I/O out of the language spec means the core of the language remains more or less intact regardless of the platform on which it is running. This also means C is a good language choice for full-fledged computers, or embedded systems that use dedicated microcontrollers.

Richie and Thompson relegated the I/O tasks to what is called the *C Standard Library*. The C Standard Library is a collection of functions, each of which is designed to perform a single task. Outside of the Standard Library, there are many additional C function libraries designed for a specific purpose. Some libraries process numeric data (for example, convert a simple number into a trigonometric value found in the math.h header file), others process textual data (string.h), and other libraries are designed for reading/writing data to disk files or sending output to a printer (streams.h). There are dozens of specialized libraries, too. Because such C libraries already are tested and debugged, writing new programs becomes that much easier. Indeed, one of the tasks of a new C programmer is learning about the existence of all of these specialized libraries. After all, no sense writing all that code from scratch when you can stand on the shoulders of so many who have gone before you.

Third, because C is a terse language with relatively few keywords, it doesn't take much in the way of program statements to get something done. I did some COBOL programming years ago and it was like writing the Great American Novel. A COBOL program might take four or five pages of programming statements for a relatively small task where the same program in C might take less than a paragraph. Also, which is easier to wrap your mind around: Five pages of statements or a half paragraph?

I don't know what languages you may have used in the past, but you're going to throw rocks at it once you know C. I honestly think you'll enjoy programming in C...if you choose to do so.

## Bad Things About C

Depending upon your programming experience, C can look very terse, especially if your experience is with COBOL or Basic. Now, flip your mindset: Terseness is a good thing since there is less typing for a given task compared to many other languages! It also means there is a lot less to learn...it's a bare-bones language.

## Aw, Man...Do I Really Have to Learn How to Program?

No. This book is written so you can simply download the program code, compile/upload it to the microcontroller, and start using it. To us, that's sort of like going to a five star restaurant, ordering the meal of a lifetime, and then just looking at it. So, while you don't have to know programming to get these projects working, you're going to enjoy each project more if you understand why we did things the way we did.

Also, a cookbook approach to software sort of misses the spirit of this book. We want you to view these projects as starting points for your ham radio projects. Many of you will want to enhance and extend the functionality of what we've presented here, and that might involve learning some programming. Like one of my students once said: "Do I really have to know the details about programming to use the device? After all, I can drive a car without being able to build one." True, but I'll bet the driver whose car broke down 300 miles north of Las Vegas in the middle of the desert at 2 AM kinda wishes he knew a little more about how cars worked. Even being able to change a program's splash screen so your name and call sign appear instead of ours is a simple, yet fun, task to take on...it personalizes your work. Also, you just might be surprised how much enjoyment there is in writing a program of your own design.

In any case, at least invest a little time in reading the rest of this chapter. We think it will prove worthwhile.

## The Five Program Steps

Virtually any program currently residing in the core of any computer on the planet can be reduced to the *Five Program Steps*. Knowing these steps provides two benefits. First, it provides a starting point for understanding how a program works. Indeed, we think this understanding removes some of the mystery that often seems to surround software development. Second, it also provides a starting point for designing your own software. It's the old "How do you eat an elephant" issue: Where do you start when you want to design your own program (or enhance an existing program). While you might be mumbling "Not gonna happen!" right now, bet'cha when you see that shiny new tool hanging from your belt, the urge to use it will become too much to resist!

**Figure 2.2** Shows the Five Program Steps. Each of these steps is discussed below.

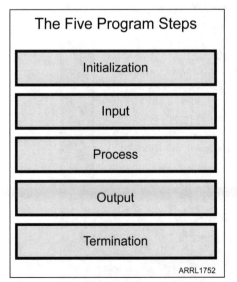

**Figure 2.2 — The Five Program Steps.**

### Step 1. Initialization

The purpose of the *Initialization Step* is to establish the environment in which the program executes. Usually this step involves a bunch of things that go on before the user sees anything on the screen. You have likely used a program such as a spreadsheet or a word processor where it keeps a list of the most-recently accessed files. **Figure 2.3** shows the Open Office word processor's recently accessed documents.

Obviously, the program is maintaining a 10-deep file list somewhere and the Initialization Step is reading that list and adding the file names to the program's recently-used list so you can eventually see them. Other programs use the Initialization Step to open printer ports, make USB connections, get database access, use your Wi-Fi connection to form an internet link, get file handles, request other resources from the operating system during program startup, and potentially a whole host of other things. All of this is taking place at the speed of light and, before you know it, a display screen lights up and you're ready to use the software. The Initialization Step has done its job; it sets the environment for running the program.

The Arduino programming environment also has an Initialization Step. It has a function (more on those below) named *setup( )* whose duty is to "set things up" and establish the environment in which your microcontroller (μC) program can run properly. We have a lot more to say about *setup( )* later in this book. For now, simply think of it as our Initialization Step.

## Step 2. Input

Virtually every program is designed to take information (data) as input in one form, crunch on it for a while, and then output that "crunched" data in a new format. (See Figure 2.1.) Getting data into your program is the purpose of the *Input Step*. While we are most familiar with programs that input data from the keyboard, input data can originate from hundreds of different places. That data might come into the program over the USB port, a fiber optic connection,

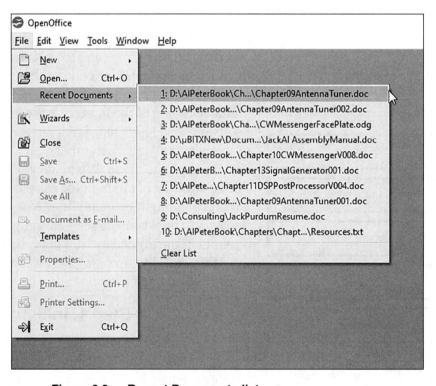

**Figure 2.3 — Recent Documents list.**

a disk data file, an RF source, or one of the bazillion sensors you can hang onto a µC. These connections can be as simple as a switch to something as complex as a fingerprint reader, retinal scanner, or GPS sensor. The point is, data comes into the µC by some means from somewhere.

Alas, data sources can screw up. Sensors fail. Power sources break down or become lost. People have been known to fill out long, complex, pregnancy heath questionnaires and later check the box that says they're male. Typing errors are the bane of programmers simply because users make mistakes. Likewise, database connections expire, batteries die rendering sensors unreliable, components overheat, and people do spill coffee onto sensors and the magic white smoke escapes and the sensor dies. The end result is that *the Input Step must validate the input data* so see if it is in a usable format and makes sense to the program. We could write an entire chapter on data validation, but we think you get the idea: data should be validated before it is used.

## Step 3. Processing

The *Processing Step* takes the validated input data and, following a specific sequence of actions, transforms that data into a new form. The nature of those specific actions is defined by the program's algorithm. As we mentioned earlier, an *algorithm* is simply a detailed plan for how the data is to be transformed. For example, you might take a sequence of characters typed into the program by the user and have an algorithm that translates those characters into Morse code. Or you might have a voltage applied to one of the analog inputs to the µC and then map that reading to another scale (for example, forward and reflected voltages to an SWR scale).

A faulty algorithm produces faulty output: Garbage In, Garbage Out (GIGO) as they say. A computer, by itself, has all the innate intelligence of a bag of hammers. It's the software that allows a computer to do interesting and useful things and a good algorithm is a must. Indeed, much of the progress in computer technology is the result of improved algorithms. When Jack was teaching programming courses at Purdue University, he was part of the Computer Technology department. He isn't smart enough to be a faculty member of the Computer Science department. That's where the *really* smart propeller heads work designing new and better algorithms.

Keep in mind that there are multiple ways to take the same data and produce the same results. One programmer I know was charged to shorten the processing time for generating morning sales reports for a very large international company. Batch sales figures came in from all over the world and a program automatically gathered that data and constructed the required reports and made inventory adjustments. The process automatically began as a "batch process" at 1 AM each morning and rarely was ready for an 8 AM sales meeting that morning. My programmer friend closely examined the Processing Step and tweaked the algorithm. When he finished, his program generated the same reports in 47 minutes, worst case. Algorithms matter.

Also, what appears to be a bad algorithm choice under one set of circumstances might actually be the best choice under different circumstances. For example, suppose you have a completely random list of a million Social Security numbers and you need to place them in sorted order. One of the simplest sorting algorithms you can use is called the Bubble Sort. While it is simple, is it notoriously slow. Indeed, someone once said continents drift apart faster than the Bubble Sort sorts data. However, suppose you have a previously sorted list of a million Social Security numbers, but want to insert one new SS number into that list. In that case, the Bubble Sort is probably the fastest sort you can use.

So, how do you pick the "right" algorithm? "It depends" is the only honest answer. Also, as was the case in the sales reports, *how* you process a body of data does matter. Both algorithms generated the required reports, but one was more than seven times faster at performing the task. Generating an effective algorithm also implies a good understanding of the task at hand.

## Code Quality

Because of this disparity in code quality, I have different classes of code. BC, or *Bad Code* doesn't even compile. Such code suffers from syntax and semantic errors that need to be fixed before you can proceed. If you're new to programming, you should expect a ton of Bad Code...it's part of the learning process. Expect to make many "flat forehead" mistakes. (You know, mistakes where you slam the heel of your hand into your forehead while asking yourself how you could be so dumb!) Most good programmers have extremely flat foreheads.

The next level of programming code is RDC — *Really Dumb Code*. This is code that compiles and may even produce the correct output, but it begs for improvement. I worked as a consultant for a bank and found a construct where their processing code checked the date and then performed a specific task for each day of the month. The way the code was written, if the day is the 31st of the month, the code performed 30 false date checks before it took any action. This means that, on average, any given date would perform 15 bogus date checks per day for more than 7 million bank customers. I changed the code to two simple statements that would only perform one date check regardless of the date. That simple change saved them more than $26,000 in CPU time each year. Alas, I referred to this section of code in a code walk-through meeting as the best example of RDC I had ever seen. Unfortunately, that section of code was written by the person who hired me as a consultant and everyone in the room, except me, knew it. I was fired the next day.

I have one more class of code called SDC — *Sorta Dumb Code*. This code works correctly and is reasonably efficient. Truth be told, most of the code I write is probably SDC. I am certain there are programmers out there who could take my code and make it faster and use less memory. I'm not the sharpest knife in the drawer, and I know that. You might reasonably ask: So why are you writing programming books? I think the reason is: Because I am a little less smart than they are, I can more readily appreciate the problems people have while learning to program. That appreciation makes me a better teacher. I had an employee named Tim who was probably the best programmer I've ever known. That said, he was a horrible teacher because he couldn't understand why something wasn't obvious to everyone, including beginners. When I had my software company, I made the mistake of assigning an intern to Tim. She came into my office at the end of the first day, crying, and was trying her best to say something. All I got was "Tim's a monster. I quit!" before she stormed out of my office.

I suppose I should have a code class called WC for *Wow Code*. Wow Code is the kind of code that you look at and simply admire its elegance. I've only seen it a few times, but it sets a target for the kind of code I'd like to write.

Obviously, we've spent some additional time in this book discussing the Processing Step because it often separates the wheat from the chaff. If you're just getting started in programming, expect a lot of BC at the outset. Getting to the SDC stage is significant progress, especially since no one else needs to see your code. If it works to your satisfaction, that's good enough!

## Step 4. Output

Spending a lot of time and energy acquiring and processing data does little good if you don't do something with it. What you do with it depends in large measure on what you designed the program to do in the first place. Often the output, or result, of the program output step is simply displayed on the Serial Monitor or perhaps an LCD display. (The *Serial Monitor* uses the USB connection to display the program output on your PC.) However, it's not uncommon for a program to read a sensor and, depending on the reading, trigger some external action. You might have a temperature sensor in a chemical vat and when the temperature reaches a specific value, a signal is sent to open a valve that adds another chemical to the vat.

Program output can be sent to thousands of different electro-mechanical devices, disk data files, databases, printer ports, EEPROM, network connections, fiber optic networks...even other programs; the list is almost endless. The point is, eventually the old data takes on a new form and you are going to do "something" useful with that new data.

## Step 5. Termination

It's fairly common to think of the Termination Step as simply shutting the program down and it ends. Usually, however, a well-designed program does much more than that. If the Initialization Step opened a printer port or a database connection, this step should close that port or otherwise break the database connection. If the program maintains a list of recently-used data files, this step should update that list with the most-recently used file(s). Simply stated, the *Termination Step* often reverses what was done in the *Initialization Step*. A program that terminates gracefully returns the resources it used during program execution to the operating system. While our µCs don't really have an operating system, all of the µCs we're using have a program called a *bootloader* that allows the Arduino IDE to communicate with your computer via its USB link. It's just good programming practice for programs to terminate programs gracefully.

That said, many µC programs are written in such a way that they are never expected to terminate. Fire alarm systems are designed to run forever, as are many other devices from traffic lights to buoy systems floating in the ocean and space probes on their way to Andromeda. Many programs are designed to terminate only when 1) power is lost, 2) a component fails, or 3) there is a program reset. Most of our ham radio projects are expected to lose power when you leave the shack, so we do have some future comments about terminating a µC application.

## Anatomy of an Arduino Program

Every program written in the Arduino environment must have two basic parts to it. (We call them Arduino programs even though many projects in this book use μCs other than those from the Arduino family. All of them, however, use the Arduino IDE for code development and have the same startup function calls.) These two program parts are functions named *setup()* and *loop()*. When you select New from the File menu (i.e., File → New), the Arduino IDE provides you with a basic programming shell as a program starting point. You can see this in **Figure 2.4**. The purpose of the *setup()* function is to provide a starting point that defines the environment for your program. In other words, the *setup()* function often holds the code that performs the Initialization Step discussed earlier. The code in *setup()* is only expected to be executed once. *setup()* is automatically called first whenever you start or restart a program. The *setup()* function is never called again, unless you restart the program.

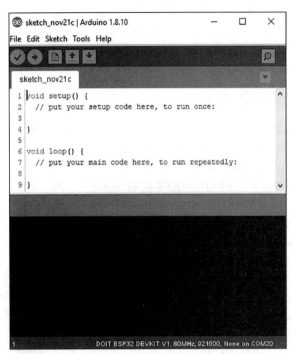

Figure 2.4 — The Arduino IDE program shell.

## Function?

So what is a function? A *function* is a small piece of program code that is designed to perform one specific task. **Figure 2.5** shows the syntactic parts of a function. The program statements you see in Figure 2.5 were borrowed from the Blink program as distributed with the Arduino IDE. You can access that code using the following menu sequence in the IDE: *File → Examples → 01. Basics → Blink*.

The *function type specifier* tells you the type of data this particular function can send back to a point in the program. For example, you might write a function that reads some data from a sensor. You can return that data item back to the point in the program where the function was called. (We explain what

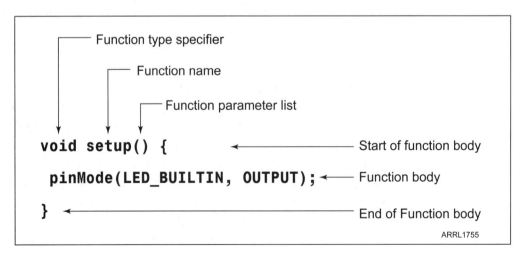

Figure 2.5 — Parts of a function.

"called" means later on.) The function type specifier tells you the type of data your function returns. The *setup()* function has a *void* type specifier, which means that this function is not written to return back any data value to whomever is calling this function.

After the function type specifier comes the *function name*. The name can be just about anything you want it to be, but it must start with a letter and cannot have any special characters, such as a question mark, comma, or similar characters. Function naming rules are the same as for C variable names. You also don't want to use function names that are already in use in the program. If two functions had the same name, the compiler wouldn't know which one to use, resulting in what's called a *name collision*.

As a programming convention, for functions that we write ourselves, we capitalize the first letter of the function. Most library and other standard functions (for example, *setup()*) use lowercase letters for the start of the function name. This makes it easier to know who wrote a given function — us or someone else.

After the function name comes a set of opening and closing parentheses, which form the starting and ending points of the function's parameter list. If your function needs outside data to accomplish its task, that data, called a *parameter list*, would be passed into the function as one or more variables that would appear between these two parentheses. The parameter list can contain zero variables, as is the case with *setup()*. An empty parameter list simply means the function can perform its task without external data being passed to it. Other functions may contain a list of one of more variables that it needs to fulfill its task. For example, *SetThermostat(72)* has a parameter that is likely the desired room temperature. If the parameter list is empty, it is often referred to as a "void parameter list" or a "null parameter list."

After the closing parenthesis of the parameter list is an opening brace. The *opening brace* marks the start of the function body. The *function body* is simply all of the program statements that appear between the opening and closing braces. Forgetting one of these braces is a common source of program errors for beginners. (Depending upon how you write the function code, the IDE usually adds the closing brace for you automatically.) It is the statements within the function body that actually accomplish the task at hand. The Blink program only has one statement as its *setup()* function body. Note that *pinMode()* is also a function and that it has a parameter list with two variables, LED_BUILTIN and OUTPUT. If a programmer were to read the code in Figure 2.5, they would verbalize it as: "The setup function calls pinMode." The word "*calls*" to a programmer means that program execution is transferred to that function momentarily to perform the specific task of the function.

Another term for you to learn is function signature. A *function signature* refers to the information about a function that starts with the function's type specifier and ends with its closing parenthesis of the parameter list. Therefore, the function signature for *setup()* is:

```
void setup()     // Function signature for setup()
```

The function signature conveys what we need to know to properly use a function. The function signature tells us the type of data that the function can give back to us (nothing for this function). The function signature tells us the function's name, and it tells us the type of information we need to pass to the function for it to work properly (again, no parameter list for the *setup()* function).

## Writing a Simple Function

Let's write a simple function that squares a number. Code **Listing 2.1** shows us how we might write a function that squares a number.

The top part of the function that begins with the /***** sequence is the start of the function header and that header ends with the *****/ sequence. Everything between these character sequences is called a *multiline comment*. The compiler ignores everything within the multiline comment. As a result, you can use multiline comments to document what a function or a statement does and it has no impact on the executable program's size, memory use, or execution speed.

Multiline comments can be added to a C program more simply by using a "/*" pair to mark the start of the comment and a "*/" to mark the comment's end. We just expand the comment sequence to make it stand out a little more. Within the comments we explain what the function does, the parameter(s) from elsewhere in the program that it may need (which can be nothing, or *void*), the type of data returned from the function, and any precautions that might be necessary when using the function. Because comments are totally ignored by the compiler, there's no reason to be terse in your explanations about the function. If the task itself is very complex, we often put a URL where the reader can find additional information on the algorithm used to perform the function's task.

The code for the *Square()* function works as advertised, but no self-respecting programmer would write the function using the RDC code in Listing 2.1. A more likely example would be the code shown in **Listing 2.2** (ignoring the function header for a moment).

The reason the code in Listing 2.2 is better is because it does away with three variables that we really don't need in the function. In both cases, if *number* is too large to square safely, –1 is returned to the caller. If the return value is –1, we can test for that value to sense an overflow condition since the square of a number cannot be negative. (The largest number a signed *int* data type can represent is limited because most Arduino μCs use only two bytes, or 16 bits, to store an *int*. A little math shows that $2^{15}$ is 32,768. The square root of 32,768 is 181. The reason we raise 2 to the 15th power rather than the 16th power is because one bit is used as a *sign bit*, telling us whether the number is positive or negative. If the *number* parameter is greater than 181, the result would overflow an integer data type and the world as we know it would end. Well...okay... that's an exaggeration, but nothing good is going to happen in your program.) If *number* is 181 or less, the square of *number* is returned. Now let's bring it all together into a simple program. The code is shown in **Listing 2.3**.

Note how we have left the *loop()* function totally empty. We did this because there's no reason to run this program again. Once we see the answer, the output won't change with a second run, so what's the point?

On the other hand, whatever code you put in *loop()*, it is run over and over

**Listing 2.1**
**An RDC Function to Square a Number**

```
/*****
   Purpose: A function that squares the number that is passed to
it and returns its square

   Parameter list:
        int number       the number to be squared

   Return value
        int              the square of the number

   CAUTION: The number cannot be larger than 181 without
overflowing an integer for a 16-bit int
*****/
int Square(int number)
{
   int a;
   int b;
   int result;

   a = number;
   b = number;
   if (number > 181)
      result = -1;                // An error
   else
      result = a * b;
   return result;
}
```

until one of the three terminating conditions is met. (Review the Termination Step above.) *loop()* is the coding equivalent of the movie Groundhog Day. The program statements in *loop()* are executed repeatedly without stop. Because the square of 9 has not changed much since math was invented, seeing the answer once seems good enough, hence putting all of the code in *setup()* instead of *loop()*. You might move the code into *loop()* as an experiment just to see what happens.

## How to Think about Functions

We want you to think of library functions as houses that have no windows, as depicted in **Figure 2.6**. Each house (function) has a front door, but there is no other way to get inside the house or see what's going on inside the house. Each function reminds you of Las Vegas — what happens in Las Vegas stays in Las Vegas. So it is with functions; what goes on in there doesn't concern you. If the function gives you the right behavior, that's all you need to know. From your perspective, each function is a mysterious black box.

**Listing 2.2**
**An SDC Function to Square a Number**

```
int Square(int number)
{

   if (number > 181)
      return -1;                    // An error
   else
      return number * number;
}
```

**Figure 2.6 — Functions as windowless black boxes.**

Why does the function name matter? For example, suppose you write a function named *ShellSort()* and use it in a program. A few weeks later, you discover that an Insertion Sort is faster in this application. So, what do you do? Do you write a new sort function called *InsertionSort()* and, after testing and debugging, go through all of your source code and change *ShellSort()* to *InsertionSort()*? Trust me, that's a train wreck waiting to happen. Now, quit looking at the trees, step back, and look at the forest. What's the purpose of the two functions from the user's point of view? Their goal is to have the data sorted...period. Do you think they're going to yell at you based on whether you use the Insertion sort over the Shell sort? They might, but only if they see the program's takes longer to execute. But that's an algorithm choice you make, not the user. In the final analysis, call your function *Sort()* and hide the implementation details inside the function. Now you can change the code in the *Sort()* function as much as you want, but the application source code remains the same—no error-prone search-and-replace.

Now let's look at the program in Listing 2.3. Because we know that *setup()* marks the Initialization Step (Step 1), we look in the function body of *setup()* and see that we define an integer variable named *value*. Next, we call the *be-*

**Listing 2.3**
**A Complete Program Using the Square() Function**

```
/*****
   Purpose: A function that squares the number that is passed to it and returns
its square

   Parameter list:
      int number    the number to be squared

   Return value
      int     the square of the number or -1 on error

   CAUTION: The number cannot be larger than 181 without overflowing an integer
*****/
int Square(int number)
{
   if (number > 181)
      return -1;       // An error
   else
      return number * number;
}

void setup()
{
  int value;

  Serial.begin(9600);
  value = 9;

  Serial.print("The square of ");
  Serial.print(value);

  value = Square(value);    // Call the function

  Serial.print(" is ");
  if (value == -1) {         // Is the value too big?
    Serial.println("too large.");
  } else {
    Serial.println(value);
  }
}

void loop()
{
```

gin() function, but because it is buried inside the *Serial* object, we are forced to call the function using the syntax *Serial.begin(9600)*. (More on this syntax structure in the next chapter.) Clearly, the *begin()* function has one parameter and we are setting it to the value 9600. As it turns out, 9600 is the baud rate for our communication link via the USB connection to the host PC. This tells your PC that the Serial object will be sending it data at a 9600 baud rate. Your PC will then adjust its serial port to that communication rate. If the PC and Serial object didn't sync the baud rate, your PC would think it's receiving gibberish for data and the program would fail.

Next, we assign the value 9 to the variable named *value* and that becomes the parameter for the function. Think of yourself walking along a street and stepping inside a house (function) named *setup()*. You take another step and define a variable in memory named *value*. You take another step and call a function named *begin()* by using a variable named *Serial* that is automatically made available to you via the IDE. You decide that all conversations between the μC and your PC take place at 9600 baud.

For the moment, we will skip over the statements that use the *Serial* variable. We explain those statements in detail later in Chapter 3.

Now we get to the program statement which is a function call to the function we wrote named *Square()*. Because *Square()* needs to know the number it is supposed to square, our *Square()* function has a parameter list that defines the number to be squared. Therefore, you take off your backpack (you always wear one when walking through a program!) and put a *copy* of the two bytes (lvalue) holding the variable *value* (rvalue = 9) into our backpack and speed off to find the black box named *Square()*. Upon finding the house named *Square()*, you knock on its door and things happen very quickly. The door to the windowless black-box house opens, a hand comes out, grabs your backpack, and withdraws with the backpack into the black box. A few milliseconds later, the door reopens, the hand shoots out of the door, replaces the backpack, retreats back into the black box and slams the door. Instantly you are transported back to the point in the program where the call to *Square()* originated.

Note: you have no clue what went on inside the *Square()* function. Again, functions are like Las Vegas: what happens inside the function is none of your business. What transpires inside the function is under the purview of the person who designed and implemented the algorithm. We want you to think of functions — especially library functions — that way. Functions are black boxes where you pass data in, some magic happens, and some data gets transformed.

The function call's syntax form is such that we are standing on the right-hand side of the equal sign in this statement:

```
value = Square(value);     // Call the function
```

Clearly we are supposed to take the value produced by the call to *Square()* and assign it back into *value*. Because the function type specifier for the *Square()* function is *int* (integer) the hand inside the black function box knew it had to put a two-byte integer result into your backpack before it sent you back to the *setup()* program code. So when you return from the function call to *Square()* and look inside your backpack, you discover that the hand had placed

> ## Another Algorithm for Squaring a Number
> 
> Quite by accident I discovered a different algorithm for squaring a number:
> 
> The sum of N odd integers equals the square of N
> 
> So, if you want to find the square of 3, sum up the first 3 odd integers:
> 
> 9 = 1 + 3 + 5
> 
> If you want to try some programming, try implementing this alternative algorithm for squaring a number. It's what I call an *oddity algorithm*: kinda interesting, but not very efficient. Still, if I wanted to I could use this algorithm in a *Square*() function and most people would never notice the difference. However, if I called the function *OddSquare*(), I'm sure someone would get nosy and wonder what's going on inside that black box.

the two-byte value 81 in your backpack. Your responsibility now is to take that new number 81 and stuff it into the two bytes in memory named *value*. That's what the equal sign means: take the value 81 returned from the call to *Square()* and assign it into *value*.

While it's a simplified explanation, that is essentially how all function calls work. If the function's signature has a type specifier named *void*, like *setup()* and *loop()* do, returning from those functions results in an empty backpack. If you call a function named *sin()*, it returns a floating point data type called *float*. Hence, *sin()* has a function type specifier of *float*. In that case, your backpack would contain four bytes of data that, collectively, form a floating point number (a number that can have a fractional value) that is the sine of whatever parameter you placed in the backpack to begin with.

Any time a function type specifier is anything but *void,* your return trip from that function is going to end up with something in your backpack. You can choose to ignore it, but that may or may not be a good idea. Also, C allows you to invent your own data types, too, and a properly-written function can return that new data type if you want it to. C is an incredibly flexible language.

## Declaring Variables in a Program?

Probably more than 90 percent of computer programmers are *very* sloppy when it comes to using programming terms (and you are *not* going to be one of them!). Perhaps the most misused word by programmers is the word "declare." It's common to hear (or read an article) where the programmer states: "I declared variable x as:

```
int x;
```

in the program."

*Wrong!*

The statement above is a *definition* of the variable *x, not* a *declaration* of variable *x*. There are very real differences between the terms "define" and "declare" in C. Writing, testing, and debugging C programs is easier if you understand the difference. Let's see why.

# What Happens When You Define a Variable?

The discussion that follows is a simplification of process that goes from program source code to an executable program. Still, it's a useful way to envision the process.

Let's pretend for a moment you are a C compiler and you come across this statement in a C program:

```
int val;
```

This is just about as simple a program statement as you can write in C, but consider what the compiler has to do to process that simple statement.

First, the compiler has to check and make sure the statement the programmer wrote obeys the syntax and semantic rules of the C language. Simply stated, C says that a *program statement* must have one or more valid C expressions which are terminated with a semicolon. Our program statement does contain a valid C expression for a data type specifier (the *int* keyword), an identifier (the name of the integer variable; *val* in this statement), followed by a terminating semicolon. Once the compiler reads the semicolon, it breaks that statement into its parts (data type specifier, id, terminating semicolon) and then checks the C language rules to see if that syntax structure is covered by the syntax rules for C. The statement passes the syntax check.

Second, it now must check to make sure you don't already have a variable named *val* defined in this program. If you do, you would get a "multiply-defined variable" error message from the compiler. The compiler keeps track of each variable in a program by using a symbol table. A *symbol table* is a compiler construct that maintains an attribute list and other information about each variable in a program.

Assume we already defined another integer variable earlier in the program named *y*. When the compiler checks the symbol table, it might look like **Table 2.1**. If there was already a variable in the symbol table with the identifier name *val* (at the same scope level), the compiler would issue a duplicate definition error. If the two variables are identical, which one should the compiler use? Since our symbol table only has variable *y* defined in it, there is no duplicate variable named *val* so the statement passes the syntax check.

In the third step, the compiler checks to see if the statement is semantically correct. Semantic errors can occur when the syntax rules are followed, but the statement is used out of context. For example, the syntax rules for a proper English sentence are for a sentence to have a noun and a verb. The sentence:

```
"The dog meowed."
```

passes the syntax rules (there is a noun and a verb), but the context is wrong — dogs don't meow. Therefore, the sentence contains a semantic error. Likewise, it is possible to create C program statements that have the wrong context (for example, trying to increment a constant). Our program statement does not contain a semantic error, so the compiler moves to the next step.

Now the compiler tries to allocate enough memory for

**Table 2.1**
**A C Symbol Table**

| Identifier | Data type | Scope | lvalue |
|---|---|---|---|
| y | int | 1 | 800 |

the variable your program is trying to define. If we think of the bootloader as a mini operating system, the compiler sends a message to the op system for a storage request. Because many microcontrollers use *int* variables that need 2 bytes of memory to store them, the compiler sends a message to the Op System saying: "Hey, Op System! My programmer needs to store a new integer variable named *val*. Have you got 2 free bytes of memory laying around?" The compiler scans its free memory list (in SRAM) and looks for 2 contiguous, unallocated, bytes of memory. If it doesn't have any free memory, it will send back an "out of memory" error message. Because microcontrollers don't have a lot of SRAM memory for variables, this happens more often than you'd like.

However, if the op system does have 2 free bytes of memory for the new variable, it sends back to the compiler the memory address of where those 2 bytes are located. Let's pretend that those 2 bytes start at memory address 625. When the compiler gets that information, it immediately places that address in the *lvalue* column of the symbol table. **Table 2.2** shows what the symbol table looks like now. Note that variable *val* now has its lvalue column filled in. The *lvalue (pronounced ell-value) of a variable is the starting memory address of a variable*. If we diagram the variable *val*, it would look like **Figure 2.7**. (A memory jogger for the diagram is to think of lvalue as the "left" leg in the diagram and the rvalue as the "right" leg.) We will explain what the rvalue leg of the diagram is in a moment.

Perhaps later in your program, you might write the statement:

```
val = 10;
```

This statement causes the compiler to form the number 10 (0b00000000 00001010 in binary as 2 bytes) and place those 2 bytes starting at address 625. Our diagram has now changed to that shown in **Figure 2.8**. The rvalue of *val* is no longer a question mark like it was in Figure 2.7. It's good programming practice to assume the rvalue of a defined, but unassigned, variable is a random sequence of bits (garbage) that contains nothing useful. In the statement above, however, the rvalue now holds the numeric value (10) assigned to it.

The *lvalue* in the symbol table tells the compiler *where* to place the binary value of *val* in memory. What you place at that memory address (10) is called the *rvalue*. Simply stated, *the lvalue of a variable is where a data item is located in memory and the rvalue is what is stored starting at that memory address*.

If you later write a program statement to copy the value of *val* into y:

```
y = val;
```

the compiler generates instructions to go to memory address 625, fetch the 2 bytes located there, and copy those 2 bytes starting at address 800 (the lvalue for *y*).

Think about it.

**Table 2.2**
**A C Symbol Table with the Memory Address Filled In**

| Identifier | Data type | Scope | lvalue |
|---|---|---|---|
| y | int | 1 | 800 |
| val | int | 1 | 625 |

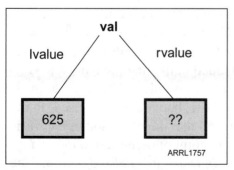

**Figure 2.7** — Diagram for defining a variable.

## The Bucket Analogy

Think of a program variable as a bucket positioned somewhere in memory. The *location* of the bucket in memory (its memory address) is that variable's *lvalue*. The *contents* of that bucket is that variable's *rvalue*. So, if you go to the lvalue 625 and peek inside the 2-byte bucket stored there, you now see the binary value 10 (its rvalue). This is illustrated in **Figure 2.9**.

Buckets come in various sizes. The C data type name *char* is a 1 byte data type, often used to hold a text character such as "A" or "B" or "?". (By the way, some people pronounce this as char, as in char a steak, while others pronounce it to rhyme with "care." Since we prefer to think of something as being kind rather than a burnt piece of meat, we prefer the "care" pronunciation.) If the symbol table defines a *char* variable, visit its lvalue (for example, memory address 700 in Figure 2.9) and you find a small 1-byte bucket holding an (rvalue of) "A". On the other hand, a *float* data type (used for data that may need a fractional value) requires 4 bytes of storage, so it has a 4-byte bucket. Note that each bucket is "sized" to fit its particular type of data.

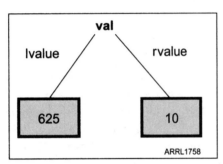

**Figure 2.8** — Determining the rvalue of val.

Note that it's usually safe to pour a 1-byte bucket (a *char*) into a 4-byte bucket (a *float* or a *long*) because you're going from a smaller bucket size into a larger bucket size. However, trying to pour the rvalue of a *float* bucket into a *char* bucket is going to slop 3 bytes of data all over the floor. (There are ways of coping with this situation, called a *cast*.) Most of the time, you want to use the same bucket sizes in C statements, especially assignment statements.

## What Happens When You Declare a Variable?

Suppose you have a project with multiple source code files. The primary source code file must have the two required Arduino IDE functions *setup()* and *loop()* and must have the .ino secondary file name (such as MyProject.ino). Suppose the second file is named MyMenus.cpp (a C Plus Plus, or C++, source code file) and a statement in that file needs to use variable *x* which you have defined in the first file named MyProject.ino. Now you have a problem. You have defined a variable in the first file, but you want to use it in the second file, too. Recall that the definition of a C variable extends from its point of definition t*o the end of the source code file in which it is defined.* Uh-oh. The Arduino compiler cannot "see" across multiple files. What do you do?

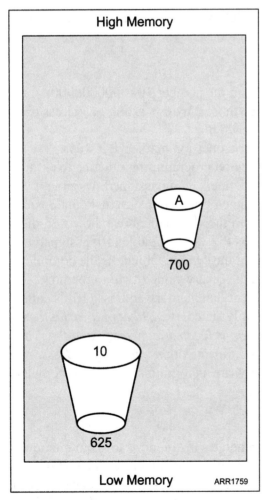

**Figure 2.9** — The Bucket Analogy diagram.

Think about it. What you really want to do is tell the compiler enough information about the variable defined in the first file so the compiler can let you use it in the second file. The "enough information" is actually the attribute list you see in the symbol table in Table 2.2. You can provide the necessary attribute list by doing a data *declaration* instead of a data definition.

A *data declaration* in C constructs an attribute list for a variable, *but does not allocate memory storage for it.* Read the preceding sentence 100 times over so it really sinks in. This statement means the entry in the symbol table has an empty lvalue column. You can construct a declaration attribute list using the *extern* keyword in the statement:

```
extern int val;
```

The *extern* keyword in the second source code file you are saying to the compiler: "This variable is defined in some other (external source code) file in this program, but let me use it in this file as an *int* variable named *val*." The symbol table entry would look like that shown in **Table 2.3**. Because *val* is not *defined* in this file, the symbol table leaves two question marks in the symbol table for the lvalue because it doesn't know where *val* has been placed in memory at this point in the compile process. So how can we find where *val* is in memory to be able to use it in this file? That is: Where is *val*'s bucket?

**Table 2.3**
**A Data Declaration**

| Identifier | Data type | Scope | lvalue |
|---|---|---|---|
| val | int | extern | ?? |

Another program element that's part of the IDE, called the *linker*, follows the compile process and resolves all of the "unknowns" (question marks) found during compilation. When the linker gets to the two question marks in the second file, it looks in its "Grand Symbol Table" and sees that *val* has an lvalue of 625. It then fills that unknown memory address in the second file with the lvalue of *val* from the first file. Therefore:

> A *data definition* always allocates storage for a variable and there can only be one data definition for each variable. A *data declaration* has the complete attribute list for a variable, but *does not* allocate storage for it. There can be multiple data declarations for a variable.

If nothing else comes from this chapter, memorize the section above. The statement above makes sense if you think about it: Why would you want the same variable located at two different places in memory in the same program? You could never be sure which one you were using. Nine-nine percent of the programmers say "Variable x is declared at the top of the .ino file." and they are just flat-out wrong. More likely the variable is defined in the file, not declared. Knowing the difference can make program testing and debugging a lot easier. (Also, you won't look like an uninformed programmer!)

If we ever meet in person and you mention you have our book, I may ask you to explain the difference between define and declare. Yeah...I'd do that. Woe to the person who doesn't know the answer! (Just kidding, but you really do need to get it right and distinguish yourself from those who don't know better.)

**An Overview of C**

## Libraries

Wait a minute! We used a variable named *Serial* in the *setup()* function, but I don't see *Serial* defined anywhere in the program code in Listing 2.3. How can we use a function we didn't even define? Very true and a good question, but the reason is easily explained. The compiler has a bunch of libraries from which it can draw code to enhance the functionality of your programs. A *library* is a collection of related, pre-written and tested, functions designed to perform specific tasks. The *Serial* variable (actually, it is properly called a class object) comes from one of those libraries and is designed to provide a communications link between your Arduino IDE program and your PC. An example will help explain how they work.

The C compiler thinks of a C library like a book. In this example, that book is named *Serial*. You (the compiler) open the book to its Table of Contents (TOC) and look down the list of chapters.

However, instead of chapter titles and page numbers, there is a collection of function names like *begin()*, *find()*, *print()*, *read()*, *readBytesUntil()* and many more. After each function name there are two sets of numbers instead of just a single page number you would find in a normal TOC. (See **Figure 2.10**.)

Consider the *print()* function. (Actually, we should call it a *method* rather than a function. That fuzziness is removed in Chapter 3.) Suppose the first number on the *print()* line in the TOC is 1050, followed by a comma, and the second number is 231. The TOC entry might look like:

```
print()      1050, 231
```

What this says is that if you skip over the first 1050 bytes in *Serial* library file (book) and drop the disk head down, you will be on the first byte of the *print()* function's library code. The second number says you must read 231 bytes to collect all of the program instructions needed for the *print()* function to work properly. The linker can now take those 231 bytes of code and stick them into your program at the appropriate place. Shazam! Your program now has the

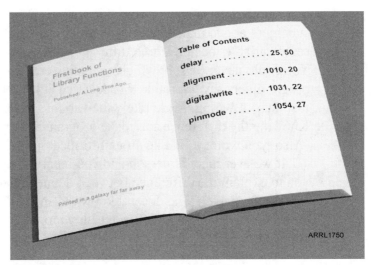

Figure 2.10 — Libraries are like a book's Table of Contents.

ability to send messages from your program to your PC without you having to write the method code to do it. How cool is that?

## How Many Libraries are There?

Part of the task of a beginning programmer is to discover all of the libraries that already exist. These libraries are fully functioning chunks of code, tested and debugged, waiting for you to use them in your own programs. The IDE is distributed with almost two dozen libraries that you can see listed in the Arduino IDE's *libraries* subdirectory. However, there are just under a bazillion other libraries that you can find on the web that can also be used in your programs. (I just looked on my system and I presently have 127 libraries in my primary library directory...there are others, too!) Most libraries written for the Arduino IDE are Open Source. *Open Source* is a collection of software donated by other programmers for you to use in your own programs free of charge. All of the code Al and I have written in this book is Open Source, so you are free to use it as you will. Pretty much all the Open Source license requires is that you give the authors credit for the code.

One of the first steps you should always do when writing a new program is an internet search on the intended purpose of the program. For example, suppose you want to develop a solar tracker using a μC to move the solar panel during the day. I just typed in "Arduino solar tracker" into a search engine and got 1.290 *million* hits. I'll bet one of those comes pretty close to what I want to do. With enough digging, you will likely find a library that makes your job just that much easier.

## How to Make a Library Part of your Program

Not all libraries are automatically included in your program just because you reference a function from that library. For example, suppose you want to use an LCD display in your program. The Arduino IDE does include a library that works with most LCD displays. The name of the library is *LiquidCrystal*. If we stick with our book analogy for a library, somehow you need to tell the compiler that you have placed the *LiquidCrystal* library on the bookshelf for use in this program. After all, the compiler does not automatically include the code in the LCD library in every program because not every program uses an LCD display.

The way to include a library in your program is to use the *#include* preprocessor directive. A *preprocessor directive* is an instruction that is processed *before* the compiler even begins its work. All preprocessor directives start with a # character. For our LCD library, the preprocessor directive is:

```
#include <LiquidCrystal.h>  // A library distributed with the IDE
```

It's called a directive rather than a statement because it's aimed at the preprocessor, not the compiler. (Note the absence of a semicolon at the end of the line and before the comment. C statements have a semicolon at the end of the statement and are used by the compiler as a statement terminator, remember? Preprocessor directives do not use a semicolon.)

The *#include* preprocessor directive causes the compiler to go to the Arduino *libraries* subdirectory on your computer and read the file named *LiquidCrystal.h*. Because the secondary file type is ".h", this type of file is called a header file. *Header files* contain details about the structure of the library code that the compiler needs to know in order to use that library in your program. Because C is case sensitive, calling the library <LiquidCrystal.H>, <liquidCrystal.h>, <liquidcrystal.h>, or <liquid crystal.h> all fail because one or more letters in the file name is incorrect. The compiler is incredibly fussy about such details. If you ever get a "library not found" error message, the first thing to check is the spelling of the header file name in the *#include* directive. Once you *#include* the library, the linker can use its TOC to locate and place the library code in your program.

Why the angle brackets (<, >) around the header file name? The brackets cause the compiler to look for the missing library file(s) in the default library directory (called *libraries*) of the IDE. Had you written the directive as:

```
#include "LiquidCrystal.h"   // A library distributed with the IDE
```

using double-quote marks instead of the brackets, the compiler would *first* look in the project directory where the program source code you are compiling is stored for the missing library. If you didn't store the library in the program's directory, it would *then* search the default library directory. Guess what happens if the compiler doesn't find the library in either place?

Why offer this library location option? Suppose you are using a graphics library and you are running out of memory. Graphics libraries tend to have a lot of code in them. The Arduino linker grabs all of the code for the library, even if you're only using a small part of that code, and puts it all into your program memory. Now suppose you realize that you could take some of the library functions you need from the standard graphics library, and strip them down to their bare essence, perhaps saving a significant amount of memory. Because you want the compiler to use your stripped down code instead of the standard graphics library, you put the stripped down code in the directory with your program and use double quotes on the header file instead of the brackets. Because the compiler finds what it needs in your directory, it doesn't bother searching the standard library files. Instead, it looks in your project directory for the Cliff Notes version of the library you edited. This allows you to use a custom version of a library, but still be able to use the non-customized version of the library with other programs.

The Arduino Reference Page (**www.arduino.cc/en/Reference/HomePage**) contains information on just some of the library functions that are available to you. Check online for available libraries. Some items that appear on that page are actually not functions, but rather objects that you can use.

Objects...what? Objects are the topic of the next chapter.

## Coding Conventions and Styles Used in this Book

Jack has taught university programming courses for almost 40 years and has picked up a few conventions and styles over the years that make the programming process a little bit easier to manage. While we are not saying these conventions are etched in stone, at least give them a try and see if they don't make your programming efforts easier.

## Variable Names

We always start variable names with a lower-case letter, then use camel notation (i.e. a capital letter on sub-terms) thereafter. Examples might be:

*myHatSize*   *printerPort*   *zipCode*   *ptrSisters*

Just like function names, variable names cannot start with a digit or punctuation character. Try to pick names that are long enough to be descriptive, but short enough not to be burdensome to type. In the book's narrative, we always print variable names in italics, for example *myHatSize*, so they stand out from the text. We also show C keywords in italics, such as *for*, to distinguish them from the narrative equivalent. Most good variable names take on a "noun-like" nature (a person, place, or thing).

## Function Names

Functions are code blocks you write with global scope and that are not defined within a class. Function names start with an upper-case letter and then use camel notation thereafter. Most function names assume a "verb-like" nature (they suggest some kind of action or process). Examples:

```
MyDelay()      ProcessMenu()      ReadEncoder()      PrintReport()
```

As a rule, when we refer to functions in the book's narrative, we include the parentheses after the function name, like *MyDelay()*, and we print them in italics. The parentheses for functions make it easy to differentiate between variable and function names. The reason for using an initial upper-case letter in the function name is so you know that these are functions that we wrote and are not part of some library (or other) pre-written code. Most authors of C programming libraries start class method names with a lower-case letter (for example, *Serial.print()* for the *print()* method). Therefore, seeing a lower-case method name tips you off that the code is found in a library or some other pre-written code (for example, *setup()* and *loop()*). Consider this data definition:

```
int (*mf[3])() = {Mine, Yours, Theirs};
```

Using our conventions, even if buried in a class definition, we can immediately tell that *mf[ ]* is an array of three pointers to functions that return an *int*. (Yeah, a little showing off here, but it does show how you can create complex data if you need to. If this is confusing, see Jack's Right-Left Rule: **jdurrett. ba.ttu.edu/3345/handouts/RL-rule.html**.) Because all of the function names begin with an upper-case letter, we know those functions were written by us and are not class methods buried in a library.

## Function Headers

Any time we write a new function, we add a function header block to provide information about that function. Consider this example:

```
/*****
  Purpose: This gives a one or two sentence description of what
  the function does.

  Parameter list:
      int myValue    value used to calculate the value (type
  specifier followed by variable ID)

  Return value:
       int          0 means an error, non-zero the answer for
  myValue (this could be void)

  CAUTION: If this function expects some precondition before
  it is called, this might say something like: "This function
  expects the Serial monitor to be open @ 115200 baud to function
  properly" Normally, functions should not depend on other
  functions or preconditions if at all possible (i.e., no coupling
  of functions).
*****/
int MyFunction(int myValue)
{
       // function body
}
```

The five asterisks after and before the slash make it possible to write a filter program that would create a DOC file for any functions we write. Also note that the opening brace ("{") for the function is on its own line. I was against this at one time, but since the IDE allows the editor to "match braces," it could be useful in debugging. Also, it makes it easier to copy the function's signature if you need a function prototype for it.

## Include Files

Any library header file not distributed with the Arduino IDE by default has the URL where you can go on the Internet to download the library as a comment on the same line:

```
#include <Rotary.h>       // https://github.com/brianlow/Rotary
```

This makes it easy for you to locate any non-standard libraries we might use. Preprocessor directives are usually placed at the top of the source code file that contains them.

### Positioning of *setup()* and *loop()*

These two functions should be the last two functions defined in the project's .ino file. Placing these two functions near the end of the .ino source code file makes it easier to quickly find the controlling functions for the project.

### Use *#define* Instead of *const*

To be truthful, some programmers disagree with our stance. Symbolic constants may be written using either of these two ways:

```
#define DISTANCE     125
const int DISTANCE = 125;
```

Both versions are seen in sample code and suit their purpose. We use *#define* in most cases for several reasons. First, symbolic constants do not take up compiler symbol table space. While this may not make any difference with controllers with large amounts of SRAM memory, it is a drawback. Second, and more importantly, symbolic constants are typeless. That is, the compiler does care if its data type is a *char, byte, int, long, float*...whatever. However, if you use *const* and you later decide that DISTANCE won't work as an *int*, but will as a *long*, you need to change its definition. With a symbolic constant, it doesn't matter. This means you don't need to cast it when it's used in a C expression.

### Multiple Code Files

Many Arduino-based projects use a single file, where are all of the project's source code is contained in a single .ino source code file (such as, MyProject.ino). We saw one project that was more than 10,000 lines of code in a single .ino file. Scrolling through that many lines of code looking for something is not a search, it's a career! Organizing that project that way is simply stupid.

The Arduino IDE supports multiple code files and makes it easy to work with them. **Figure 2.11** shows an example. The project's .ino file (the one that contains *setup()* and *loop()*) is *always* presented in the left-most tab of the project. All of the remaining project source code files are presented in alphabetical order in the tab list, going from left to right. This project happens to have two header files (files that end in "h"). Header files should not contain executable program code, but rather data declarations or library support information. All of the rest of the files are .cpp (C Plus Plus) files that contain program source code.

Note that all of the project's code files *must* be stored in the same project directory to appear in its own tab in the IDE. The Tab Control shown in Figure 2.11 can be used to add a new tab to the project. After you've named the new tab (*.cpp or *.h), you can then use the IDE to add the desired program statements to that file. You can use the *Five Program Steps* discussed earlier in this chapter as one way of chopping a large program into manageable chunks with each chunk having its own file and tab.

There is another powerful reason for using multiple project files. Recently, we were developing a project that had more than 11,000 lines of source code spread over 19 files. In the morning when we performed our first compile, it would take close to a minute to compile the project, and that's on a 8-core

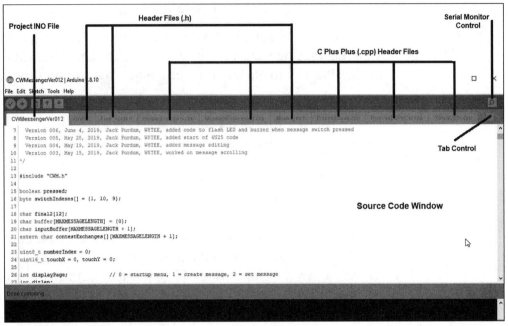

**Figure 2.11 — A project with multiple source files.**

system with a 3.8 GHz clock and 32 GB of memory. Now suppose that I'm debugging one of the 19 files and recompile the project. Now the second (and subsequent) compile only takes about 5 seconds. Why?

The reason for the shortened compile time is because the compiler is capable of incremental compiles. That is, if you've only changed one source code file, the compiler is smart enough to only compile the changed file and then link it together with all 18 of the other pre-compiled files. True, it only saves about 40 seconds, but if you do 50 compiles a day, that saves you over a half hour of thumb-twiddling each day!

Moral: On anything but trivial programs, it makes sense to use multiple source files. Use the Five Program Steps as a guideline.

### Write Cohesive Functions

A *cohesive* function is a function that is designed to perform a single task and do it well. Many beginning programmers try to write a function that performs like a Swiss Army knife; it does a lot of things, but none of them well. If you cannot explain to someone what your function does in two simple sentences or less, it's not cohesive. The function is too complicated and needs to be broken into multiple functions and simplified. As a bonus benefit, you will quickly find that cohesive functions more readily lend themselves to re-use in other programs because you're not dragging all this extra (code) baggage along.

Cohesive functions rarely depend upon sequencing or help from other functions you've written. In other words, avoid writing functions that require a specific sequence of function calls to get a task done. (This dependency on function call sequencing is called *coupling*. As a general rule, coupling functions is a bad thing.) True, there are some tasks that cannot be performed without sequencing (fir example, writing to a disk file requires a previous function call to open or create the disk file). If that's the case, make an entry in the CAUTION field of the function header discussed earlier. Use the CAUTION to inform the reader that a certain sequence of function calls is necessary.

### Group Similar Code Statements Together

If you have a header file with a bunch of *#define*s for symbolic constants and *#include*s for library files, place those near the top of the header file with the *#include*s first. Also group *extern* data declarations and function prototypes as groups in header files (we tend to alphabetize them within groups).

Within a code (.cpp) file, try to sequence the functions in the order that the programmer might call them (for example, *CreateMenuList( )*, *InitMenuList( )*, *SelectMenuItem( )*, *EditMenuItem( )*, *StoreMenuItem( )*). Some form of logical order should make it easier for the programmer to find things within the source code file. Alphabetical order by function name also works.

### Cascading *if* Statement Blocks Versus Switch/Case

A cascading *if* statement block looks like:

```
if (day == 1) {
   DoMondaysStuff();
} else {
   if (day == 2) {
      DoTuesdaysStuff();
   } else {
      if (day == 3) {
         DoWednesdaysStuff();
      }
            // ...and so on to Sunday...
```

This is a prime example of RDC. I already mentioned some bank code that used a cascading *if* for the 31 days of the month. Recall that the reason it's bad is because at the end of the month, you'll perform up to 30 unnecessary *if* tests before finding the one you want. Not good.

With a *switch/case* statement block, the same code looks like:

```
switch (day) {
   case 1:
     DoMondaysStuff();
     break;
   case 2:
     DoTuesdaysStuff();
     break;
   case 3:
     DoWednesdaysStuff();
     break;
            // ...and so on to Sunday...
```

When the compiler sees a *switch/case* statement block, it determines the value of *day* and then performs a jump to the memory location of the correct *case* statement block. In other words, the compiler creates a jump table of memory addresses and uses *day* to index into the jump table. In terms of the generated code, it uses one compare, a load instruction, and a jump. No RDC here.

## Interrupts Versus Polling

Anyone who's been around a young child knows what an interrupt is. An *interrupt* is a notification by the program that some aspect of the program needs immediate attention. Polling, on the other hand, means that the program is written to periodically check something to see if it needs attention. An example will help explain this.

Suppose you are charged with writing a fire alarm system for the Empire State building. You decide that each floor needs 100 fire sensors. The program sits in the basement and visits each sensor on each floor to check for a fire. If you visit a sensor and a fire is present, the sensor returns a 1 which causes the program in the basement to turn on the sprinklers, sound the fire alarm, and auto-dial 911 and the closest fire station. We'll assume a "sensor visit" takes one second to read the sensor and to return 1 (fire) or 0 (no fire). After all 100 sensors are visited on the first floor, the program increments *floorCounter* and starts with sensor 1 on the second floor. This process repeats for all floors, at which time it starts over on the first floor. (Here's a bug...it should start in the basement!)

Now, being the unlucky person you are, you read sensor 100 on the first floor and, just as you move to sensor 1 on the second floor, sensor 100 on the first floor bursts into flame! By the time you read the last sensor on the 108th floor, the fire has had almost a three *hour* head start! You'd be amazed how cranky people get when they have to wade through 10 floors of fire to go to lunch.

Suppose you had decided to install a different type of sensor. Suppose each sensor has the ability to sense the presence of a fire, but it can also issue its own pants-on-fire message to the main program in the basement. In other words, each sensor can interrupt the normal program flow with an urgent fire message. Now the worse-case delay is one second before the alarm sequence begins. This is the way interrupts work.

As we pointed out in Table 1.1 of Chapter 1, all of the microcontrollers we use have interrupt and polling capability, so which one should you use? If you're writing code that processes human reactions (such as turning a knob), from the point of view of the microcontroller, continents drift apart faster than humans react. However, if your program is auto-tuning where an out-of-tune condition will vaporize a set of $200 transistors within the span of a few milliseconds, it would make sense to use interrupts. There is no definitive or "right" answer. We use both approaches in our projects.

## Program Comments

As you know, you can use /* and */ to start and end a multi-line comment, or the // pair to make a one-line comment in a program. Because comments are ignored by the compiler, their presence has absolutely no impact on the size or execution speed of the program. This gives rise to the question: When should you use a comment?

We've seen programs that have comments on every line, even one like this:

```
b++;                    // increase the value of b by 1
```

Really? The person who wrote this thinks the reader is stupid enough not to know what the statement does? That is not a comment...it's clutter. Code filled with stupid comments like this makes it harder to read the code than if the code has no comments at all.

Good comments are short, informative, and to the point:

```
if (sensorValue == FIRE) {  // This code block turns on
                            //  sprinklers, sounds alarm, dials 911
```

and there are no more comments in the next dozen lines of code. Good comments point to code blocks that perform some task, or single out a complex section of code to explain what it does. I've even seen complex code (such as complex code doing some kind of FFT transform) that included a URL where the reader could go for more complete information.

While we're here, it's a good idea to place a comment at the top of the .ino file that presents a version number for the code, who wrote the code, the date, and any changes made to the code since the last entry. This is also the standard place for the Open Source comment requirement of naming who originally wrote the code. If you make a new version of the project each time you make major changes to it, it makes it easier to revert to an earlier (working?) version if need be without using a source code control program. It also makes it easier to blame someone else when things go south. (You did a backup save for each earlier version of the program is a safe place, right?)

As a rule, use enough comments to make it easy to find and read (and maybe debug?) the code by someone else. After all, six months from now when you read the code, *you* will be that someone else!

All of the above are suggestions, and you are free to adopt or ignore them all. Programming is a very personal endeavor and you should feel free to pursue what works for you.

## Conclusion

We covered a lot of ground in this chapter, but you should have at least some idea of what happens when you write a program and use libraries to make your life easier. As an experiment, use the following menu sequence to load the ASCII Table program:

*File → Examples → 4. Communications → ASCII Table*

Once the program appears in the IDE's source code window and *before* you run it, walk through the code line by line and try to verbally state out loud what you think that line does. If your significant other walks in while you're talking to yourself and gives you a weird look, get used to it. It happens to us all the time. Regardless, verbalize every line in the program.

Once you are finished verbalizing the program, compile, upload, and run the program. You can activate the *Serial* monitor by clicking on the "magnifying glass" icon you see on the upper-right side of the IDE. (See Figure 1.6 in the previous chapter.) A dialog box opens and the output of the program is displayed. Did the program do what you expected? If so, you are doomed...you're a programmer!

If you want additional details about C programming, there are a bazillion sources on the web for you to consider. That said, we think the book *Beginning C for Arduino* is one of the better ones, but we may be biased.

# CHAPTER 3

# A Gentle Introduction to C++ and Object Oriented Programming

## Why C++?

In this chapter we delve into C++ language, its classes, and C++ objects. Why? After all, everything you've read up to this point discusses programming using C within the Arduino IDE. True, but as you gain some experience using microcontrollers, you'll discover that the heavy lifting in program development is done by the programming libraries that support the microcontroller. Quite often, the code you write is really little more than syntactic glue that holds the library chunks together. Indeed, a major reason you don't hear more about non-Arduino IDE microcontrollers that have vastly superior performance characteristics is because they don't have the library depth the Arduino IDE microcontrollers have. Programmers are smart enough to realize that a skinny library base means more work for themselves.

Okay, but why make us even touch on C++? The reason is because virtually all of the libraries written for the Arduino IDE are written in C++. There are features inherent in C++ that can lead to more bulletproof code than straight C can offer. Once again, we find ourselves in the "build-a-car, drive-a-car" discussion we had earlier. If you're just going to *use* the library, why do you need to know how to dig around in it?

It's similar to you putting a bandage on a small cut on your finger. You're not a neurosurgeon, but you do know how to "fix yourself" within certain limits. Likewise, it's beneficial to understand how a library works with a "forest-for-the-trees" perspective. Even a limited understanding of the innards of a library will make you a better programmer. An appreciation of C++ helps you understand how the projects in this book actually work. Also, there may come a time when a program you're working on is right on the cusp of the resources afforded by the microcontroller. However, because you understand how libraries work, you also know there are features in the library that you don't need. With a little understanding, you can surgically remove the unnecessary parts, perhaps freeing up enough resources to complete your project.

We think the material presented in this chapter will help you to understand better how libraries work and result in you writing better programs. Indeed, some of the design goals that C++ programmers seek can be implement in "pure" C, provided you understand those goals.

## Object Oriented Programming (OOP)

You probably already know that C++ is a beefed-up version of C. If you've compiled and run at least one Arduino IDE sample program, you've already used C++. The Arduino IDE uses the Open Source GNU C++ Compiler (aka *GCC*) to convert your English-like C source code program statements into binary machine code instructions the core processors understand. The Arduino IDE convention is that the Arduino support libraries are almost always written in C++ rather than C. C++ has some very powerful features built into the language. However, with that enhanced power comes enhanced responsibilities. Someone once said that C gives you the power to shoot yourself in the foot, but C++ gives you the power to blow your whole leg off. Still, mastering that power is worth the effort.

C++ is based on the concept of objects. Indeed, that's what Object Oriented Programming (OOP) is all about. *C++ objects are just simplifications of things we see around us*. There are model airplanes based upon the Boeing 747 aircraft that a 10 year old child can assemble. We can look at that model and get some appreciation and understanding of what a 747 plane is. However, instead of assembling the 6 million parts that actually make up the 747, our understanding is derived from a few dozen parts. So it is with C++: C++ objects are just simplifications of things around us.

Essentially, our understanding of a C++ object can be reduced to two things:
1) Understand the attributes, or properties, that make up the object.
2) Understand the actions that the object does or can support.

Sticking with our 747 example, we know there are really more than 6 million parts in a 747, but we might simplify it to a fuselage, wings, engines, landing gear, tail assembly, number of passengers and crew, and cockpit. These simplified parts are the attributes, or properties, of a 747. As to what the 747 can do and support: it can take off, fly, land, cook meals, transport people and cargo, consume fuel, emit exhaust, and crash. Collectively, the attributes and actions that describe an object are comprise what is called a C++ class.

## What is a C++ Class?

We want you to think of a class as a set of blueprints. Those blueprints can show you how to build everything from a 747, a 100 story skyscraper, or a simple shed. Objects can be less specific, too, like a book, an invoice, or a paycheck. Let's consider a set of blueprints used to build a class named *clsHouse*. (It's fairly common to begin a class name with *cls*.) Our class needs to provide details on many things, like:

```
houseColor,
houseSidingType      // e.g., brick, stucco, vinyl, wood, etc
numberBedroom
numberBathrooms
numberFloors
kitchenCounterTopType
```

## Class Attributes

If you've ever built a house, you know there are thousands of things in the house from siding to wall switch plates that need to be determined by the builder. These items of the *clsHouse* class are call *class properties*, or *class attributes*. Collectively, the class properties, or class attributes, are what makes it possible to have houses that look completely different from one another, both inside and outside, even though all of them are still called houses. The properties that describe a Boeing 747 are vastly different than those that describe the Lockheed Martin F-35, even though both objects are aircraft.

## Class Methods

Sticking with our house example, houses have to be able to do and support things, too. For example:

```
toggleWaterMain()
toggleElectricMain()
toggleGasMain()
toggleKitchenSinkLight()
toggleBathroomVentFan()
adjustShowerHotWaterFlow()
toggleWaterHeater()
toggleGarageDoor()
heatTheHouse()
coolTheHouse()
```

Each of these "toggle" methods would have a parameter passed to it telling whether to turn the item on or off (for example, *toggleKitchenSinkLight(ON)*). Some are passed a parameter that sets a target state (such as, *heatTheHouse(72)*) for the item. These are the actions, or capabilities, that are provided by the *clsHouse* class. They have parentheses at the end of their names, making them look like the functions you studied earlier in Chapter 2. However, most OOP purists prefer to call these capabilities that are only available inside a class as *methods* rather than functions. Think of methods as being associated with C++ classes and functions being associated with "pure" C. (In terms of purpose, the distinction between C functions and C++ methods is somewhat artificial.) While our hypothetical list of class methods don't show any parameter lists, most of them probably have parameter lists. If the house has more than one bathroom, parameters to define which bathroom fan to turn on or off would make sense (for example, *toggleBathroomVentFan(MASTERBATH, ON)*).

Regardless of how you refer to them, *the purpose of a class method* remains the same: *To perform a single task and do it well.* Recall that this is exactly the same design goal for a pure-C function, too. We try to be consistent in using the term method for class actions and functions for straight-C functions.

If you assign values to all of the class properties and write the code for all of the class methods, you end up with a custom set of blue prints for your dream home. Alas, you can't start moving your furniture into a set of blue prints. You need to hand the blue prints to a builder who actually constructs the house from those blue prints. A class, therefore, is a simplification, or abstraction, that describes "the real thing." The class blue prints, therefore, are a *declaration* of a class object.

## Object Instantiation

The process of moving from the set of class blue prints to a "real" home you can actually live in is called *instantiation* of a class object. Until you instantiate an object of a class, you do not have anything that you can actually use in your program. You have *declared* the class (if it has a list of attributes and methods) with your set of blue prints, but you haven't *defined* an object of the class (no memory has been allocated for it, hence no lvalue that can be used) that you can actually use in your program.

If you look at the Display.ino program that is an example program for the *LiquidCrystal* library, you see the statement:

```
LiquidCrystal lcd(rs, en, d4, d5, d6, d7);
```

In light of what we just discussed, you might verbalize this statement as: "Go find the *LiquidCrystal* set of blue prints and instantiate a *LiquidCrystal* object named *lcd* and set the display's reset property to *rs*, set the enable property to *en*, and set the four data line class properties to *d4, d5, d6,* and *d7*." Once the compiler finishes processing that statement, you now have instantiated a *Liquid-Crystal* object name *lcd* that you can use in your programs. Stated differently, you now have an lvalue where the *lcd* object "lives" in your program...there is a "bucket" at that memory address waiting for you to fill it with data. This also means that you can use the *setCursor()*, *print()*, *clear()* and other methods of the *LiquidCrystal* object in your program.

Notice how *class properties are similar to nouns* in a sentence and *class methods are like action-based verbs*. The nouns-verbs concepts are worth keeping in mind when you start creating names for your class properties and methods. In fact, the noun-verb concept applies to "pure C" variable and function names, too.

## Class Constructors

Let's take a look at what happens behind the scenes when you instantiate (define) a class object. Let's consider the statement where we instantiated the *lcd* class object:

```
LiquidCrystal lcd(rs, en, d4, d5, d6, d7);
```

We explained how this instantiated the object and initialized six class attributes (or properties) with specific values. However, you do *not* have to initialize properties when you instantiate an object. For example, there is a class library that creates touch screen buttons called *MyButtons*. To instantiate a button object, you can use:

```
MyButtons myButton;
```

Anytime you instantiate a C++ object, you implicitly call a class constructor. It is the responsibility of the *class constructor* to define all of the properties and methods that are contained within the class. What's really interesting is that, even if you don't explicitly write a constructor, the C++ compiler does it for you! However, if you want to initialize some variables to specific values, then you need to write your own class constructor. If you don't use a parameterized constructor such as we did for the myButton object, the class constructor initializes the class attributes to a default value that is appropriate for the data type (usually 0 for numeric variables and NULL for textual data items).

The *lcd* object we instantiated earlier used four data lines as part of its parameter list. Because microcontroller I/O pins are often a scarce item, communication protocol libraries have been developed that use fewer I/O lines. One of those protocols is the *Inter-Integrated Circuit* (I2C) communications protocol. An LCD library supporting the I2C protocol for the Arduino IDE is named LiquidCrystal_I2C.

If you look in the LiquidCrystal_I2C.cpp source code file for the I2C class libraries for LCD displays, you'll find:

```
LiquidCrystal_I2C::LiquidCrystal_I2C(uint8_t lcd_addr,
   uint8_t lcd_cols, uint8_t lcd_rows)
```

The statement above is the class constructor for an object instantiated using the LCD-I2C library. (The two colons together (::) are called a *scope qualifier*. You can verbalize the scope qualifier as "belongs to" or "is a member of." That is, the *LiquidCrystal_IC2()* method belongs to, or is a member of, the *LiquidCrystal_IC2* class.) How can you tell that *LiquidCrystal_IC2()* is the class constructor? *A class constructor always shares the same constructor method name as the class name.* Notice the shaded names in the statement above; they are the same. If you look at the HelloWorld.ino program example in the *LiquidCrystal_I2C* library, you'll see:

```
LiquidCrystal_I2C lcd(0x27, 16, 2);
```

This statement instantiates a LCD object that uses the I2C interface. You might verbalize the statement as: "Call the *LiquidCrystal_I2C* class constructor to instantiate a *LiquidCrystal_I2C* object named *lcd* and assign it the I2C device number 0x27, which is an LCD display with 16 columns and 2 rows." That statement then calls the *LiquidCrystal_I2C* constructor described earlier (**LiquidCrystal_I2C**(*uint8_t lcd_addr, uint8_t lcd_cols, uint8_t lcd_rows*)). The end result is you have defined a *LiquidCrystal_I2C* class object named *lcd* that now exists in memory and you can use in your program.

Alas, we should mention that there are multiple libraries available that use this same library name. In those cases where a library is *not* part of the Arduino IDE default installation, we give you the URL where you can download it. The URLs are written directly into the program source code so you always have access to the proper URL.

We don't need to go into greater detail about class constructors, but you should at least know what they are and the role they play when instantiating an object. If you're interested, look online for "C++ class constructors." You'll find plenty of reference material.

## Using a Class Object

Let's examine our *LiquidCrystal* class object named *lcd* in a little more detail. Suppose you see this statement in a program:

```
lcd.print(value);
```

What's going on here?

We want you to think of the program you are writing as though it is a city street. You're at the top of a hill and you start walking down the street. The street has a bunch of houses on either side. The odd thing is: None of the houses have any windows. They are all painted black with a black roof. The only point of entry into each house is a single, stout, steel door on the front of the house. Above that door is a sign with some writing on it. That's it. Your program is like a street filled black house-shaped boxes, each of which is a class. It might look similar to **Figure 3.1**. Notice the *lcd* object on the left side of the street.

Now you walk up to one of the black houses and read the sign over its door. It reads: *lcd*. From outside, you can't see anything inside the *lcd* object house. All you can see is the word *lcd* painted above the front door of this windowless house. The front door has a massive lock on it and knocking on the door seems to do no good. Simply stated, all of the information and capabilities of the *lcd* object are locked and hidden away inside this windowless house and we have no clue what's going on inside. Everything about this object is hidden (encapsulated) inside that windowless house. What we need is a key that can unlock the massive door of the *lcd* object so we can get inside the house.

## The Dot Operator

Look closely at the C++ statement above again. See that little dot between the object name *lcd* and the method name *print* in the statement? That's called the dot operator. The *dot operator* is the key that opens the front door of a C++ object. *You use the dot operator like a key to open the door of a C++ object to gain access to the attributes and methods that are inside the object.* The dot operator is a universal key. That is, the dot operator is used to gain access to *any* C++ object. The expression on the left side of the dot operator (*lcd*) tells

**Figure 3.1 — Program classes and their objects are like black-box houses.**

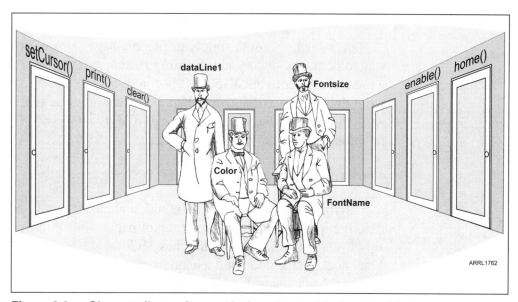

**Figure 3.2 — Class attributes (properties) resting inside the lcd object.**

you the name of the specific class object you are about to unlock (*lcd*). The expression on the right side of the dot operator (*print()*) tells you which attribute or method you wish to access. Read this paragraph until you are certain you understand what it says. That understanding makes the rest of the chapter easier to understand.

The first thing you see after walking through the door is a bunch of people sitting in a large living room. Each person has a name tag. Closer inspection of some of the name tags reveals names like *fontSize, color, fontName, charactersPerLine, numberOfLines, dataLine1, dataLine2*, and so on. These "people" are the properties (attributes) that "live" inside the class. It is their values that make this *lcd* object different from some other *lcd* object. See **Figure 3.2**.

You look around some more and you see a bunch of closed doors labeled *setCursor(), print(), clear(), home()* and many other doors with other labels on them. Each of these doors leads to a room that holds the library code for the method whose name appears above the door. These class methods work the same as functions did in your backpacking days when we discussed C functions in Chapter 2. However, unlike C functions, C++ methods are hidden inside a black house (a class object) and can only be accessed through the front door using the dot operator.

## Serial Object

The Arduino IDE automatically creates a Serial object each time you compile a program. That object is used to download the compiled code from your PC to the microcontroller via the USB cable. You can also use the Serial object to print things on your PC's display using the Serial Monitor that's part of the IDE. (See Figure 1.6 in Chapter 1.) Indeed, it is this Serial Monitor that serves as a primary means by which you can examine the values of variables in your program.

We have more to say about the Serial object in the next section.

## C Functions versus C++ Methods

Suppose you want to simply print a message on your display screen. One way to do this is study your operating system manuals, and find out how to write a character to the display. It becomes a simple task to extend this character write routine to one that can write a complete message to the display. So, you write the code for the function and call it *MyPrintMessage()*. To use your new function, you might write the code as:

```
charsWritten = MyPrintMessage("Microcontrollers rule!");
```

As you learned earlier in Chapter 2, code like this causes you to write the message in quotation marks on a slip of paper, stuff it into your backpack, and zip off to the function house that has *MyPrintMessage()* above the front door. You knock on the door, and a hand immediately opens the door, grabs your backpack, and slams the door. A few milliseconds later, it opens the door and returns your backpack to you. You peek inside and see the number 22 inside your backpack. You then run back to the point in the program that caused you to call on the *MyPrintMessage()* function, open the backpack, and hand the number 22 over to the variable named *charsWritten*. Then, *charsWritten* dutifully takes that number, jumps to its lvalue in memory, and sticks the number 22 inside its bucket. The rvalue of *charsWritten* is now 22. All of this happens in a few milliseconds.

As it turns out, you wrote your *MyPrintMessage()* function in such a way that it returns the number of characters it wrote to the display. (You just went back and counted the letters, didn't you?) Perhaps you wrote it that way so you could verify that entire message was sent.

That's what happens with a C function. Now let's perform the same task, but do it using C++.

The first thing you should do with any programming task is find out if someone has already written code for whatever it is you're trying to do. After a little internet digging, you discover something called the *Serial* object: **www.arduino.cc/reference/en/language/functions/communication/serial/print/**. You learn that the *Serial* object can send human-readable (ASCII) data a serial port. Because you know that the USB (serial) cable connects your Arduino IDE to your PC, this looks promising! A little more digging and you see a print method can be used with the *Serial* object, so you try:

```
charsWritten = print("Microcontrollers rule!");
```

which yields nothing on your screen other than an error message. Then you remember that the *print()* method is buried within the black, windowless, house named *Serial* and you have to use the dot operator to get inside. So, you try:

```
charsWritten = Serial.print("Microcontrollers rule!");
```

Alas, still no joy. Then you remember the Five Programming Steps and you wonder: Perhaps the serial port needs some information to properly talk from the Arduino IDE to the PC. A little more research unveils a *Serial* method named *begin()*, whose purpose is to establish the rate at which data bounce back and forth between the IDE and the PC on a serial communications link. So you try:

```
Serial.begin(9600);

// Probably a bunch more program statements...

charsWritten = Serial.print("Microcontrollers rule!");
```

You read the first program statement and realize you need to stick the number 9600 in your backpack. Then you run down the street looking for that house named *Serial* again. (Figure 3.1 has a *Serial* black house in it. The Arduino IDE creates a default *Serial* object for you that you can reference using the name *Serial*.) Once you find the Serial black house, you use your dot operator key to go inside the windowless *Serial* house and you look for a door with *begin()* written on it. You knock on the *begin()* door and immediately the door swings open, a hand shoots and grabs your backpack, and slams the door. A few milliseconds later, the door opens, hands you an empty backpack, and sends you back to the next program statement.

A little while later, you see you need to go back to the *Serial* house after writing the message on a piece of paper and placing it in your backpack. Once you're back at the *Serial* object house, you use the dot operator to open the door, walk in, but this time you look for a door with *print()* on it instead of *begin()*. Once you find the door, you knock, the door opens, grabs your backpack with the message in it. Almost immediately, the door reopens and shoves your backpack to you. You then return to that program statement that sent you to the *Serial* object, take the number 22 out of your backpack, find *charsWritten's* lvalue, go to that memory address, and place the number 22 in the bucket you find there. The *charsWritten* variable now has an rvalue of 22.

Taa-daa! The message now appears on the PC's display screen. (By the way, do you think the *Serial* object has a property named something like *portNumber*? Could that be the same as the port you need to set using the *Tools* → *Port* menu selection? Do you think that might be part of the Step 1 program process?)

The most important thing to take away from this discussion it that objects provide a means of organizing a program so it is easy to write, test, and debug. (The term "debugging" is the process of removing program errors. The term came about when one of the earliest computers suddenly stopped working. It was later found that a moth had crawled into the computer and shorted out some of the connections. Hence, debugging became the term for fixing errors in a computer and its programs.)

## Again...Why C++?

You may be saying to yourself: "I still don't see what C++ objects bring to the table that is any different than what plain old C functions can do." You know, I asked myself the same question back in 1987 when Bjarne Stroustrup, the inventor of C++, was the keynote speaker at the Software Development Conference in San Francisco. His talk was titled *C With Classes* and 750 of us were packed into the auditorium that afternoon to hear his keynote address. About an hour later, 749 of us filed out and said: "What the hell was that all about?" I didn't have a clue.

For almost five years, I avoided C++ because, despite talking with some of the brightest programmers on the planet, I still didn't understand what OOP was all about. About this time, I was doing some consulting and needed to work with C++ as part of the project. I was studying some C++ listings late one night and, suddenly, there was a flash in my brain and everything fell into place. Seriously, it was my first (and only?) epiphany and, at that instant, I understood was OOP was all about. I can only hope that everyone gets to experience such a moment in their lifetime.

But then the question arises: I'm not a stupid person, so why did it take almost five years to understand what OOP was all about? After all, I had some of the smartest programmers in the world explaining it to me, but...nothing. And therein, friends, is one of the main reasons why some people can't teach: They're too smart.

I've mentioned before how Tim, one of the brightest programmers I've ever met, reduced his intern to tears in less than 8 hours. The reason is because to Tim, and many really smart people in general, everything is obvious. Indeed, really smart people cannot understand or appreciate why it is difficult to understand something new. If I'm honest about it, one of the reasons I've been called a good teacher is because I'm dumb enough to have experienced the same difficulties as others in understanding new concepts. But, at the same time, I am smart enough to think of ways to understand and explain those concepts to someone else.

## The OOP Trilogy

The entire philosophy of OOP (and the benefits it brings to the programmer's table) can be summed up in what's called the *Object Oriented Programming Trilogy*. The Trilogy is Polymorphism, Inheritance, and Encapsulation. While all three concepts form the heart of the OOP Trilogy, it was Encapsulation that made the penny drop for me. Let's take a quick look at each element of the Trilogy.

### Polymorphism and Function Signatures

The word *polymorphism* roughly translates to "many shapes." In Chapter 2, we discussed the concept of a function's signature. In C++, method signatures have the same general interpretation as function signatures. The only difference is that methods exist within objects while functions exist by themselves outside any object. However, the parts of a signature remain the same and convey the same information.

#### Function Signatures

In Chapter 2, we discussed a sample function used to square a number with a function signature that looked like this:

```
int Square(int number)      // C function signature
```

The *int* keyword in the signature is the function type specifier. Recall that the *function type specifier* defines the type of data that is returned to the caller by the function. (The term *caller* actually refers to a point in a program where a specific function is activated, or "called.") What it says here is that the value returned from the S*quare()* function is an *int* data type. The function type specifier determines the data type of any data that gets stuffed into your backpack.

The word S*quare* is the name, or ID, of this function. Recall that function IDs follow the same naming rules as variable names. You must keep function names unique. That is, if you try to name a function of your own with the same name as a library function used in the program, you get a "duplicate definition" error. Because we are writing the function ourselves, we start the function name with a capital letter (such as, "S").

After the function ID is an opening parenthesis followed by zero or more function parameters. The purpose of a function parameter is to supply the function with whatever outside information it needs to perform its task. As we stated earlier, some functions don't need any additional information to perform the task, in which case the opening parenthesis is immediately followed by a closing parenthesis. Examples of an empty parameter list are the *setup()* and *loop()* functions you saw in Chapter 1. Those functions have *void* (empty) parameter lists. The S*quare()* function, however, needs to know the number that is to be squared, so that value is passed in as an *int* parameter named *number*. The closing parenthesis marks the end of the function signature.

So, if you pass this function the value of 5 for *number*, this function is going to return the value 25 to the caller as an *int*. Because the Arduino C compiler uses 2 bytes for each *int* variable, your backpack enters the function with 2 bytes that form a binary *int* with the value of 5. When you leave the function, your backpack still has 2 bytes in it, but the value of those 2 bytes has been changed to 25.

## What Size *int*?

In Chapter 2 we mentioned that the Arduino microcontrollers use an 8-bit architecture. However, the Teensy, STM32, and ESP32 all use 32-bit architectures. As a result, while a Nano *int* is 16 bits, the other three microcontrollers all use 4 byte, or 32-bit, *int*s.

How can you tell what the byte size is for an int for the microcontroller you are using? Assuming you have the *Serial* object instantiated in your program, just add this line at some point after the *Serial.begin()* statement:

```
Serial.println(sizeof(int));
```

While the *sizeof* operator looks like a function call, it really isn't. (Formally, *sizeof* is actually a parameterized macro.) The *sizeof* operator can be used to determine how many bytes have been allocated to a defined data variable or object. In the example statement above, we are using a *Serial* object to print out the number of bytes allocated to any *int* data type. Note that *println()* method of the Serial object is a little different from plain old *print()* in that *println()* causes the next call to print its output on a new line. Without the addition of the "*ln*" in the method name, the next *print()* call prints its output on the same line.

## Method Signatures in C++

If we were writing the same square function for a C++ program, the method signature might look like:

```
int math::square(int number);    // A C++ function signature
```

In this case, we see that the *square()* method is a *member method* of the *math* class. Someone has already written a bunch of math methods (*sin()*, *cos()*, *log()*, *pow()*, and so on) for you, so there's no reason for you to write, test, and debug your own version (other than as a learning exercise.) You might verbalize the *two colons* as: "class has a member method named." Verbalizing the entire method signature becomes: "The *math* class has a member method named *square()* that is sent an *int* parameter named *number* and returns an *int* value to the caller."

Polymorphism adds a new wrinkle to the way methods can be used in a program. Simply stated, *polymorphism lets you reuse the same method name, but forces you to change the parameter list of the method*! C won't let you do that. Consider these program statement lines:

```
char message[] = "Microcontrollers rule!";
int number = 5;
int answer;
float pi = 3.14;

Serial.print(message);
answer = square(number);
Serial.print(answer);
Serial.print(pi);
```

Think about it. The first call to the *print()* method of the *Serial* object is passed a *char* array named *message[]* as its parameter. The second call to the *print()* method is passed an *int* as its parameter. Finally, the last *print()* method

## Function Names in C

The astute reader is probably wondering why the function named *Square()* in Chapter 2 starts with an uppercase "*S*" whereas the class method named *square()* discussed here uses a lowercase "*s*". (We touched on this distinction in Chapter 2.) I'm sure someone has a good explanation, but I have yet to find one. However, C got its start in the early 1970s, but there was no working language standard for C until 1989. That's plenty of time for habits to form. From that date on, however, C and C++ had very formal language definitions as laid out by the American National Standards Institute (ANSI). One statement in the C++ standard is that C++ method names (identifiers) will start with a lowercase letter. My guess is that someone said: "Well, if that's the case for C++ methods, let's make plain old C functions start with an uppercase letter."

We try to remind ourselves to start C function names with an uppercase letter and C++ methods with a lowercase letters, but the best laid plans of mice and men...

call is passed a *float* data type as its parameter. Think what this means: All three calls have the same method name, *print()*, but each must have a different parameter list. Those methods must be written with the following method signatures:

```
int Serial::print(char msg[]);
int Serial::print(int val);
int Serial::print(float num);
```

The method type specifier is the same data type (*int*) for all of them, the method name is the same, but the parameter lists are different, as shown by the shaded areas. Polymorphism: Many shapes (different parameter lists), one name (same method ID). While you may not appreciate it right now, polymorphism makes it *much* easier to write and use program libraries. As long as the method signatures are different (i.e., different parameters) methods can share a common name and type specifier. This means you have fewer method names to learn!

## Inheritance

*Inheritance* is a feature of C++ that allows you to declare a set of common denominator properties and methods for one class and use those properties and methods to create new classes. For example, some time ago I was asked by a real estate investor to write a program that would track their real estate holdings. The investor had three basic types of buildings: residential, commercial, and apartments. Each type of rental property had its own special considerations. While the number of bedrooms affects both residential and apartment properties, it has no impact on the commercial properties. Likewise, commercial properties had to have so many parking places, of which some fraction had to be for handicap parking. Also, bathroom facilities in commercial properties were affected by the square footage of the building. There were even snow removal restrictions that varied by property type as well as township location. So, how do you minimize the complexity of the software?

You can reduce the complexity by looking for common features for all property types, and worry about the details a little later. For example, each property has a number of attributes in common: an address, property taxes, purchase price, insurance cost, mortgage lender, mortgage amount, and so on. We could create variables for these attributes of a property and place them in something called a *building* class. We can take all of the attributes that are common to all properties and hide them in the black, windowless, house named *building* as shown in **Figure 3.3**. These class properties become the "people" living within the black house. In the figure, *building* holds all of the properties and methods that are common to all three property types.

We now create three new classes, one for each property type (residential, commercial, apartment). Each new class holds the properties and methods that make it a *unique* building and differentiates it from the other

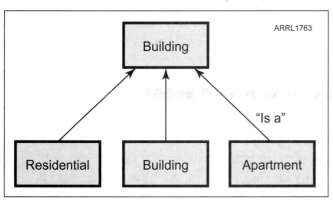

**Figure 3.3 — Building types.**

building types. For example, the residential class might have an attribute named *squareFeetInFinishedBasement*, which likely does not apply to the other two classes. Likewise, a commercial property might have an attribute named *handicappedParkingSlots* that doesn't apply to the other classes.

What Figure 3.3 illustrates is that a residential, commercial, and apartment buildings are all a special type of a base class called *building*. In other words, these three special types of building "inherit" all of the basic elements shared in common for all buildings. The arrows pointing from the three specific building types to the common building type is called an "is a" relationship and says that each of those special building types inherits all of the traits (attributes and methods) of *building* class. That is, a *residence* "is a" type of *building,* as are *commercial* and *apartment* buildings.

### Base and Sub Classes

OOP jargon often refers to *building* as the *base class* and the three building types as *subclasses* of the base class. (Some programmers prefer to call the base class the *parent class* and the subclasses as *child classes*. Pick whatever makes sense to you...the meaning is the same.)

So, what does inheritance bring to the party? Note that, instead of writing all of the program code needed to track property taxes, mortgages, addresses, and other common attributes for each type of building, we can push all of the code for those common attributes and methods into a "common denominator" class (the base class) and simply let each subclass inherit and use those members variables from the base class. The same is true for many of the methods that are common to all properties (*payMortgage()*, *payInsurance()*, *payPropertyTaxes()*, *payBribes()*, and so on).

Perhaps another example would help. You could have a base class named *clsCookieCutter* and subclasses named *clsStar, clsWreath, clsChristmasTree,* and *clsCandyCane*. Each cookie cutter has a metal frame, sharp metal edges, and a handle of some sort affixed to it. While each subclass shares these attributes, the angles of the metal are different for each on and they are constructed slightly differently.

However, until you take the desired class cookie cutter (such as *clsStar*) and push it into the cookie dough (the memory), only then have you instantiated a cookie that you can bake and, ultimately, eat. It is through the act of instantiation that you define something you can use (or eat!).

By using inheritance from the base class, you've reduced the number of lines of program code for tracking those variables and methods by two-thirds what they would be otherwise. Reducing the lines of code you have to write also means fewer lines to test, debug, and maintain...that's a good thing.

## Encapsulation and Scope

To us, encapsulation alone makes learning OOP worth the effort. Indeed, it is the feature that led to Jack's epiphany moment. Also, encapsulation is the one aspect of the OOP Trilogy that you can practice using plain old C. Practicing the tenets of encapsulation in *any* language is going to make you a better programmer.

Simply stated, *encapsulation is the practice of data hiding*. Data hiding

**Listing 3.1**
**Illustrating the Concept of Scope**

```
int pervasive;              // Defined outside all functions

void setup() {
  int count;                // Defined within a function
  for (int index = 0; index < 10; index++) {   // Defined within a statement block
        // code lines for loop body
  }        // Point C
}          // Point B

void loop() {
}
// Point A
```

makes it more difficult for other parts of the program to "contaminate" the data. To appreciate the concept of encapsulation, you need to know about program scope. The *scope of an program entity (a variable, function, or method) refers to both its visibility and lifetime in a program.* We can best illustrate this with a skeletal program listing. Consider **Listing 3.1**.

### Global Scope

First, note how the variable named *pervasive* at the top of the program in Listing 3.1 is defined outside of any function. Any variable defined outside of a function is said to have global scope. *A program element with global scope "lives" and is visible from its point of definition to the end of the source code file in which it is defined.* This means that the scope for *pervasive* extends from the semicolon of the first line in Listing 3.1 to last line in the program (point A in Listing 3.1, just after the closing parenthesis for the *loop()* function). This also means that *any statement* between these two limits can "see and has full access" to the variable named *pervasive*.

When program execution reaches Point A in Listing 3.1, the variable named *pervasive* "dies." The variable *pervasive* at Point A is dead and no longer accessible. Any expression that tries to access *pervasive* after point A will draw an error message from the compiler.

I can hear you saying: "Well, its death is kinda obvious since Point A is the last line in the program." Well, that's true and most of the program examples have a single (.ino) file that defines the entire program. However, the Arduino IDE lets you build programs that may be sufficiently complex that it makes sense to spread the program out over multiple source code (such as .cpp) files. We talked about multiple source code files in Chapter 2, remember?

Notice near the top of **Figure 3.4** there is a tab named JackAlV068. The left-most tab in the Arduino IDE is always the "main" program file. That is, that file always has the secondary file extension .ino (JackAlV068.ino) and it always contains the two functions every Arduino program must have: *setup()* and *loop()*. The JackAlV068 program, however, is fairly complex consisting

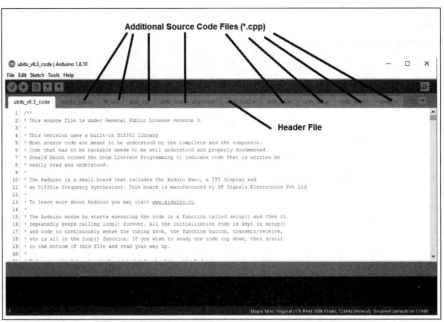

**Figure 3.4 — The JackAl source code.**

of about 11,000+ lines of C source code which compiled to more than 210,000 bytes of program code. Because of this complexity, we spread those 11,000+ lines of source code over 18 additional source code files. The names of each source code file suggests its purpose in the program (for example, *AddAllButtons*) and the 19 files each have their own program tab in Figure 3.4. (You can't see them all without scrolling the Source Code Window of the IDE.) Also note that each additional file has the secondary file name of .cpp (C-Plus-Plus, *AddAllButtons.cpp*) or .h (a header file, *JackAl.h*).

Okay, so what's the point? The point is *pervasive* dies at the end of the main .ino file, but what if you need to use *pervasive* in source code file number 18? The scoping rules of C says that the visibility (accessiblity) and life of a global variable only applies *from its point of definition to the end of the source code file in which that definition occurs.* More formally in the terms used in Chapter 2, the lvalue and rvalue of *pervasive* are unknown to the other 18 source code files.

How can you fix that? There are several ways. First, you could pass *pervasive* as a function parameter to whatever function in source code file 18 needs it. That is, throw *pervasive* in your backpack and trundle off to file 18 and let the function use your backpack to get the contents of *pervasive*. If that function in file 18 uses *pervasive* to generate a new number, the function can stick that new value into your backpack and send you back to use that variable in the main .ino file.

The second way is to add a new line to the top of any file that needs to use the global variable:

```
extern int pervasive;
```

The C keyword *extern* essentially is saying to the compiler: "There is an *int* variable named *pervasive* which is defined in some other source code file, but please let me use it in this file as an *int* data type named *pervasive*." In other words, this is *not* a data definition of *pervasive*, it is a data *declaration* for *pervasive*. As you learned in Chapter 2, a data declaration creates an attribute list for a variable, but does not allocate storage for it. It is the attribute list (*int* and its ID, *pervasive*) that lets you use *pervasive* properly in source code file 18.

Wait a minute! If file 18 doesn't have an lvalue for *pervasive*, how can it have access to it? After all, it's the lvalue that lets you know where a variable resides in memory and gives you access to its rvalue. Without knowing a variable's lvalue, there is no way to use its rvalue. Actually, the compiler solves this problem by working in cahoots with another program called *the linker* that is part of the compile process. (We touched on this in Chapter 2.)

When the compiler see the *extern* declaration for *pervasive*, it makes a note to itself to allow program statements to use *pervasive* as an *int* in this source file. However, instead of filling in the lvalue for *pervasive*, the compiler leaves two question marks in the compiled file everywhere *pervasive* is referenced. (Yeah, it's a simplification of the process, but conceptually correct.) The reason the compiler uses two question marks is because memory addresses on most Arduino microcontrollers use two-byte memory addresses. (Most 32-bit processors use 4-byte addresses for variables so the compiler would leave four question marks in those cases.)

Now the linker program comes along and sees the two question marks for *pervasive*. The linker then looks in its own "super" symbol table for *pervasive* and, upon finding it, puts its lvalue where the two question marks used to be. Now file 18 has complete access to *pervasive*. If the linker did not find *pervasive* in its symbol table (perhaps you misspelled it), it would issue an undefined variable error message.

Therefore, we can make a variable have global scope even across multiple files if we need to. As we will see in a moment, global scope may not be such a good idea.

**Function Scope**

Look back at Listing 3.1 and find the definition of *count*. Because *count* is defined within the *setup()* function, it can't have global scope. Instead, *count* is defined with function scope. *Function scope for a variable extends from its point of definition to the end of the function in which it is defined.* For variable *count*, its scope extends from semicolon at the end of the first line in *setup()* to the closing brace of the *setup()* function, or Point B in Listing 3.1. In other words, the scope of a function scope variable extends from its point of definition to the end of the function in which it is defined. After Point B, *count* dies and is no longer visible in the program.

Note what this means: No other file or function even knows that *count* exists in the program. That is, we have encapsulated (hidden) the variable named *count* inside of the function named *setup()*. No other part of the program other than the code in the *setup()* function can access *count* directly.

The implications of this are huge! If *count* suddenly takes on some unexpected value, we know *exactly* where to go to figure out what went wrong: we look in the *setup()* function because no other part of the program has access to *count*. If *count* were defined with global scope, we really have no clue where to start looking for the problem. By limiting the scope of *count* to the *setup()* function, we have substantially reduced the number of source code lines to search if a program bug is causing mischief with *count*.

**Statement Block Scope**

Statement block scope is even more restrictive than function scope. *Statement block scope extends from the variable's point of definition to the end of its encompassing statement block.* In Listing 3.1, variable *index* is defined immediately after the opening parenthesis of the *for* statement block:

```
for (int index = 0; index < 10; index++) {   // Defined within a
                                             // statement block
    // code lines for loop body
}   // Point C
```

as indicated by the shaded part of the *for* statement block. The scope of *index* extends from its point of definition to the closing brace of the *for* statement block, or Point C above. Suppose we write this:

```
r (int index = 0; index < 10; index++) {   // Defined within a
                                           // statement block
    // code lines for loop body
}   // Point C
Serial.print(index);
```

where we try to print the value of *index* on the *Serial* object. What would the compiler do? It would issue an undefined variable error at the *Serial.print()* statement call because *index* died when the closing brace of the *for* statement block (Point C) was reached. To the compiler, *index* doesn't even exist anymore by the time execution reaches the last statement.

Statement block scope is encapsulation with a vengeance. However, it makes it really easy to determine why *index* might have a bogus value, since any change in the value of *index* could only occur within the confines of the *for* statement block.

## Scope and Debugging

As you might guess, debugging a program is more difficult when you use a lot of variables defined with global scope. Suppose we had kept the JackAl program as a single file program instead of breaking it out into 19 source code files as we did. Further assume that the definition of *pervasive* appears at the very top of the file. Suppose that, suddenly, *pervasive* takes on some really weird value as the program executes. Because *pervasive* has global scope, any one of the 11,000 program statements could be the cause of the bogus value. Not good! Reducing the scope level from global to function (or statement) block level substantially reduces the search time necessary to isolate and fix the bug.

C++ simply takes this one step further by hiding all relevant data inside an object. You must be very deliberate when accessing object data because you must use the dot operator to gain access to the data. In gaining that access, you must at least consider which object is being use to access a member property.

Limiting the scope of your data is a good thing because it makes it easier to test and debug your programs. You can often avoid globally scoped variables by using function scope and passing the variable as a parameter to whatever function needs to have access to it. While that may seem like a little more work, the time you save in testing, debugging, and maintaining the code more than offsets that additional effort. Studies suggest that, when developing a new program, 20% of the time is devoted to the design and writing of the program and 80% of the time is used to test and debug the program. Anything you can do to hide (encapsulate) your data cuts into that 80% figure.

### Assigning and Retrieving Values of Attributes

It does little good if the class attributes and methods are buried within the class object in such a way that you can't access them. For example, suppose you have instantiated a graphics display object named *myDisplay* and the class has a bunch of attributes and methods that you can use. Let's further assume that you want to set the text font size (for example, *fontSize*) to 4. To assign the value of 4 to the graphics class member named *fontSize*, you would use the following statement:

```
myDisplay.fontSize = 4;
```

You should be able to explain the statement using the backpack analogy: Take the value 4 and stuff it into your backpack, look for the black, windowless, house named *myDisplay* and insert the (dot operator) key into the lock and open the door. Look around the room for the person wearing the *fontSize* name tag and hand him your backpack. He will reach into the backpack, pull out the value 4 (probably as an *int* data type) and place it on his lap. You retrieve your backpack, go through the front door, closing it securely behind you, and you jump back to the beginning of the next program statement. Job done.

Now suppose you want to find out the current font size that is being used on your display and assign it into a variable named *currentFontSize*. The statement to do that is:

```
currentFontSize = myDisplay.fontSize;
```

Once again, you should be able to "backpack-explain" what's going on: Go to the *myDisplay* object and use my (dot operator) key to gain access to the house, look around the room inside the class object for Mr. *fontSize*, approach him and say: "My programmer needs to know your rvalue." This question causes Mr. *fontSize* to rise out of his chair, walk over to the copy machine, and make a copy of the value that is in his lap (4). He then grabs your backpack and stuffs the *copy* of the value into it. You shoulder the backpack, exit through the door, locking it behind you, and are immediately transferred back to the assignment operator ("=") point in the code. You then take out the contents of your backpack (4), and hand it to *currentFontSize*. He then takes that value, goes to

**Table 3.1**
**Assigning or Retrieving an Attribute Value**

| Location of object expression | Effect on Attribute | Example |
|---|---|---|
| Left side of equal sign | rvalue changes | `myDisplay.fontSize = 4;` |
| Right side of equal sign | rvalue unchanged | `currentFontSize = myDisplay.fontSize;` |

his lvalue (where his bucket is located in memory) and places the value (4) at that memory location so his rvalue is now 4. That is, *currentFontSize*'s bucket now contains the value 4.

Note that we said that, when you leave the *myDisplay* object, your backpack contains a *copy* of the rvalue of *fontSize*. While it may be obvious to everyone, it's important to note that assignment statements are not "destructive". That is, we can fetch a class attribute value, but that attribute (or variable if it's a C function) still retains that value. Only when the object expression (for example, *myDisplay.fontSize*) is on the left-hand side of the assignment statement is the value of the attribute changed. **Table 3.1** shows the general syntax of assigning or retrieving a class attribute.

Although it's probably obvious, it's good programming practice for the data types in the statements to be the same. That is, the data type of *currentFontSize* should be the same as *fontSize*.

## An Explanation of Function and Method Calls... Sort Of

In earlier discussions, we often used the terms "function call" and "method call." What does it mean "to call a function or a method"? We've sort of explained how a function or method call works, but a little more detail should be useful in helping you understand how a program works. We've taken a few liberties with the actual mechanics, but this section should help you understand how method calls work. (We frame the discussion in terms of methods, since many of your program calls are to C++ libraries that hold the object's methods. However, the process is much the same for functions, too.)

### Program Flow

Think of yourself at the start of a sidewalk. Each concrete slab that forms the sidewalk has a piece of paper on it that is a program instruction. Looking down the street you see an endless stretch of sidewalk slabs with a program instruction on each slab. A bell rings and you pick up the first instruction and read it. Perhaps it says *#include <Wire.h>* so you dutifully read the contents of the Wire.h header file into the program at that point. Since that instruction is done, you move to the second sidewalk slab, pick up the instruction and do what that instruction tells you to do. The process repeats itself, advancing down the street one slab at a time.

Now you're at slab 99. You pick up the instruction and it says to call the *Serial.print()* method with the function argument "Type in your name:". You place the message in your backpack and start walking down the street looking for a house with the name *Serial* above the door. Actually...we lied to you in the earlier discussion, mainly because we didn't think you were ready for what really happens.

What really happens is that the compiler can't stand to see you waste time walking down the street looking for the *Serial* object's house. Instead, when the compiler saw the *print()* method call, it put two "address question marks" in the code. It had to do this because it did not know where the code for the *print()* method would end up in memory. However, later on the compiler does generate the program code for the *print()* method. However, it doesn't "backup" and fill in the address question marks. Instead, it's the linker's responsibility to change those two question marks into the actual memory address where the *print()* method code begins (its lvalue).

So there you are, standing on slab 99 with "Type in your name:" in your backpack and the instructions tell you to "jump" to slab 234 to find the *Serial* object's house. There's no time wasted waddling down the street to slab 234. Nope. You're able to make that jump "in a single bound" and land right on slab 234. But, just before you perform your jump, the program code tucks away the next slab number of your next program instruction, slab 100 in this example. It tucks this value into something often called the *instruction pointer*. In an actual program, what we have been calling the slab number is actually the memory address of the very next program instruction after the method call.

So, you "call the *Serial.print()* method" which is actually a jump instruction to slab (memory address) 234. You go inside the black house using your dot operator "key" to gain access, look for the door with *print()* above it and knock on the door. The door opens, grabs your backpack, and slams the door. A few milliseconds later, the door opens and a hand shoves your backpack in your hands. Being a nosy person, you look inside and see the number 18, but don't think much about it. You put the backpack on your back and head for the front door.

You walk outside the *Serial* object's house and immediately jump to... where? Aha! The instruction pointer automatically teleports you back to slab 100. You have just "returned from the method call to *Serial.print()*". You now pick up the instruction paper on slab 100 and read what to do next.

Wait a minute! What happened to the value 18 that was placed in the backpack by the method call? Nothing. Your programmer chose not to use it. Many people don't know it, but any time you use the S*erial.print()* method, it returns the number of characters that were processed successfully by the *Serial* object's *print()* method. (You just went back and counted the number of characters in the message again, didn't you?)

At some point in your walk down the sidewalk, you're going to come to the very last slab. You bend over and pick up the instruction and what does it say? You should already know the answer. Because microcontroller programs are not designed to "end," that last instruction is going to tell you to jump to the very first slab that appeared in the *loop()* function and start the journey all over again. Think about it.

## Conclusion

You do not need to know or use C++ to write your own Arduino programs. However, understanding the benefits OOP and what C++ brings to the table can help you write better program code, even if its only "pure" C. Encapsulation is a concept that applies to all languages and the more you encapsulate your data, the less time you will have to spend fixing things.

## References

www.arduino.cc/en/Hacking/LibraryTutorial
www.arduino.cc/en/Reference/APIStyleGuide
www.arduino.cc/en/Hacking/HomePage
www.arduino.cc/en/Guide/Libraries
www.alanzucconi.com/2016/05/11/libraries-for-arduino/

# CHAPTER 4

# Displays

In Chapter 2, we discussed the Five Program Steps. Step 4 of that series was program output. After all, it does little good for a program to grind away on some data input if you can never use the program's output. Sometimes, the output of the program is passed along to another program and the output is not directly visible (for example, a fire sensor program that sends a signal to another program that sounds an alarm and automatically dials the fire department). However, the programs presented in this book need a way for you to sense the output. In other words, in many projects you need a device of some sort to display the output generated by your program.

The good news is that you have a lot of display options from which to choose. The bad news is that you have a lot of display options from which to choose. Even more good news is that you have a variety of common display devices available to you at widely diverging costs. The bad news is that each of these display devices is programmed in a slightly different way to accomplish a given task.

The purpose of this chapter is to show you some of the display options available, provide details on the cost of each display type, explain their functional differences, and give you enough information to make an informed display choice for a given task.

## Display Options

**Figure 4.1** shows a few of the display options that are available to you. One of the first display types that most microcontroller programmers use is a 16×2 (16 columns by 2 rows) LCD as seen in the upper-left corner of Figure 4.1. (Note that most of the displays still have their protective plastic wrap on them to protect their surfaces from being scratched.) Characters on the displays are generated by the appropriate illumination of one or more pixels. The "row" LCD displays have each character element formed by a 5×8 pixel block. The small (0.96 inch) OLED display has a 128×64 pixel resolution and is surprisingly easy to read. The TFT (Thin Film Transistor) displays have much higher resolutions (for example, 800×480 for the 5-inch display and 320×240 for the smaller TFTs), which allows for considerable flexibility in terms of font and graphics choices.

The costs vary from less than $5 for the "row" LCD displays to $35 for the

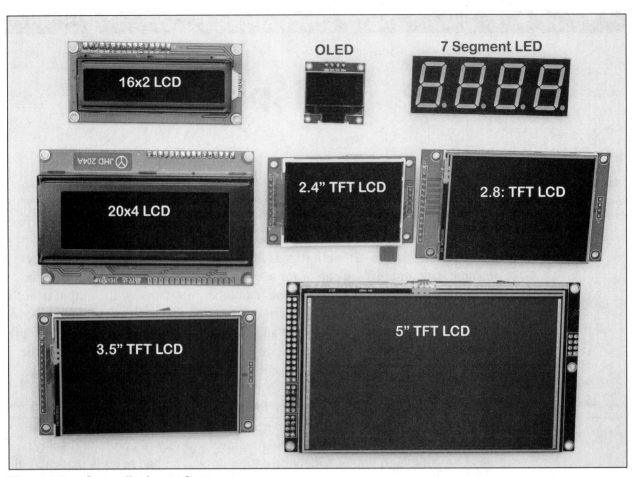

Figure 4.1 — Some display options.

5-inch touch screen display. The costs, however, are not linear. For example, a 16×2 LCD display using the I2C interface (explained below) is just under $6. Yet, with a little careful internet shopping, you can buy a 2.4-inch TFT LCD touch screen color display for less than $5! Why so cheap for the color display? In a nutshell, supply and demand. Most program designers now prefer a color TFT display over the 16×2 "blocks" display, so the demand for TFT color displays is high. However, the initial high prices for the TFT color displays attracted lots of producers, so the supply increased dramatically. Result: relatively low prices for a graphics-capable color display.

The Nokia (5110) display was used extensively in cell phones, but lost favor with the advent of the smart phone with color displays. The Nokia display can be purchased for about $2.50. Despite the Nokia 5110's low cost, not many projects use it due to its relatively low resolution (84×48). Another issue is relatively poor visibility in bright sunlight.

## Which Display to Choose?

Actually, that's a very difficult question. The reason display selection is difficult is because there are so many factors to consider. If the project results in a product that is used outdoors most of the time, some of the LCD displays are difficult to read in bright sunlight. While a large TFT display is easier to read in bright sunlight, TFTs also consume considerably more power, making them a less obvious choice if the device is likely to be battery powered. Another factor that complicates the choice is the options within a given variety of display. For example, the 16×2 LCD display can have blue, red, green, yellow, white, or no backlighting. The pixels can be black, white, or amber. Also, there are all kinds of different interface options with respect to the way your microcontroller communicates with the display. There are a host of other complications that you may need to consider, too (for example, weight, robustness — does it break easily, and other factors).

As seen in the project details throughout this book, we favor the TFT LCD approach, primarily because of its flexibility in showing outputs of various types. The cost is slightly higher, but the results are miles ahead of conventional LCD displays. Note that we tend to use a display suited to the type of information to be output. For instance, the Mini Dummy Load (MDL) only requires simple text, so the low-power OLED is ideal. On the other hand, both the DSP Post Processor and Signal Generator show lots of data in different forms, so a 3.5-inch high-resolution TFT LCD is appropriate.

**Figure 4.2** shows a common 16×2 LCD display. However, the top two displays in the photo are versions that use a standard parallel data access. Notice the small knot of wires at the top of the display — these wires connect to the microcontroller. (Actually, there could be more wires because the display in Figure 4.2 is wired to use four data lines rather than eight.) The middle display in the picture shows what the back of a 16×2 display looks like. The two displays shown toward the top of Figure 4.2 are "standard" in that they use a pin-to-pin parallel interface with the microcontroller.

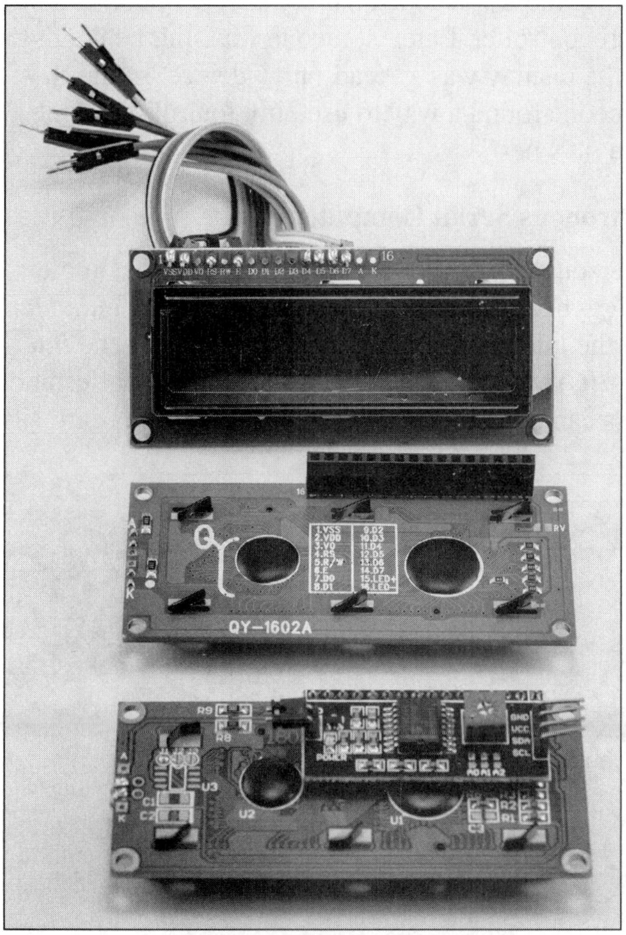

The display at the bottom of Figure 4.2 has an interface board attached to the display. While it does not show up too well in the photo, the board is connected to the solder pads you can see on the top two displays. However, the small "piggyback" board contains the electronics that allow you to control the display with only two data lines. (The other two wires are the positive and negative [GND] power connections.) The point is that there are different flavors of LCD displays, both in terms of colors and interfaces to the microcontroller.

The next section presents a discussion of the interface options offered by the various displays. This is important as it can affect other project design elements.

Another consideration is the voltage of the display. Some displays can work with only 5 V, while others can only use 3.3 V. There are many displays, however, that can use either 5 V or 3.3 V. The key is to make

**Figure 4.2 — Common 16×2 LCD displays.**

sure that your μC uses the same voltage as the display. For example, if you use a display with 5 V but hook that display up to a μC with pins that are only 3.3 V tolerant, you run the risk of burning out the μC I/O pins since the display is feeding 5 V signals into a chip that only can handle 3.3 V. (If you find that you must use a display and μC combination that use different logic levels, you can buy bi-directional level shifters online for a few dollars. However, that seems unnecessary given the wide variety of display choices available.)

Another important factor is the chip that is used to control the display. Some libraries are written specifically for one controller (for example, the ILI9431) and would not work with a display that uses a different controller (such as the ILI9488). Always make sure there is a library for your μC and your display's controller chip before you buy the display.

## Interfacing with a Display

Let's start off with a display that is often included as part of an "experimenter's" microcontroller kit: the common 16×2 LCD display. If you look closely at the 16×2 displays in Figure 4.2, you can see 16 solder pads in the upper-left corner of the display. Each pad has a specific use for its connection: ground, supply voltage, contrast, chip select, read/write, enable, D1-D8 data lines, and backlight power. This suggests that we need a rat's nest of wires between the microcontroller and the display. One of the first things experimenters did to control the ugliness of this wiring mess was to use 16-conductor ribbon cable to connect the display to the microcontroller. Later, someone was smart enough to notice that 99.9% of the time the display was a "read-only" device, so the R/W wire was redundant. Then someone found a way to use only four of the eight data lines, further reducing the rat's nest.

### I2C Synchronous Serial Computer Bus

Eventually, some clever person determined they could reduce the complexity even further with a little piggyback board and relegate all of the communications between the display and the microcontroller to something called the *Inter-Integrated Circuit* (aka *I2C* or *IIC*) serial communication bus. The manifestation of this little piece of cleverness can be seen in **Figure 4.3**. The I2C bus card sits

**Figure 4.3 — The I2C bus connection for the LCD display.**

on the back of the LCD display with the display's 16 pins soldered directly to the I2C card. The four wires you see going off to the right side of the figure are routed to the microcontroller — a reduction from an ugly tangle of 16 lines to a more comely four lines. Not only does this reduce the interfacing complexity, but more importantly it frees up a potential 12 microcontroller pins that can be used for other purposes.

To us, the freed pins represent the real win with the I2C bus. As you can see in Figure 4.3, you only need to supply power (usually 5 V) and ground connections, plus synchronized clock (SCL) and data (SDA) lines between the microcontroller and the display. Once you determine which pins of your microcontroller are dedicated to the I2C bus clock and data lines, you're home free. And it gets better! The microcontroller (the master device) can assigned a unique device number to each I2C device (the slave device), which (in theory) allows a single microcontroller to control up to 128 external devices! (One bit is reserved for R/W messaging. If you look closely at Figure 4.3, you see six solder pads just below the blue trimmer pot labeled A0, A1, and A2. By soldering jumper wires across those pads, you can set the device number if the default device number conflicts with other I2C device numbers.)

True, you pay a little extra for that I2C board you see hanging on the back of the display in Figure 4.3. The display in Figure 4.3 came with the I2C board in place for $5.50.[1] Whenever possible, we take advantage of devices that implement the I2C bus. However, you can find cheaper displays on the internet, ranging from $1.50 for a standard 16×2 display, to $2.00 for one that has an I2C interface. If your project uses relatively few I/O pins, there's nothing wrong with using a standard display.

The I2C interface, however, is not the only game in town.

### Serial Peripheral Interface Bus

An alternative interface is the *Serial Peripheral Interface*, or *SPI bus*. SPI is also a synchronous serial communications interface, but has the advantage that its master-slave architecture is full duplex. **Figure 4.4** shows how the control lines are configured for the SPI interface. The serial clock line (SCLK) controls

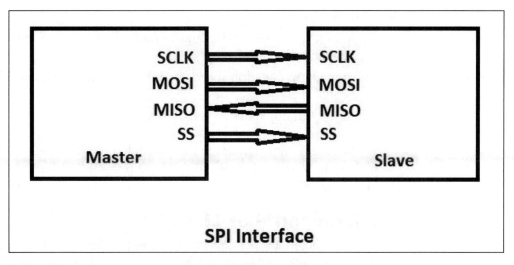

**Figure 4.4** — The SPI interface control lines.

the signals that flow between the master and slave devices. The MOSI signal is the Master Out, Slave In signal line, while the MISO is the Master In, Slave Out line. Note that both the slave and master have an input/output line. The Slave Select (SS) line determines which slave device is "talking" to the master device. (The SS line serves a function that is similar to the I2C device number for the slave device.)

Without getting into too much detail, the SPI interface has the advantage that it can operate in full duplex. Also, the clock rate does not have to be limited by the microcontroller clock, which can result in higher data rates. (Think of the clock rate as the rate at which a traffic light changes from red to green, thus controlling the traffic flow. The SPI clock serves a similar purpose in that it controls the flow of data between the microcontroller and the SPI device.) From our point of view, however, the major disadvantage of the SPI interface is that it requires more control pins than the I2C interface does. Some devices may provide only an SPI interface. While there are other advantages/disadvantages (see below), when given a choice, we usually opt for the SPI interface. (We show four control lines in Figure 4.4, but did not bother to show the voltage and ground lines.)

There are other computer buses (PCI, SCI, CAN, SAS, and so on), but the I2C and the SPI buses are the most popular for the devices we use. Also, all of the microcontrollers we use in this book provide these two interfaces. However, the more we use these different interfaces, the more we favor the SPI bus.

## I2C versus SPI

So, which interface is best? As always, the answer is: It depends. Here's a list of the pros and cons of these two popular interfaces:

**I2C:**
- Needs fewer connections; only two required
- Uses an open collector bus, which often allows you to use either 3.3 V or 5 V
- Can use multiple master devices

**SPI:**
- Simpler protocol
- Use of a device select line allows multiple devices of the same kind to be connected
- Faster communication speeds over longer ranges (i.e., not limited to processor clock speed)
- Can run full duplex (I2C is half duplex)

We use both interfaces in this book, but the SPI seems to be becoming more pervasive. Most of the color displays support the SPI interface and are fairly inexpensive.

## Microcontroller Libraries

For a beginner, perhaps one of most perplexing aspects of writing code for microcontrollers is the proper use of program libraries. As we pointed out in

Chapter 2, there are dozens, if not hundreds, of libraries available that can be used within the Arduino IDE and the microcontrollers we plan on using. That's the good news because it means that a lot of work has already been done for you.

The bad news is that the Arduino family of libraries is not very well organized. By that, we mean that most of the libraries were written by individuals, probably for their own use, but who were kind enough to share them with us. However, there is no precise format or style standard for submissions, which yields a hodgepodge of libraries. For example, one of the project areas we do in this chapter is write programs for all three microcontrollers using the liquid crystal displays (LCDs). We just did an internet search on LCD libraries and got 2.4 *million* hits! That doesn't mean that there are more than two million LCD libraries, but there are a lot of them. One reason for the large number is that there are a lot of different chipsets (for example, HD44780, RTD2485D, PCA8530, and so on) used to drive the displays. Even worse, many library programmers use the same name (for example, LiquidCrystal) for their library even though a different library written for a different chipset might use the same library name. This can lead to a lot of teeth gnashing and hair pulling.

Library code is made available to a program using the *#include* C preprocessor directive. The syntax looks like this:

```
#include <SPI.h>    // Standard with IDE
```

Verbalizing the directive would be: Search the default Arduino library path for a library named SPI, look in its directory and read the file named SPI.h. Disk files ending in "h" are called *header files* and they are much like the table of contents of a book because they tell the compiler something about the content of the library itself. Usually, if there is a SPI.h header file, there is also going to be a SPI.cpp (SPI.C-Plus-Plus) source code file. As a general rule, the primary file name of the header and source code files (SPI) is also the library directory name that holds the files. Therefore, looking in the SPI *library* subdirectory, we find another subdirectory named *src* that contains:

```
SPI.cpp
SPI.h
```

(By the way, the SPI library is kind of hidden relative to the main library path. Usually you can find it in off the IDE's path by following the path: \hardware\arduino\avr\libraries\.) If the header file functions like a book's table of contents, the .cpp source file holds the book's narrative details.

Most libraries have a subdirectory named *Examples* that shows you common ways the library is used. The examples found in the library are great learning tools.

## Finding Different Libraries

You should know, however, there are lots of libraries that are *not* distributed as part of the Arduino IDE installation and we need to use some of them. For example, Adafruit has an excellent graphics library that we use in later chapters, but it's not distributed with the Arduino IDE. So, how do you know where to go to download the library for your use? We use the following convention (and wish everyone did):

```
#include <Adafruit_GFX.h>   // https://github.com/adafruit/Adafruit-GFX-Library
```

Anytime a sketch uses a library that is not distributed with the Arduino IDE, we follow the *#include* preprocessor directive with a comment that gives the URL where you can go on the internet to download the library. Once it's downloaded, you need to install the library in the Arduino *library* directory. (There are plenty of online instructions on how to download and install a library.) Most libraries are also Open Source, which means you can download and use them without charge.

It should be obvious that, anytime you want to compile and upload one of our programs, you need to make sure you have installed any libraries that might be included in the program but are not part of the IDE. Also, adding a new library to the IDE requires that you restart the IDE after it's been installed. The Arduino IDE then reconfigures its startup code (part of the Step 1 Initialization Step detailed later in this chapter) to register the new library.

## Connecting From Your Microcontroller to the LCD Display

This section shows you how to connect the microcontroller you selected to the LCD display. For these examples, we assume that you have purchased a 16×2 LCD display using an I2C interface like that shown in Figure 4.3. Obviously, the I2C interface needs the power supply lines (Vcc and GND in Figure 4.3) and the two control lines, SCLK and SDAT. Most displays use +5 V for power, but there are exceptions. The current demands for an LCD likely exceed the power that can be supplied from a single I/O pin. On the other hand, using the 5 V pin on the microcontroller usually has sufficient power to drive the LCD display. Be mindful of the total current capability of the microcontroller.

### I2C/SPI and Arduino Nano

**Figure 4.5** shows the pins used to connect the I2C display to an Arduino Nano microcontroller. You can see that physical pins 18 and 19 are associated with analog pins A4 and A5. (These are circled in Figure 4.5.) You can also tell from Figure 4.5 that those same pins are defined as SCL and SDA, which matches the I2C interface pins on the I2C display board. Therefore, you would need to connect the A4 (aka SDA, pin 18) and A5 (aka SCL, pin 19) to the SCL and SDA pins on the I2C interface attached to the back of the LCD display.

But, what if you want to use the SPI interface instead of the I2C? Look at pin numbers 10 through 13 and note they carry these labels: SS, MOSI, MISO, and SCK. Sound familiar? Most pinout images for a given μC like that shown in Figure 4.5 provide labels for the SPI interface, too.

## I2C/SPI and Teensy 3.6

**Figure 4.6** shows the pins for the Teensy 3.6 microcontroller. (We are only showing the pins available on the top side of the Teensy. There are also I/O lines that can be accessed from the bottom side as well.) If you look on the right side of the Teensy, you can see that pins 18 and 19 align with A4 and A5. However, if you look a little to the right of that, you can see that those pins are also used for SCL0 and SDA0. These pins are the I2C clock and data lines for the I2C interface. These two pins from the Teensy should be connected to the clock (SCL on the LCD to SCL0/A5 on the Teensy) and data (SDA on the LCD to the

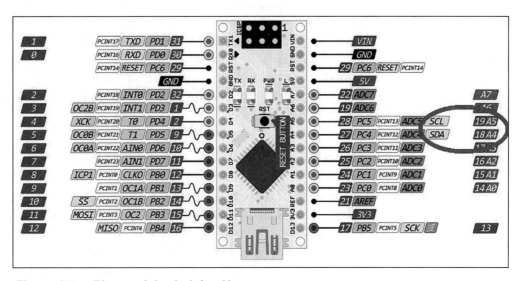

Figure 4.5 — Pinout of the Arduino Nano.

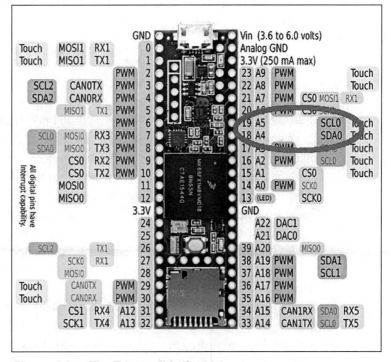

Figure 4.6 — The Teensy 3.6 pinout.

**Displays    4-9**

SDA0/A4 on the Teensy) lines. The Teensy supports two independent I2C interfaces. For testing purposes, the USB cable supplies the voltage to the Teensy.

**Note**: The Teensy is a 3.3 V device and most LCD displays use 5 V. For that reason, we use a separate 5 V supply to power the LED portion of the LCD display. Even if you can find a 3.3 V LCD display, make sure you know how much current it draws, as the Teensy can only supply a maximum of 250 mA. Likewise, you can see a 5 V tap point on the Nano in Figure 4.5, but it should not be used to power the LCD display as the Nano would also be overtaxed by the load. Also note that care must be used mixing 3.3 V and 5 V devices. Unless the microcontroller explicitly states that some of its pins are 5 V tolerant, the only safe way to go is to use a level shifter.

The wiring couldn't be simpler. Connect your 5 V dc power source to the Vcc and GND connections on the I2C module on the back of the LCD display. Connect two wires from A4 (pin 18) and A5 (pin 19) of the Teensy (or Nano) to the SCL (clock) and SDA (data) lines on the I2C module.

Can you find the SPI pins in Figure 4.6? (Hint: the Teensy supports two SPI interfaces, so you might look for MISO1, MOSI1, CS1, and SCK1.)

## I2C/SPI and Blue Pill

**Figure 4.7** shows the pinouts for the Blue Pill (STM32F103). In Figure 4.7, we have circled the I2C interface pins, which correspond to the pins labeled B6 and B7. For the I2C interface, connect B6 to SCL pin on the LCD's I2C interface board and B7 to the SDA pin. Again, the Blue Pill is a 3.3 V device, so you should power the LED portion of the LCD display from a separate 5 V source, but be careful of display/microcontrollers with different voltages.

The legend on the left side of Figure 4.7 indicates that the image also shows the SPI interface pins. If you find an image similar to Figure 4.7 on the internet, the legend is color coded, making it very easy to identify the interface (and other) pins.

## I2C/SPI and the ESP32

The ESP32 also makes the I2C interface available, as can be seen in **Figure 4.8**. As seen in the figure, the general purpose I/O pins (GPIO) numbers 21 (GPIO21) and 22 (GPIO22) form the data and clock lines of the I2C interface.

As was pointed out in Chapter 1, the ESP32 is a little different than the other microcontrollers in that different vendors seem to place their I/O pins in a manner that suggests very little attempt to make the various boards interchangeable. Indeed, we also mentioned that some ESP32 boards have 30 pins, some 36, and some 38. The board in Figure 4.8 shows all 38 pins, some of which can only function as input pins. (See Table 1.2 in Chapter 1.) If you decide to use the I2C interface with an ESP32 board, you'll need to check a pinout diagram for the ESP32 board you are using to determine which pins are used for the I2C interface.

From a survey of the literature about various microcontrollers, it appears that the I2C interface is less popular than the SPI interface. Also, we see more SPI displays, and programs using those displays, than we do for the I2C interface. A quick check on eBay showed about 2100 I2C displays for sale (most

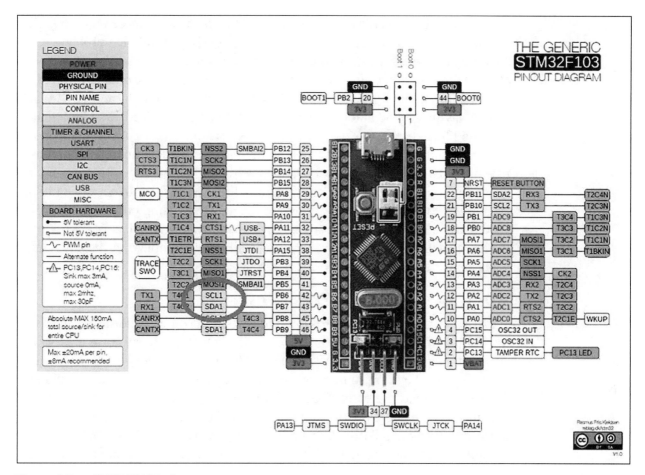

Figure 4.7 — STM32F103 pinouts.

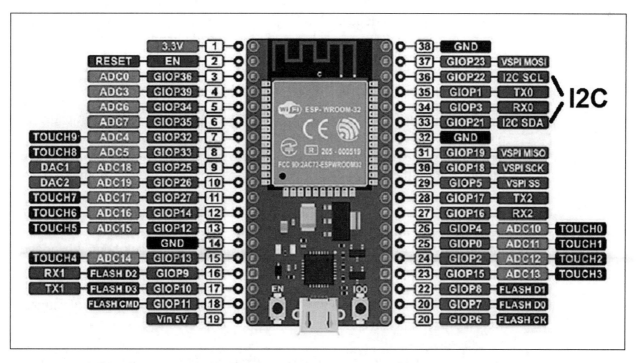

Figure 4.8 — ESP32 pinout and the I2C interface pins.

Displays    4-11

were small OLED displays). On the other hand, there were more than 3600 SPI displays and in a much greater variety of display sizes. While it's probably a "don't care" in most projects, the SPI interface is not limited by the clock speed of the microcontroller. For those reasons, we concentrate our coding efforts in this chapter on the SPI interface. Still, some of you probably have a 16×2 LCD display laying around and there's no sense not using it. Indeed, it's likely that it uses the I2C interface.

Once again, you shouldn't have too much difficulty locating the SPI interface pins. (Hint: 29, 30, 31, and 37.) The actual pin numbers will likely vary depending upon the ESP32 board that you purchase. In all instances, we find it very useful to have a full color image of the µC we're using on any given project close at hand. Keeping the image in a clear plastic sleeve lets you write on the image with a marker pen, recording details (for example, pin-display wire) about how the pin is connected to the circuit.

## A Simple I2C LCD Program

Now for the really cool part. Load the program source code shown in **Listing 4.1** into the Arduino IDE as discussed in Chapter 2. Because we are using a standard interface (for example, I2C), *you don't have to change anything in the source code!* The same source code can be used regardless of the actual microcontroller you have selected. Because we are using a standard interface with a defined device address (for example, 0×27), the compiler knows everything it needs to know to talk to the external device (the LCD display). (Well, it is possible to get a nonstandard display that might use a device number other than 0×27. We show you how to fix that problem below.)

If we had decided not to use a standard I/O interface, we would have had to contend with changing the source code to match power and control data lines. True, we still need to "wire up" the microcontroller correctly, but not having to change the source code itself is a huge win. Now let's examine what the code in Listing 4.1 does. (We repeat subsections of the code listing to make it easier for you to follow the discussion.)

## Code Walkthrough for Listing 4.1

Code walkthrough? What's that? When Jack had his software company, the programmers were divided into small teams. Usually, teams were assigned a task that involved a two or three week timeline. On the Friday at the end of the timeline, *all* of the programmers were required to meet in the morning to put a "fresh set of eyes" on the code. Copies of the team's code were handed out to each programmer and they literally "walked through" each line of code the team had written. As an added incentive for the team to write error-free code, if no bugs were discovered during the code walkthrough, Jack bought beer and pizza for lunch and everyone got Friday afternoon off. Even with a fairly small programming staff, buying beer and pizzas for the programmers plus losing all of your programmers for a half a day probably sounds like code walkthroughs were a pretty expensive activity.

Not really.

I can't tell you how many times I walked into the office late at night before

**Listing 4.1**
**Display a Message on LCD display, Teensy Version**

```
// Teensy 3.6: Write message to LCD using I2C interface, Jack Purdum, W8TEE,
// Dec. 13, 2018.

#include <Wire.h>
#include <LiquidCrystal_I2C.h>           // https://bitbucket.org/fmalpartida/
                                         //   new-liquidcrystal/downloads/

LiquidCrystal_I2C lcd(0x27, 16, 2);      // Create LCD object: I2C device address
                                         // 0x27, 16x2 display

void setup()
{
  lcd.begin();                           // initialize the LCD

  lcd.backlight();                       // Turn on the backlight and print a message.
  lcd.print(" CQ CQ DE W8TEE");
}
void loop()
{
  char spaces[17] = "                "; // 16 spaces with room for null
  char message[]  = " CQ CQ DE W8TEE";
  static int row  = 0;

  if (row != 0) {
    lcd.setCursor(0, 0);                 // Clear out row 0
    lcd.print(spaces);
    lcd.setCursor(0, row);               // show message on row = 1
    lcd.print(message);
    row = 0;
  } else {
    lcd.setCursor(0, 1);                 // Clear out row 1
    lcd.print(spaces);
    lcd.setCursor(0, row);               // Show message on row = 0
    lcd.print(message);
    row = 1;
  }
  delay(500UL);                          // Pause for a half second...
}
```

a walkthrough and saw *everyone* there helping the team with their code. Not only did the carrot-on-a-stick forge the programmers into a cohesive team, it also made all of the programmers somewhat familiar with all elements of the project's code. Also, with a small company, if one programmer left for vacation or was sick, it made it easier to temporarily fill in the gap with someone else. Nope...code walkthroughs are a good thing on many levels. I can't tell you how many times Al did a code walkthrough on my code after I spent hours beating my head against the wall. I'm still not sure how to feel when, after looking at the code for a few minutes, Al would say: "Ah...here it is!" Yep...a fresh set of eyes often are a big help.

## Preprocessor Directives and Global Data

Let's call the following few lines Listing 4.1, Part A.

```
// Teensy 3.6: Write message to LCD using I2C interface, Jack Purdum, W8TEE,
// Dec. 13, 2018.
#include <Wire.h>
#include <LiquidCrystal_I2C.h>         // https://bitbucket.org/fmalpartida/
                                       //   new-liquidcrystal/downloads/
LiquidCrystal_I2C lcd(0x27, 16, 2);    // Create LCD object: I2C device address
                                       // 0x27, 16x2 display
```

Note how we have two *#include* preprocessor directives near the top of the program. The code associated with these two libraries allows you to instantiate a *LiquidCrystal_I2C* object named *lcd* in the third program line. Note that the I2C display we are using requires the use of the *LiquidCrystal_I2C* library that can be downloaded at the URL given on the include line; it is not part of the standard IDE install. We always provide you with the download URL anytime a nonstandard library (one not supplied with the Arduino IDE by default) is used in the program.

Recall that the I2C interface needs to know the address associated with the I2C device. In this particular example, the bottom line above is used to instantiate an I2C LCD object named *lcd*. (We covered class instantiation in Chapter 3.) The first argument tells the compiler that the *lcd* object is to be addressed as device 0×27 and that it is configured as a 16×2 display device. This is a parameterized constructor that passes the device information at the time the object is instantiated. If you don't know the address for the device you are using, you can use Nick Gammon's *I2C Scanner* program to determine the device number (**playground.arduino.cc/Main/I2cScanner**).

### *setup()*, Program Step 1 — Initialization

The next statement block is the *setup()* function that we discussed earlier in Chapters 1 and 2. Its purpose is to establish the environment in which this program is to be run. The first statement in *setup()* is a method call to *begin()* of the lcd object. We know it's an object because of the dot operator between the object and method names. The *begin()* method initializes the necessary data to make the lcd object capable of displaying data.

```
void setup()
{
  lcd.begin();              // initialize the LCD
  lcd.backlight();          // Turn on the blacklight and print a message.
  lcd.print(" CQ CQ DE W8TEE");
}
```

The second statement is a call to the *backlight()* method, which turns on the LCD display's backlight. (Sometimes the backlight is not turned on to conserve power.) The third and final statement simply prints a CQ and Jack's call on the display. Since the display was not given any directions where to place the message on the display, it starts on the first of the two rows in the first of 16 columns. (However, since everything in C is zero-based, the first letter printed is at the reference location "0,0". The first character display is a blank space...not a "C"!)

In this case, the only data used in this program is the quoted text message. Therefore, Step 2, Input, is also taking place within the *setup()* function. It's not uncommon to find default data values assigned to variables in the *setup()* function.

## *loop()*, Step 3, Process and Step 4, Output

Most of the work being done in the program takes place within the *loop()* function. The code is repeated here:

```
void loop()
{
  char spaces[17] = "                "; // 16 spaces with room for null
  char message[]  = " CQ CQ DE W8TEE";
  static int row  = 0;
  if (row != 0) {
    lcd.setCursor(0, 0);     // Clear out row 0
    lcd.print(spaces);
    lcd.setCursor(0, row);   // show message on row = 1
    lcd.print(message);
    row = 0;
  } else {
    lcd.setCursor(0, 1);     // Clear out row 1
    lcd.print(spaces);
    lcd.setCursor(0, row);   // Show message on row = 0
    lcd.print(message);
    row = 1;
  }
  delay(500UL);              // Pause for a half second...
}
```

The *loop()* statement body begins by defining three variables. The *spaces[]* array is nothing but 16 blank spaces. As the comment suggests, we must define enough memory for 17 characters, because all string data ends with a null ("\0") character. The *message[]* array duplicates the string that you just saw in *setup()*. (This is a little bit of RDC, but we did it to make the process easier to follow.) The third line defines an integer variable name *row*, which is used to determine which row should be used to display the message.

On the first pass through the code, *row* equals 0 because that the value we gave it when we defined it. The *if* statement checks to see if *row* in not equal to 0. However, because it is equal to 0, we skip over the next five statements and execute the *else* clause of the *if* statement block. This means we execute a call to the *lcd* object's *setCursor()* method. The two arguments (column is 0 and row is 1) mean the cursor is placed on the second row (line 1, remember?) and the first character (column 0).

The next line calls the lcd's *print()* method to print out 16 spaces. This has the effect of erasing anything that used to be on the (*row*) line.

The next statement is another call to *setCursor()* which places the cursor at column 0, and *row*. We have to do this because the previous *print()* call leaves the cursor at the end of the second line (i.e., the 1). So this second call to *setCursor()* moves the cursor back to column 0 for *row*.

The next statement is another call to the *print()* method, which prints out the contents of the *message[]* array. This means that the CQ message appears on the second row of the display. (Think about it.) The call to the *delay(500UL)* call gives you a half a second to view the message. (The argument to *delay()* expects an <u>u</u>nsigned <u>l</u>ong data type, so the "UL" at the end of the numeric constant 500 documents its data type.)

As you learned in Chapters 1 and 2, when the last statement in *loop()* has been executed, control returns to the very first statement in *loop()* which redefines the *spaces[]* array. Now look at the third statement that defines *row*. If we redefined *row* to be initialized to 0 again, the first five statements of the *if* statement block would never be executed. This means the message would always appear on the second row. Read this paragraph again 100 times or until you're sure you understand why...whichever comes first.

The *static* keyword at the start of *row*'s definition statement tells the compiler to treat row differently. In essence, it says define *row* and initialize it to zero on the very first pass through *loop()*, *but never redefine it a second time.* What this means is that *row* can retain its current value on each pass through the loop. Because we set *row* to equal 1 on the first pass through the *else* statement block, *row* keeps that value for the second pass through *loop()*. If you have followed this discussion closely, you should now understand what the program does.

Once you have your program working, it should look something like that shown in **Figure 4.9** after the first pass through *loop()*.

**Figure 4.9 — Listing 4.1 in action.**

You can see the four I2C wires running from the display to the Nano. (An expansion board like that shown in Figure 4.9 really makes connections easy!) The I2C data line (usually abbreviated SDA) is connected to pin A4 on the Nano. The I2C clock line (SCK) is connected to Nano pin A5. (These are the standard I2C pins for the Arduino family.) The other two I2C interface wires go to +5 V (Vcc) and GND on the Nano. That's it.

## Naming Conventions for Methods, Core and Custom Functions

The call to *lcd.begin()* in used to initialize the LCD device. If you want to know exactly what the method does and how it works, look in the *LiquidCrystal_I2C* class definition for the *begin()* method. As mentioned in Chapter 3, *most C++ class method names begin with a lowercase letter and then use camel notation from that point on* (for example, *lcd.setCursor()*). We also mentioned that many programmers *start function (not method) names created by the project programmer with an uppercase letter* (for example, *MyDelay()*). This is probably not really necessary, since function names are not preceded by the dot operator like method names are. However, it does make it easy to see which functions you've written versus functions provided by the core of the IDE (for example, *delay()*, *millis()*, and so on). Core IDE functions tend to begin with a lowercase letter, similar to library method names, but without the C++ object name and dot operator.

How are these conventions helpful? If you write perfect code every time, they aren't helpful. On the other hand, if a hiccup appears in your code every so often, if the function or method (nothing to prevent you from writing your own classes!) starts with a capital letter, you know that the source code for that function or method is part of the current project and that you (or a team member) wrote it.

## Using SPI with TFT Color Displays

In this section we write a simple menu program that illustrates how to connect your μC to a TFT color display, but using the SPI interface this time. The code that accompanies this book has separate project source code files for the ESP32, the STM32, and the Teensy μCs. The core of the code is the same for all three μCs, but the pins used for the SPI, encoder, and encoder switch varies according to the μC you wish to use. Therefore, when you implement the demo code, look in the MenuXXXXX.h header file for the specific pin assignments. (The XXXXX is filled in for your μC, such as MenuSTM32.h.)

The menu project has four source code files:

- MenuXXXXX.ino      // main project file with setup() and loop()
- MenuXXXXX.h       // the header file for the project
- clsMenu.h         // the menu class header file
- clsMenu.cpp       // the source code file for the menu class

For the most part, all of the files with the exception of MenuXXXXX.h are the same for each μC. After all, the project's task is the same regardless of the process you decide to use. We can "re-use" most of the code without changing it, and that's a real strength of object-oriented programming (OOP) code.

## Simple Menu Example Using the SPI Interface

The output of the program is shown in **Figure 4.10**. What you cannot see in the figure is that the first menu choice ("Band") is black lettering on a white background while the other two choices are green text on a black background. Therefore, the currently-active menu choice is Band. Some graphics libraries don't allow you to change the background color of the text field, in which case we would change the active menu choice to use white letters on a black background. The other menu options are displayed as green text on a black background. The white and green letters provide enough contrast to see easily which is the currently-active menu choice even without the background coloring. However, if your display/μC support background colors, menu selection is more effective if you use them. Sometimes, the background approach makes the text harder to view. Check other options if this is the case, such as changing the text color when selected.

Most of the code for the menu demo program does not change when the processor is changed. Indeed, the output looks the same regardless of the μC used. However, because we are using the SPI interface, some of the pin assignments do change. These pin assignments for the various μCs are presented in **Table 4.1**.

Figure 4.10 — Simple menu program.

**Table 4.1**
**Pin Assignments**

| Description | Teensy | STM32 | ESP32 |
|---|---|---|---|
| **TFT Display:** | | | |
|     SDO (MISO) | 12 | N/C[1] | 19[1] |
|     LED | 5 V | 5 V | 5 V |
|     CK | 13 | PA5 (SCK1) | 18 |
|     SDI (MOSI) | 11 | PA7 (MOSI1) | 23 |
|     DC | 9 | PA0 | 2 |
|     RESET (through 100 Ω resistor) | 8 | Vcc | 4 |
|     CS | 10 | PA1 | 15 |
|     GND | GND | GND | GND |
|     Vcc (Some may use 5 V) | 3.3 V | 3.3 V | 3.3 V |
| **Encoder:** | | | |
|     ENCODER1PINA | 4 | PB12 | 12 |
|     ENCODER1PINB | 5 | PB13 | 13 |
|     ENCODERSWITCH | 6 | PB14 | 14 |

[1] N/C – No Connection – Slave does not output in this example. Added for ESP32 for completeness.

## Making a Menu

Our menu only has three choices: Band, Mode, and Config. In this case, the default selection starts with the Band menu option highlighted. So, how to we get to the Mode option? We use a rotary encoder. By turning the encoder shaft clockwise (CW), the highlighted menu option moves to the right. If we rotate the shaft counterclockwise (CCW), it moves to the left.

What happens if Band is highlighted and we turn the shaft CCW? It depends upon how you want things to work. We like the highlight to "wrap around" the endpoints. This means that Config would now be highlighted. Likewise, if Config is highlighted, a CW rotation moves the highlighted option back to Band. Some programmers prefer to "stall" the menu at the endpoints. Because we don't like this, our menus "wrap." (As Mel Brooks once said: "It's nice to be King!")

## Rotary Encoders

An easy way to implement a menu system in a project is via the use of rotary encoders. **Figure 4.11** shows a common rotary encoder that can be purchased online in small quantities for less than $1. This particular example is a KY-040 encoder and includes a built-in switch. Be forewarned that many internet companies sell encoders without switches and that do not have the threaded mounting shaft. As a general rule, you'll want to buy encoders with a built-in switch and threaded shaft, as it makes panel mounting them much easier.

Mechanical rotary encoders work by rotating a contact arm in a circular motion and counting the contact closures as the contact arm rotates. Think of the contact arm rotating in a circle that has speed bumps on the surface. A second arm is following behind the arm, but is 90 degrees out of phase with the first arm. When the contact arms rise over the speed bump, they make contact with a surface that is above the contact arm. These speed bumps are more properly called *detents* and the KY-040 encoders have 20 contact points per full revolution. By measuring the nature of these armature contacts as the encoder shaft turns (called the "pulse chain"), software can determine whether the contact arm is rotating in a clockwise (CW) or counterclockwise (CCW) direction. Because the KY-040 encoder has 20 contact points per revolution, this means that every 18 degrees of angular rotation, the encoder sends out a pulse chain that the microcontroller uses to determine rotation direction. As a general rule, the more detents, the more expensive the encoder. Some encoders have 200 detents per revolution and optical encoders can have even higher detent counts, but they are pretty expensive.

You'll note that the KY-040 encoder in Figure 4.11 has a small circuit board with five pins. Going from top to bottom in Figure 4.11, the pins are: 1) clock, 2) data, 3) switch, 4) positive voltage, and 5) ground. The encoder switch is activated by pushing on the encoder shaft. Not all encoders have switches,

**Figure 4.11 — KY-040 rotary encoder.**

but we find it useful to only buy those that do have switches. Our code examples assume you are using encoders that have a built-in switch that is activated by pushing the encoder shaft.

Okay, but how do you read the encoder pulse chain? Basically, there are two ways to read the data from a rotary encoder: 1) polling, or 2) interrupts. We'll discuss polling first.

### Polling

Let's revisit the example used in Chapter 2. Suppose you're designing a fire alarm system for the Empire State Building. Let's further assume that each floor has 100 fire sensors. With polling, you would write the code so that it would start on the ground floor, read fire sensor number 1, which would in turn send a signal back that there is a fire (for example, perhaps a 1) or no fire (a 0). The program would then visit fire sensor number 2 on the first floor, read the sensor, and report back fire (1) or no fire (0). Let's assume that it takes 1 second to communicate an accurate fire sensor reading.

After reading sensor 100, your program increments the *floorCounter* variable to floor number 2 and visits fire sensor 101. Sensor 101 sends back a 1 or a 0, and then fire sensor 102 is read. This process continues to repeat itself for all 103 floors in the building. We'll assume that if a fire is sensed, the sensor sends back a fire signal (a 1), and the program turns on the sprinkler system, sounds an alarm, and calls 911.

But, there's a problem with polling. Suppose your luck is such that just after you read sensor 100, a fire breaks out on the first floor. Even at a rate of one sensor read per second, it will be almost three *hours* before the alarm will be sounded! You'd be surprised how cranky people can get when they have to wade through 10 floors of fire on their way to lunch. We need a better way.

### Interrupts

An interrupt is a signal within a system that demands immediate attention. In our fire alarm system, suppose we replace the sensors with devices that can not only sense a fire, but also build in the ability for that device to send a message to the host microcontroller. Once that fire message is sent by the sensor device, an interrupt causes the microcontroller to stop whatever it's doing and process that fire message immediately. As a result, the interrupt-driven fire alarm system can sense and raise an alarm within a second for any sensor throughout the entire building. The result is a much more responsive alarm system than one based on polling.

In truth, writing a simple menu system like we're talking about here would be sufficiently fast that a human would not know if we were using interrupts or polling in the code to read the rotary encoder. By comparison, human movement is like watching something erode when compared to computer signal processing.

## Menus Using Interrupts

Interrupts are used when you want to make sure that your program responds as quickly as possible to some event or state change. For example, if a sensor is attached to a specific pin and the value on that pin changes, you want the program to take some specific action immediately.

Presenting all of the menu code here isn't really necessary. You can download the code and understand most of it even if you're not a programmer. However, the interrupt service routine is worthy of additional comment. There are interrupt libraries, but we wrote a very simple one so you can see how it works.

## Interrupt Service Routines

The action you want to take place when that interrupt occurs is called an *Interrupt Service Routine*, or ISR. There are some guidelines you should follow when writing an ISR:

1) Keep the ISR code as short and as fast as possible
2) No parameters can be passed into the ISR (i.e., a *void* parameter list)
3) No value can be returned from the ISR (i.e., a *void* function type specifier)
4) Do not use any "blocking" function or method calls in the ISR (for example, no *Serial.print()* calls)
5) Any global data used in the ISR should defined with the *volatile* type specifier

The last two conditions may need some clarification.

Any ISR you write should not call a function that itself may call an ISR. For example, the *print()* method of the *Serial* object uses its own set of interrupts. Because that method may be processing its own interrupt via its ISR, you run the risk of having it "block" your interrupt. Microcontrollers can only service one ISR at a time so for yours to be assured of being processed in a timely fashion, your code should not call any other function or method that uses an ISR. Another common function that uses its own ISR is *delay()*. If you need a program delay for some reason, write your own, perhaps using *millis()* instead of *delay()* since *millis()* does not block. (Our *MyDelay()* function is non-blocking.)

Because you cannot pass data into the ISR as a parameter, nor return a value from the ISR, any external data needed from outside the ISR must have global scope. Recall that global scope means the data item is defined outside of any function, so it is visible to all parts of the source file from its point of definition to the end of that file.

Optimizing compilers often keep heavily-used variables in a microcontroller register to avoid having to reload them from memory. (This process of holding data in registers is called *caching*.) The bad news is that a cached variable runs the risk of being "out of sync" with the current program state until it is used in a program statement. If your interrupt occurs before the current value gets reloaded, the code may use the cached (out of date) value for the variable. The keyword *volatile* tells the compiler to *always* reload a fresh copy of this variable from memory before using it. This guarantees that your ISR uses the most current value of the variable.

**Listing 4.2** shows an ISR instead of a polling method of using the encoder. The ISR routine simply reads the two pins you've assigned to the encoder. If you download the ISR routine, you'll see that ENCODER1PINA and ENCODER1PINB are defined as input pins.

**Listing 4.2**
**An ISR for the Encoder**

```
/*****
  Purpose: A simple ISR for the rotary encoder.
  Paramter list:
    void

  Return value:
    void
*****/
void interruptServiceRoutine()
{
  if (digitalRead(ENCODER1PINA))
    state = digitalRead(ENCODER1PINB);
  else
    state = !digitalRead(ENCODER1PINB);
  fired = true;
}
```

All of the interrupt magic takes place in *setup()* because of one function call:

```
attachInterrupt (digitalPinToInterrupt(ENCODER1PINA),
   interruptServiceRoutine, CHANGE);
```

The *attachInterrupt()* function is a standard library function and has three arguments. The first argument is a macro that holds the microcontroller pin that you want to monitor to trigger an interrupt event (for example, *ENCODER1PINA*). If we were writing our fire alarm system, this would be the pin the sensors should send the fire message to. The second argument is the function name (*interruptServiceRoutine*) you've given to the ISR that you want to execute when the interrupt occurs. (Anytime you use a function name by itself with no parentheses, the compiler uses the address in memory where the code for that function starts — its lvalue. The compiler can then use that memory address and "jump" to it to execute that function's code.) The last argument is the condition on the pin that triggers the interrupt. You can have the interrupt occur on the rising or falling edge of a pin change, or on either, which is what we are using. Therefore, for example, anytime there is any kind of state *CHANGE* on the *ENCODER1PINA* pin, the *RotateISR()* service routine is called and executed. In this program, we used the same state change and ISR for both pins because rotary encoders use two pins to sense encoder movement. The *state* variable tells us whether the encoder was moved in the CW (1) or CCW (–1) direction. *state* is defined near the top of the program using the *volatile* keyword.

If you look at the code for any of the menu programs, there is no explicit call to *interruptServiceRoutine()*. There's no need to call it directly. The compiler sets up the code to call the ISR anytime there is a state change on the *ENCODER1PINA* pin.

Actually, the difference in running the code using a polling method versus the interrupt method in the program probably will not be noticeable for such a simple program. As you add function calls to *loop()*, and perhaps those functions call other functions in complex programs, you begin to notice the responsiveness of the ISR approach over polling. In such a simple program, you can use either polling or interrupts to sense encoder movement. We chose to use an interrupt as that's the method we use most often in later chapters.

## Scaffolding

If you look in the MenuXXXXX.h file, near the top of that file is the line:

```
#define DEBUG       // To turn off debug statements, comment this line out
```

which simply defines a symbolic constant named *DEBUG*. Later, in the *MyMenuSystem()* function you see the following lines of code:

```
#ifdef DEBUG
  Serial.begin(115200);
  delay(500L);
#endif
```

The expression *#ifdef* is a preprocessor directive that encompasses two program statements between itself and its associated *#endif*. You can interpret this directive as saying: "If the symbolic constant named *DEBUG* is defined at this point in the program, compile all of the statements from this point to the associated *#endif* into the program." This means that the *Serial.begin(115200)* and *delay(500L)* statements are compiled into the program. However, if you comment out the *#define* directive for DEBUG:

```
//#define DEBUG       // To turn off debug statements, comment this line out
```

then *DEBUG* is not defined and the *#ifdef* block statements are ignored. In this simple example, not defining *DEBUG* removes the *Serial* object from the program. However, reinstating the definition of *DEBUG* reactivates the *Serial* object. This gives you a simple way of toggling the debug code out of your compiled program without actually removing the statements from the source code. Then, when some bug appears, you can turn on all of the debug print statements by simply reinstating the definition of *DEBUG*. This can be a real time saver since you're not retyping a bunch of debug statements over and over.

Note how we defined the menu options:

```
const char *menuLevel[] = {"  Band    ",
                           "  Mode    ",
                           "  Config  "};    // Top level Menu
```

We defined it this way so the background highlight would all be the same width for all menu choices. It's not a requirement, but we think it looks better.

The only difficult part of the menu system is to remember you must "erase" the current menu option before moving to a new one. All three of the µCs can use the same code, except for the MENUXXXXX.h file. The pin assignments often must differ across µCs and things like that properly belong in a header file. Note how the two "class" files don't change. "Write once, use a bunch" applies to all good classes and you should be able to reuse them as often as needed without major changes.

We encourage you to read through the program, starting with the *setup()* function and then the *loop()* function and state out loud what each statement in the program does. When you think you understand what the code does, compile/upload/run the program to see if it behaves as you expected. If not, why not? If it does, try adding a fourth menu option to see if you handle its processing correctly. If you're really ambitious, add a sub-menu (for example, if the selection is "bands", then make a new sub-menu appear under it that says "80", "40", "20") that is arranged vertically so the encoder now scrolls the selected option vertically instead of horizontally.

## Conclusion

Virtually any program you might write is going to need some form of output, be it visual (showing an S-meter reading or a battery voltage) or audio (the audio from a DSP filter). In this chapter we showed you how to use a simple LCD display using a simple four-wire I2C interface, as well as using the more versatile TFT LCD displays that use the SPI interface. You can implement a parallel LCD interface, too, but it will use more I/O pins that the I2C interface uses.

One of the main reasons the 16×2 LCD display became so popular was its cost. They typically cost less than $5 and with careful shopping, you can find them for even less. However, TFT color displays offer the ability to not only display more text in various sizes, but to do graphics as well. Small TFT color displays can also be found under $5. However, consider spending a little more (still under $10) for a display that supports the I2C or SPI interface. They both will save you some I/O pins.

From this point forward in the book, we start building projects. However, each one of them will draw on the materials discussed in these first four chapters. While you may have been tempted to skirt around these early chapters, reading them and using the information contained therein should enhance what you are about to read in the remaining chapters.

## Notes

[1] As of late 2019, the display shown in Figure 4.2 can be found at **Yourduino.com**. The site sells 20×4 displays, too, and other support products. Terry King runs the company and his site has sample code and library sources that he knows work with his products. He knows his stuff and is a good person who will bend over backward to help you.

## CHAPTER 5

# Projects Power Supply

While all of you probably have a nice 13.8 V power supply for your rig, the builders of our projects might like to have the option of a separate supply not tied to the rig. Our projects all assume that the builder has access to +12 to 13.8 V dc at a few amps. To make that easy, we have included a simple, high-capacity power supply project that is powered by the ac mains but won't break the bank. Our Projects Power Supply (PPS) also has a few features that are over and above most inexpensive power supplies.

The key to this project is to use a discarded computer power supply that most of us have sitting in the corner gathering dust. These supplies are very rugged, have a number of output voltages and typically deliver considerable power. Brand new computer supplies typically have capacities up to 500 to 650 W. We don't require anything like that for our project needs, so an older 200 to 400 W unit will do just fine and is probably smaller to boot. These supplies can also be found at any flea market or hamfest, in your neighbor's attic, or at a thrift store. The key here is to find one at the least possible cost.

We have several old computers sitting around, but to serve as an illustration, we went to a nearby surplus store and found a nice older unit for about $5. It probably came from an entry-level desktop computer. This unit is shown in **Figure 5.1**. The specs are as follows:

- 3.3 V fixed at 10 A
- 5 V fixed at 16 A
- 12 V fixed at 15 A
- –12 V at 200 mV
- 225 W total power

While looking for a suitable power supply, there are a few things you should look for. Avoid any units that are excessively dirty or have rust or corrosion. Opt for an ATX style supply that has the characteristic 20 or 24 pin white connectors. (See Figure 5.4 later in this chapter.) If possible, get an ac power cord with the unit, unless you have lots of them sitting around. (Typically, the power cord is molded hard rubber with the pins in a sort of triangular shape.) Check the fan and make sure it spins. By the way, plugging it in at the store won't do any good – there is a special turn-on connection you need to make it run. This is explained later.

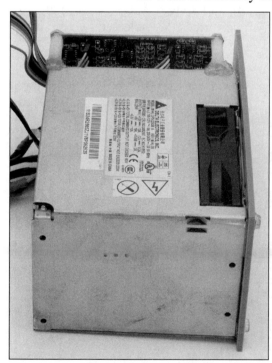

**Figure 5.1 — Computer power supply.**

## Power Supply Specifications

Because every project builder has different needs, we decided to provide you with some construction options that range from really basic and cheap to more capable and a bit more expensive. The least expensive way to go is to use the computer supply as-is, with no refinements. Our approach is to add some value to the project by adding some more ambitious specs. A bare bones PS should cost around $15 to build, while a fully-loaded PS should cost less than $40, depending on your junk box. (Table 5.1 later in this chapter gives more cost details.)

Project specs:

- Include common fixed voltages — 3.3 V, 5 V, and 12 V.
- Variable output voltage, 1 V to 30 V at 5 or 6 A maximum
- Adjustable current limiting
- Low noise output, with less than 50 mV RMS noise
- Voltage readout on front panel
- Front panel controls to set voltage and current limit
- Short-circuit protection
- Overload protection (limited at the rated power level)
- Easy to construct

Now there is a boatload of ways to accomplish these specs, but by far the easiest is to add a boost/buck dc converter to the computer supply.

A boost/buck what?

A boost/buck converter is basically two coordinated adjustable switching supplies that take the input dc voltage and either step it up to a higher voltage (boost) or step it down to a lower voltage. There are numerous available ICs that can do the job, but a more convenient way to go is to use one of the many modules that has everything needed on its own PC board. These modules come with output current ratings of a couple of amps to up to 6 A. Larger capacity units (10+ A) are available, but at much higher prices. We selected two modules to give you some choices. These options are shown in **Figure 5.2**.

Figure 5.2 — Boost/buck power supply modules.

The smaller boost/buck (B/B from now on) on the left side in Figure 5.2 has a maximum current output of 5 A, but only 80 W of total power rating. This means that the amperage is fairly limited at higher voltages. It has voltage and current control by means of adjustment pots. The on-board trimmers are replaced in the final version by front-panel pots for more convenient adjustment.

The second B/B unit, on the right side in Figure 5.2, is rated for 6 A or about 100 W, which means that it has greater current output at higher voltages. Voltage and current limits are set using the pushbutton switches. In addition, the display can be set to show voltage, current, or delivered power. Ten preset voltage and current limit values can also be stored for easy setting. The display unit can be detached from the PC board and relocated to the front panel. While we did not program this unit, it does use a microcontroller to set parameters.

## Circuit Description

A few additional components are required to complete the supply, such as filter capacitors, connectors and an on-off switch. **Figure 5.3** shows the power supply circuit diagram. The first thing to note is the on-off switch connection. This switch is connected to one of the PS wires (the green one). This green wire needs to be grounded to turn the unit on. Note that there is no ac line switch in this configuration. If you prefer to use a line switch, just permanently ground the green wire and the unit will come on when it is plugged in and the line switch set to on. **Figure 5.4** shows the color coding and connections to the ATX multi-pin connector. We prefer to cut off the connector and direct wire everything. Figure 5.4 shows both the 20 and 24 pin ATX color codes for the connectors.

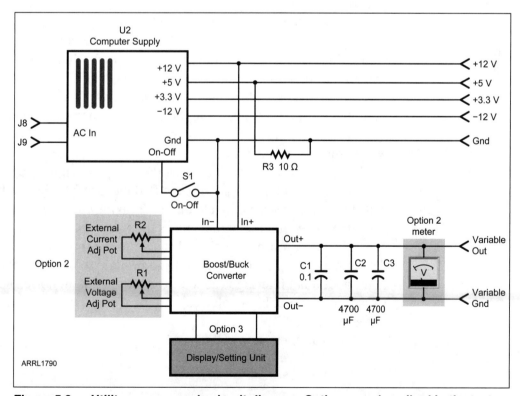

Figure 5.3 — Utility power supply circuit diagram. Options are described in the text and Table 5.1.

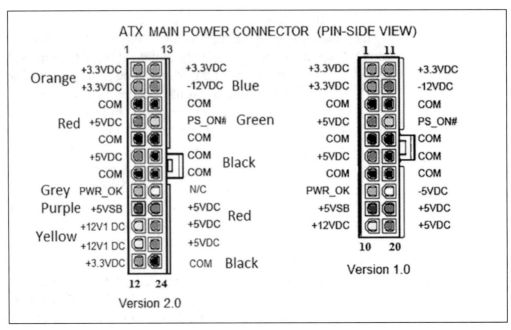

**Figure 5.4 — ATX color codes.**

The three main fixed voltages are brought out to the front panel. There is a fourth voltage, –12, but it is limited to 200 mA and we did not bother with a front panel connection. There is a 10 Ω, 10 W resistor connected across the +5 V computer supply output. This is to satisfy the need of some supplies to have a load to operate properly. Not all units require this, but just to be sure, we included it in the circuit.

Note that there are numerous articles on the internet describing how these computer supplies may be modified to change the voltages or and/or amperage rating. Because there are almost never any circuit diagrams available for our "cannibalized" PS units and the circuits vary considerably, we elected to leave the computer supply alone and use other means. By the way, we have modified a number of these computer supplies, but the range of obtainable output voltages is typically only a few volts above 12 V and we are shooting for a much wider adjustment capability.

The B/B converter is added after the computer supply to give a wider adjustable voltage range. Typically, the B/B can accept input voltages from a few volts to 25 V or more and yield outputs ranging from less than 1 V to greater than 30 V. The circuit diagram shows the two options previously outlined. In one case, the adjustment pots are brought to the front panel. In the other case the display/setting unit is detached and brought to the front panel. For both options, additional filtering is added to lower the output noise. The filtering consists of a couple of high-value capacitors. Noise levels are less than 50 mV with this arrangement.

That is all there is to the supply!

**Table 5.1**
**Cost Comparison**

| Option | Component | Cost |
|---|---|---|
| 1) Computer supply alone | Digital Panel Meter | $3.00 |
| | Computer Supply | $0.00 |
| | 8 Banana Connectors | $5.00 |
| | On-off switch | $1.00 |
| | **Total** | **$9.00** |
| 2) 80 W Boost/Buck Converter | B/B converter | $11.00 |
| | Computer Supply | $0.00 |
| | Digital Panel Meter | $3.00 |
| | 8 Banana Connectors | $5.00 |
| | On-off switch | $1.00 |
| | Two adjustment pots | $4.00 |
| | Caps | $2.00 |
| | **Total** | **$26.00** |
| 3) 100 W Boost/Buck Converter | B/B converter | $26.00 |
| | Computer Supply | $0.00 |
| | 8 Banana Connectors | $5.00 |
| | On-off switch | $1.00 |
| | Caps | $2.00 |
| | **Total** | **$34.00** |

## Cost Comparison

**Table 5.1** shows the typical costs for each of the PS configuration options. Mostly, the cost depends on how successful you are in finding a good computer supply at little or no cost. We assume you can find a free one somewhere. The interpretation of the numbers in Table 5.1 means that, if you get your computer supply at no cost and build Option 1, the project should cost around $9. (Case cost is not included.) Naturally, a well-stocked junk box can reduce these costs.

Note that the most capable option is only a few dollars more than Option 2. By the way a similar commercial unit costs upwards of $80 to $100.

We elected to build Option 3 and that is shown in the illustrations.

## Performance

The chart in **Figure 5.5** shows the performance under load for the Option 3 B/B unit. Note that even though the unit can output more than 6 A at low voltages, it is power limited, so higher voltages are at lower current levels only.

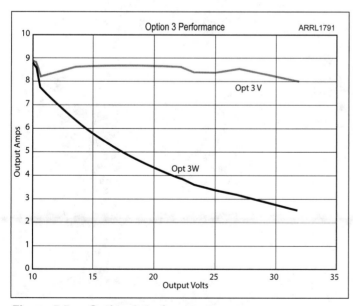

Figure 5.5 — Option 3 performance.

## Building the PS

We made up a 3D printed case for our PS, but any available case will do as long as it is large enough. (Chapter 16 details some of the construction techniques we use, and Appendix A includes information about parts sources.) Some construction notes to be aware of:

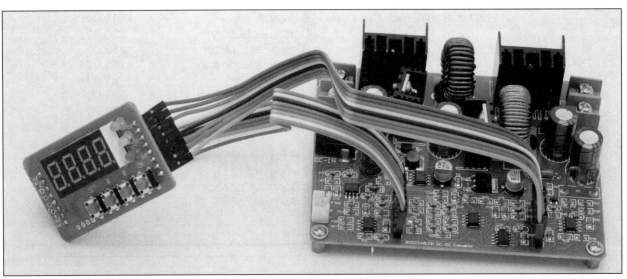

Figure 5.6 — Option 3 100-W boost/buck unit with display detached.

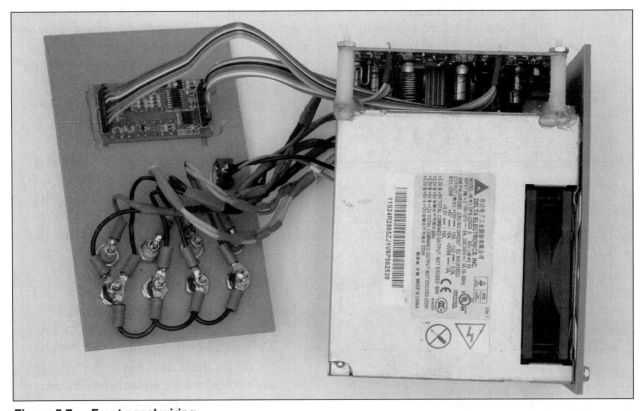

Figure 5.7 — Front panel wiring.

- Computer power supplies generate some heat, so air flow provisions for the fan must be made. The case should have adequate holes to allow circulation of air. We added vent holes at the sides, so the computer supply fan can cool the both the case interior and the B/B as well.
- Use adequate, heavy gauge wire for the main supply connections. We used #18 AWG wire, doubled up for the +12 V, +5 V and ground connections.
- The computer supply is heavy and should be mounted at the bottom of the case.
- Leave enough room around the B/B for air circulation.

**Figure 5.6** shows the B/B with the display unit detached. **Figure 5.7** shows the front panel wiring, and **Figure 5.8** is a view of the interior from the front. Note the multiple parallel wires going to the front panel banana jacks. In Figure 5.7, you can see the B/B board attached to the PS unit using threaded nylon rods. **Figure 5.9** contains a view of the unit in its case and in **Figure 5.10** we see the PPS being used with a couple of the projects from this book.

Figure 5.8 — Interior from front.

Figure 5.9 — Utility power supply in case.

## Conclusion

We elected to present this power supply project first for several reasons. First, it is a simple project to build. There are relatively few parts and those parts are easily assembled. Second, the cost of the project is fairly low... cost probably isn't a reason not to build this project. Finally, the project is useful with virtually every other project in this book. Each project needs a power source and this is going to be a more robust power supply than the wall wart you were going to use.

The power supply project shows how to use readily available computer supplies to build a very capable utility supply for the projects in this book, for less than half the cost of a commercial unit with equivalent specs.

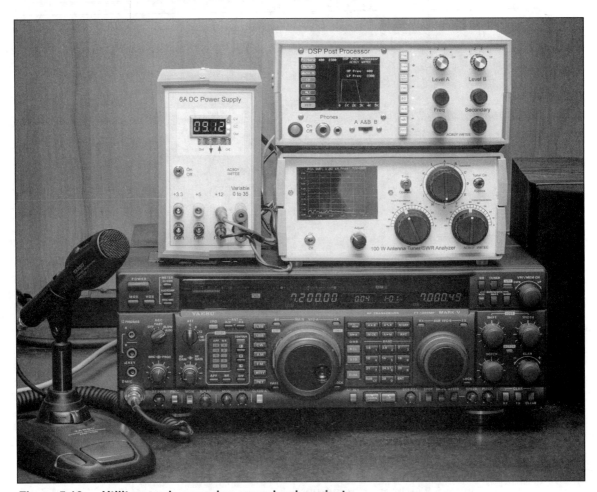

Figure 5.10 — Utility supply powering some book projects.

## CHAPTER 6

# Mini Dummy Load

One indicator of "thoughtful" hams is that they don't tune/adjust their rigs on the air. Instead, any knob-fiddling they need to do is done while directing the RF power into a dummy load instead of an antenna. All of us have heard the constant CW note that varies in intensity or tone for up to a minute while the operator is adjusting something. We've also heard others doing their SSB "Yaalll-llooww" on the air, presumably while they're also adjusting something.

That's not polite, people.

To that end, we wrote an article on building a 150 W dummy load, which was published in the November 2018 issue of *QST*. The finished dummy load ("DL" from now on) was featured on the cover of that issue as pictured in **Figure 6.1**. As you can probably tell, Figure 6.1 only shows the top of the DL. That top fits onto a quart can filled with mineral oil. It's a pretty bulky device and not exactly light to carry into the field, either. While portable DLs may fit nicely into a backpack while you're hiking up a mountain for your next Summits on the Air (SOTA) activation, most suffer from two limitations: 1) a maximum power limit of 10 W (or less), and 2) no indication of the power output you are achieving.

Figure 6.1 — The full-size DL.

## Power Limits

Essentially, the DL discussed here is very similar to the DL pictured in Figure 6.1, but without the quart can that housed the resistor network and mineral oil. The purpose of the mineral oil was to serve as a heat sink for the power going into the DL. Because the DL in the *QST* article was built for high-power applications (around 100 W), a fairly large heat sink is needed. In that article, it was the quart of mineral oil that absorbed the power. The DL discussed in this chapter is designed for portable use. The goal was to provide a small DL aimed at the QRP operator who also may need to monitor their power levels. The Mini Dummy Load (MDL) provides one solution to those design goals.

Rather than using a resistor network bathed in mineral oil, we substituted a high-power RF resistor as the load for the RF power. A typical high-power RF resistor looks similar to the one shown in **Figure 6.2**. These resistors are very small and fairly inexpensive. A 150 W, 50 Ω unit like the one shown in Figure 6.2 costs around $5 from domestic suppliers, but as little as $1.50 from online vendors. It

Figure 6.2 — A 150 W RF resistor.

Mini Dummy Load    6-1

measures less than 0.75 × 0.50 inches. The 250 W version isn't much bigger (1.0 × 0.50 inches) and sells for around $3 online. We have also seen a 650 W resistor that sells for around $20. Lower power resistors are also available and are smaller, less expensive, and would require a smaller heat sink.

While your resistor choice may affect the size of your heat sink and case, the electronics presented here remain the same regardless of power considerations. These power resistors *cannot* safely dissipate their rated power without the aid of a (large) heat sink. For this project, we added a small (4 × 2 × 1 inch) aluminum heat sink to the back of the case, with the RF resistor mounted to the heat sink. **Figure 6.3** shows the size of the heat sink and case that we used.

The project goal here was to provide a dummy load with wattmeter, but without the size and weight of the dummy load featured in the *QST* article. The unit here is quite small, portable, and can easily handle QRP power levels. Building costs are competitive with DLs with less than half the power rating.

Figure 6.3 — Heat sink and case.

### Power Tests

Obviously, there are trade offs between the size of the MDL and its ability to absorb the RF energy. The smaller the heat sink (and case), the lower the power the MDL can handle. We mounted a 150 W resistor to the heat sink shown in Figure 6.3 and applied power to it. We started out at 50 W, measuring the temperature as we applied power. Immediately the temperature of the heat sink started to rise. We raised the power to 100 W and the temperature rose faster, at which time the resistor got angry and blew the top half completely off the mounting tab to which it was secured. Clearly, even a relatively large heat sink wasn't going to even come close to the resistor's power rating.

**Table 6.1**
**Safe Power Rating for DL**

| Resistor Power Rating | Safe Operating Power Limit | Time[1] |
|---|---|---|
| 150 W | 35 W | 10 seconds |
| 250 W | 50 W | 10 seconds |

[1]Time assumes the heatsink starts at room temperature.

We also tested a 250 W version, using the same heat sink. We managed this test without destroying the resistor. Our results are presented in **Table 6.1**.

As you can see from Table 6.1, safe operating power levels are substantially less than the resistor's power rating. The tests were done with a constant power source, so the power ratings are a little on the conservative side. That is, actual use with SSB or CW as the power source is probably pretty safe for the numbers presented in Table 6.1. If you need a DL with a high power rating, you might want to consider the "full-sized" DL we describe in *QST*. However, most portable operations are at QRP levels (CW at 5 W and SSB at 10 W), which makes the MDL perfect for the QRP operator.

## Construction

The first thing to do is decide on the power handling capabilities you need. With the heat sink we used during our tests, it appears that you can expect to safely dissipate about 20% of the resistor's power rating. However, there are many larger heat sinks available which would increase the power that can be dissipated. (Indeed, we saw one heat sink for sale that was almost 3 square feet and an inch thick!) The MDL is not a good choice to accommodate the typical 100 W transceiver. The DL in the *QST* article would work in that case. For QRP operation where size and weight are an important consideration, we suggest using the 250 W power resistor with the heat sink we used.

The case should be large enough to hold the heat sink and the components. **Figure 6.4** shows the case interior of the MDL we built. As shown in Figure 6.3, the heat sink spans almost the entire width of the case. We drilled and tapped four holes in the heat sink: the two outer holes for securing the heat sink to the case, and two more to secure the power resistor to the heat sink. We also applied a film of thermal grease to the back of the power resistor before bolting it to the heat sink. (After the picture was taken, we added a ground lug to the left-most mounting bolt to accept the ground connection from the BNC connector.) We offset the heat sink toward the top of the case. That way, when the case is placed on a flat surface, the display is tilted slightly toward the user. (The OLED is sufficiently bright that viewing it in full sunlight is not an issue. Still, the slight tilt makes it easier to view the display.)

Because we opted for a fairly small case, the final placement of the components was dictated by the case size rather than electrical considerations. **Figure 6.5** shows the interior of the case with the components in place. The Arduino Nano and a few components are fixed to a small prototype board. Perf board would work fine, too. (This chapter's project is sufficiently simple that we felt a PC board was unnecessary.) You might be able to read that we used a 250 W RF power resistor for the MDL shown in Figure 6.6. Things are fairly tight in the case and we had to mount the OLED to accommodate the four Dupont connectors that protrude out from the bottom of the OLED. Also, make sure the Nano's USB connector points toward the middle of the case. That position makes it easier

**Figure 6.4 — Mounting the power resistor.**

**Figure 6.5 — The interior of the MDL.**

to connect the USB cable to your PC when you want to make a code change. Mounted the other way, the left case wall blocks the Nano's USB connector.

Near the top-center of Figure 6.5 you can see the cutout in the case for the power resistor. The BNC connector is on the left side of the case and the OFF/ON switch is on the right. A 9 V battery powers the unit, which is controlled by an Arduino Nano. The OLED is a small (0.96 inch) 128 × 64 display and uses the SPI interface, which reduces the pins needed to use the display to four. (Chapter 4 has information on the SPI interface with a display.) The OLED needs a 5 V power source, which we take from the Nano's 5 V pin. Analog pins A4 and A5 of the Nano are used for the data and clock lines of the SPI interface. The display should cost around $5, perhaps a dollar more for a two-color (blue and yellow) display like we used.

We used a prefab 4-conductor cable to attach the OLED display to the Nano, but it could also be hard-wired to the display. However, because we hot-glued the OLED display to the lid of the case, make sure you make the leads between the Nano and the display long enough to provide easy access to the interior of the case. Due to the small case size, we did not socket the Nano — a practice we rarely encourage. However, we could remove the Nano if need be because there are relatively few connections to it. You could use a larger case and socket the Nano, but that increased size makes it fit less easily into a shirt pocket.

## The Circuit

The schematic for the MDL is shown in **Figure 6.6**. R3 is the power resistor you've selected. The center pin of the BNC connector is connected to the small tab on the power resistor. The ground pin of the BNC connector is wired to the heat sink via a ground lug we placed on the threaded mounting screw of the heat sink.

D1 is a BAV21 diode. While the exact diode type is not critical, but it should have a voltage rating of 250 V or more. The purpose of the diode is to rectify the RF coming from the power resistor, which is then fed into the voltage divider formed by R1 and R2. The junction of these two resistors is connected to the Arduino Nano analog pin, A1. The Nano can only tolerate voltages between 0 V and 5 V on an analog pin. Because the voltage coming from D1 may exceed 5 V, the voltage divider created by R1 and R2 keeps the voltage going into A1 within safe limits. The current coming into the voltage divider is fairly small, so ¼ W resistors are fine for R1 and R2. If you are considering a different processor in place of the Nano, keep in mind that their pins may not be 5 V tolerant. If that is the case, you need to adjust the voltage divider accordingly. The purpose of C1 is to help average the rectified RF voltage from D1 to produce a clean dc voltage. We used a 250 V disc ceramic for C1; probably

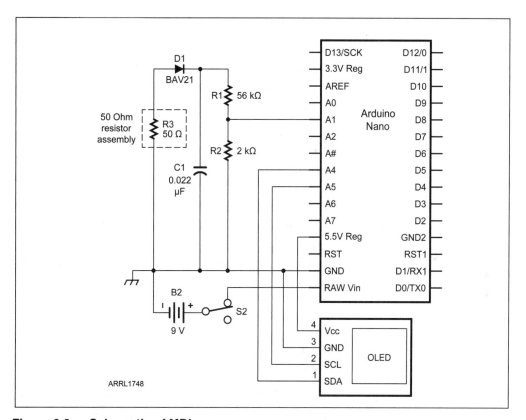

**Figure 6.6 — Schematic of MDL.**

## Safely Handling the Power Resistor

Note that the small tab from the power resistor is a little on the frail side, so try not to flex it too much when soldering to it. In our build, the connections to the power resistor's tab (the cathode of D1 and the center of the BNC connector) were the last connections we made. On an earlier build, we made the mistake of soldering a wire to the tab, which was moved around as we fit other components in the case. The tab broke off at the joint to the resistor base, rendering the resistor unusable.

> ## Safely Handling the Nano
> We've seen builders brick a Nano and then struggle with various desoldering tools to try and remove it. Folks...it's not worth the effort. If you value your time at more than a few cents per hour and the Magic White Smoke has already escaped, it makes economic sense to euthanize the $3 Nano. If you want to keep the circuit board, but add a new Nano, we suggest you "clip the legs" of the Nano as close to the chip's body as you can. When you are finished clipping the Nano's pins, remove the Nano body and use a pair of needle nose pliers to straighten the pins left behind by the surgery to a more-or-less vertical position. Use those clipped legs as solder points for the new Nano. Note that you do not need to solder all pins to the mounting pins. The schematic diagram in Figure 6.6 indicates which Nano pins are actually used by the circuit.

overkill for QRP use, but they were cheap and available. The rest of the circuit shows how the 9 V battery power is routed through the SPDT switch and to the Nano and OLED.

## Calibration

The MDL uses an analog pin to measure the RF voltage passing through D1. The analog pins on a Nano map the voltage read on the pin to a value between 0 (0 V) and 1023 (5 V). These 1024 possible values map to 0.0049 V (4.9 mV) per unit read on the analog pin. To properly calculate the power, we need to use RMS voltages rather than peak-to-peak (p-p) voltages like we're reading from the analog pin. Therefore, we take the average p-p voltage and multiply it by the square root of 2:

$$V_{DC} = V_{RMS} \times \frac{\sqrt{2}}{2}$$

For example, if the sensor returns a reading of 100, we are reading a $V_{RMS}$ of 0.49 V, and applying our equation:

$$\begin{aligned} V_{DC} &= 0.49 \times 1.4142 / 2 \\ &= 0.49 \times 0.707 \\ &= 0.3464 \end{aligned}$$

We need to add back the voltage drop across the diode, D1, which we measured at about 0.608 V for a total of 0.9544 V. However, the reading on analog pin A1 is the result of reading the voltage *after* the voltage divider network formed by R1 and R2. In a circuit, we can express the working of the voltage divider in the equation:

$$V_{out} = R2 \times \frac{V_{source}}{(R1 + R2)}$$

In our example, we know the output voltage is 0.9544 V and R2 is 2 kΩ and R1 is 56 kΩ. Substituting into the equation:

$0.9544 \text{ V} = 2000 \times V_{source} / (56000 + 2000)$
$\qquad = 2000 V_{source} / (58000)$

$0.9544 \times 58000 = 2000 \, V_{source}$

$55356.2 = 2000 \, V_{source}$

$V_{source} = 27.6776$

Applying the standard power formula:

$P = V^2 / R$
$\quad = 27.6776 \times 27.6776 / 50$

$P = 15.321$

or about 15 W.

As it turns out, our 250 W RF resistor actually measured 51.1 Ω and the values of R1 and R2 were also slightly off. In an effort to assess the accuracy of the MDL, we measured the power output from a commercial transceiver at various power settings. Those measurements were made with a commercial wattmeter and the MDL, and then we also read the actual RF voltages in the circuit so we could calculate the power, too. **Table 6.2** shows the results.

The tests show that our MDL measured a little less that the commercial wattmeter in two instances and the same or higher in the other two measurements. For the power levels calculated from the measured voltages, MDL was lower on the first two measures and higher on the last two. Still, the results were better than expected given we did not expect any linearity because of the diode. On average, the MDL is a little lower on measured power than the commercial wattmeter (19.7 W versus 19.62 W). Because we would like to approach the readings from the commercial power meter, the code multiplies the results of the calculations by *CALIBRATIONOFFSET*, which we set at the top of the code to:

```
#define CALIBRATIONOFFSET   1.00408
```

That is, taking $19.62 \times 1.00408 = 19.7$ W, which is the value of the commercial power meter. This should yield a power reading that is close enough for government work. If you expect to only use a limited range of power levels (for example, 1 to 5 W), you might discard any "higher wattage readings" (for example, over 10 W) and derive a value for *CALIBRATIONOFFSET* using only the lower power levels.

### Table 6.2
**Power Measurements**

| Transceiver Power Setting (Watts) | Commercial Wattmeter | RF Voltage In | Watts Calculated (51 Ω) | MDL Displayed Watts |
|---|---|---|---|---|
| 1 | 1.1 | 7.27 | 1.03633 | 0.98 |
| 5 | 4.7 | 16.83 | 4.91350 | 4.7 |
| 20 | 20.0 | 31.2 | 19.0870 | 19.4 |
| 50 | 53.0 | 51.4 | 51.8031 | 53.4 |
| Average | 19.7 | | | 19.62 |

## MDL Software

The software to drive the MDL is pretty simple, but there are a number of changes you need to make to insure a reasonably accurate power reading. These are symbolic constants defined at the top of the program:

```
#define MYDUMMYLOADOHMS         51       // Your RF power resistor value
#define MYR1VALUE            56000       // Exact value of resistor R1 in voltage divider circuit
#define MYR2VALUE             2000       //                "              R2              "
#define CALIBRATIONOFFSET  1.00408       // This, too. See book narrative for calculation.
#define DIODEVOLTAGEDROP     0.608       // Voltage drop for diode you selected.
```

Our power resistors, for example, varied from 49.8 Ω to 51.1 Ω. We actually measure the voltage drop for the diode we used, but you might be satisfied with the rating from its spec sheet. You can also change the values for R1 and R2, too. Each of these constants should help hone your measurements in on the "correct" value.

The first thing that the program does in *setup()* is determine if you are compiling the program for debugging purposes. If you are compiling with debugging in mind, the program not only compiles in the Serial object, but it also causes the *TestSensorRead()* function to be called in *setup()*. The *TestSensorRead()* function spins through *ITERATIONS* reads of pin A1 and displays the minimum and maximum values that it finds. This simple test is designed to give you an idea of the range of values the sensor pin is reading.

If the code is not compiled with the *DEBUG* flag set, the Serial monitor object is not compiled into the program. The program then initializes the OLED display object, shows the *Splash()* screen and then starts executing the *loop()* function.

The *loop()* function samples the power being fed into the MDL *ITERATIONS* times by using *analogRead()* on pin A1 (SENSORPIN). After *ITERATIONS* samples are taken, the average reading is passed to the *dtostrf()* function (double-to-string-formatted) to format the string for display on the OLED. The call is:

```
dtostrf(CalculateWatts(sum), 5, 2, buff); // Calculate the power
```

The first argument is a nested function call to *CalculateWatts()* that passes the average analog value read from pin A1, *sum*, converts it to watts, and passes that value back where it is used as the first argument in *dtostrf()*. The second parameter, 5, tells the function we wish to format the watts in a field of 5 digit characters. The third parameter, 2, says we wish to have 2 decimal places displayed for the watts number. The last argument, *buff[ ]*, is the character array that holds the now-formatted number of watts. Note that we defined *buff[ ]* to hold 6 characters even though 5 are the maximum digits to be displayed because we must have room for the null ('\0') string termination character. The display object then displays the contents of the *buff[ ]* character array on the OLED display.

The call to *delay()* pauses the program for one second, and then the loop begins again. The *delay()* call helps to prevent flickering when the display content is updated. One improvement would be to call *CalculateWatts()* outside of *dtostrf()*, store the return value (*watts*) and then compare that value to the previously-calculated value (*previousWatts*) and, if they are the same, bypass the display code.

Figure 6.7— The finished Mini Dummy Load.

This would cause the OLED to only repaint itself if the measured wattage has changed. You might be asking: "If this is an improvement, why didn't you do it in the first place?" Our response: "We wouldn't dream of taking all the fun away from you!" (Actually, we did do that and it works nicely that way.)

The MDL source code is fairly short and is presented in **Listing 6.1**.

Our finished Mini Dummy Load is shown in **Figure 6.7**. The faceplate for the MDL (and many other projects in this book) is made by using a graphics editor to arrange the cutout(s) and then print the faceplate on heavy photo paper. We let the faceplate dry completely and then use a very sharp knife (X-Acto) and straight edge to cut out the OLED (and any other controls that might be on the faceplate) opening. In most cases, this approach removes the need for a bezel around the display opening. More details on this type of construction is presented in Chapter 16.

## Conclusion

The goal of the MDL was to take the large DL and shrink it down so you could still easily take a DL into the field with you. The MDL is small enough to fit in your shirt pocket, but can still take up to 35 to 50 W depending on the power resistor you selected. However, it is also useful when you're building any RF circuit that needs to be terminated with a 50 Ω load. Finally, some QRP contests have fairly clear rules about power levels that may be used during the contest. The MDL makes it easy to insure that your power falls within whatever those power constraints may be.

One idea for an enhancement would be a switch that bypasses the MDL, routing power directly to the antenna. That way you wouldn't have to mess around with connecting/disconnecting each time you want to use the MDL.

We think you'll find the MDL a very useful addition to your bench and your GO bag!

**Listing 6.1**
**The MDL Source Code**

```c
/*
   Release 2.01  April 27, 2019, Jack Purdum, W8TEE, adjusted constants for power calculation
                 and modified test routine.
   Release 2.00  April 24, 2019, Jack Purdum, W8TEE, Al Peter, AC8GY Project start from
                 original DL
*/
#include <SPI.h>
#include <Wire.h>
#include <Adafruit_GFX.h>         // https://github.com/adafruit/Adafruit-GFX-Library
#include <Adafruit_SSD1306.h>     // https://github.com/adafruit/Adafruit_SSD1306

//#define DEBUG                   // Uncomment to calibrate and debug

/*
   The following symbolic constants will vary somewhat depending upon how you build
   your version of the MDL.
*/
#define MYDUMMYLOADOHMS         51          // Measure your RF power resistor resistance and
                                            //    place its value here
#define MYR1VALUE               56000       // Resistor R1 in voltage divider circuit
#define MYR2VALUE               2000        //     "     R2             "
#define CALIBRATIONOFFSET       1.00408     // This, too.
#define DIODEVOLTAGEDROP        0.608       // Voltage drop from diode you selected.

#define MILLIVOLTSPERUNIT       0.0049      // 10-bit ADC is 1024 units with a max of 5V on the pin
#define OLED_RESET              4           // Used by the library
#define SENSORPIN               A1          // Analog pin that samples the voltage
#define ITERATONS               30          // We use a sample average for display
#define SQRT2DIVIDEDBY2         0.707       // RMS to DC conversion constant

#define SENSORMAX               1023        // Max-min values for analog pin A1
#define SENSORMIN               0

Adafruit_SSD1306 display(OLED_RESET);
float CalculateWatts();

int sensorValue = 0;         // the sensor value




/*****
  Purpose: To show sign on at start up

  Parameter list:
    void

  Return value:
    void

    CAUTION:
*****/

void Splash()
{
```

```
    display.clearDisplay();        // Clear the graphics display buffer.
    display.setTextSize(1);
    display.setTextColor(WHITE);
    display.setCursor(0, 0);
    display.println(" Mini DUMMY LOAD");
    display.println("        by");
    display.println("Jack Purdum, W8TEE");
    display.println(" Al Peter, AC8GY");
    display.display();
}

void setup()   {
#ifdef DEBUG
    Serial.begin(115200);                             // Don't need the serial object unless debugging
#endif

    display.begin(SSD1306_SWITCHCAPVCC, 0x3C);        // initialize with the I2C addr 0x3C (for the 128x32)

    Splash();                                         // Show start up screen
#ifdef DEBUG                                          // Probably only need to call once to tweak things
    TestSensorRead();                                 // Done when first testing
    delay(500);
#endif
}


void loop() {
    char buff[6];
    char pad[] = "       ";    // 7 spaces
    int where;
    int i;
    float sum;

    sum = 0.0;
    i = 0;
    while (i < ITERATONS) {                           // Do a bunch of reads
        sum += (float)analogRead(SENSORPIN);          // Input from voltage divider
        delay(10);                                    // Let pin settle
        i++;
    }
    sum /= ITERATONS;                                 // Average reading after ITERATIONS analog reads
#ifdef DEBUG
    Serial.print("sum= ");Serial.println(sum);
#endif
    dtostrf(CalculateWatts(sum), 5, 2, buff);         // Calculate the power

    where = strlen(buff);                             // How big is the result?
    strcpy(&pad[6 - where], buff);                    // Update only the new number.
    display.clearDisplay();
    display.setTextSize(1);
    display.setTextColor(WHITE);
    display.setCursor(0, 0);
    display.println("     WATTS IN");
    display.setTextSize(3);
    where = (6 - where) * 10;
    display.setCursor(where, 10);
```

```cpp
      display.println(buff);
      display.display();
      delay(1000);                                  // This is here to lessen display flicker

}

/*****
    Purpose: To convert the value read from the MDL/s analog pin and convert it to watts. The
             circuit uses a voltage divider to make sure the Nano's 5V limit is not exceeded.
             This is based on the formula:

                   Vrms = R2 * Vin / (R1 + R2)

             Solving for the input voltage:

                   Vin = (Vrmst * (R1 + R2)) /  R2

             Knowing the input voltage allows us to determine the power: Pw =  Vin * Vin / R

    Parameter list:
      float averageValue     the value read on the analog input pin

    Return value:
      float                  the calculated power in watts

       CAUTION:
*****/
float CalculateWatts(float averageValue)
{
  float powerWatts;
  float voltsRMS;
  float voltsIN;

  voltsRMS = .707 * (averageValue * 6.0 / 1024.0) * (58.0 / 2.0);   // Convert mapped sensor
                                                                   // value to RMS volts
  voltsRMS += DIODEVOLTAGEDROP;                                    // Add back diode D1
                                                                   // voltage drop
  powerWatts = (voltsRMS * voltsRMS ) / MYDUMMYLOADOHMS;           // Get the power

  return powerWatts * CALIBRATIONOFFSET;                           // Apply your factor
}

/*****
    Purpose: To measure and display min/max readings from A1 voltage divider pin

    Parameter list:
       void

    Return value:
       void

       CAUTION:
*****/

void TestSensorRead()
{
```

```
  int i = ITERATONS;                    // We're going to get a series of readings
  int sensorMin = SENSORMAX;            // minimum sensor value
  int sensorMax = SENSORMIN;            // maximum sensor value

  while (i--) {
    sensorValue = analogRead(SENSORPIN);
    delay(100);                         // Settle between readings...
    if (sensorValue > sensorMax) {      // record the maximum sensor value
      sensorMax = sensorValue;
    }
    if (sensorValue < sensorMin) {      // record the minimum sensor value
      sensorMin = sensorValue;
    }
  }
#ifdef DEBUG                            // Only show if DEBUG mode set
    Serial.print("sensorMax = ");
    Serial.print(sensorMax);
    Serial.print("   sensorMin = ");
    Serial.println(sensorMin);
#endif
}
```

# CHAPTER 7

# Morse Code Tutor

Been there, done that...I get it. I got my Novice license in 1954 and sat for my General a year later. On the way to the Federal Building in Cleveland to take the exam, I told myself: "If I pass, I'll never send another letter of CW again... ever!" Truth be told, I pretty much stuck to that until I retired and subsequently discovered the fun and challenge of working the world using QRP power levels into a not-so-great wire antenna. To be effective with the limited antenna system I have, CW pretty much dictated the mode of choice. Even though I was horribly rusty, I quickly found out that I really enjoyed CW. I also found that other CW operators are extremely tolerant of a sloppy fist.

At the time I was taking my General class exam, I dreaded the code test and wished I didn't have to go through it. But now, after a period of inactivity, QRP operating and CW has breathed a new energy into me that I wish everyone could experience. There's a feeling of excitement and joy in making DX contacts with 1 W and it's satisfying to listen to a good fist.

I surveyed my club and found that, of the members who currently do not know Morse code, 87% said they would like to learn it. Thus encouraged, at the next meeting I offered to teach a Morse code course and set a date and time during the week when I figured most members would be available. No matter what date/time combination I picked, that was the precise date and time everyone was rearranging their sock drawer. For whatever reason, I simply cannot sell learning Morse code to the members. It is our hope that this Morse Code Tutor (MCT) may encourage some of you to learn Morse code and see what you've been missing.

## Morse Code Tutor

But first, let's see what our MCT looks like. **Figure 7.1** shows the prototype I built. Some of you may think the case looks like the Antenna Analyzer case from the November 2017 issue of *QST*. You'd be right, since I had a few cases

### Morse Code Tutor PC Boards

We should mention that, since Jack presented this project in a paper given at the Four Days In May conference in 2019, there are already two PC boards that others have made using his FDIM design. If you plan to build the MCT, do an internet search for the various boards that are available.

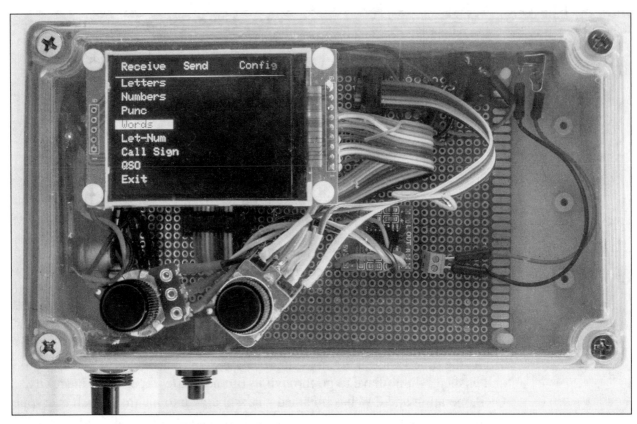

**Figure 7.1** — The Morse Code Tutor in its case.

left over from a club build. The case is roughly 7 × 4 inches and, as you can see, has lots of extra room left over.

The MCT is based upon the STM32F103 microcontroller, aka the "Blue Pill" (BP from now on). So far, I've yet to find a library that doesn't work with the BP. (It appears that some kind soul(s) took the time to redo the core libraries for the BP.) By comparison, the BP has 128 KB of flash memory (Nano is 32 KB), 20 KB of SRAM (versus 2 KB), and is clocked at 74 MHz (versus 16 MHz). The BP is slightly larger than a Nano, and can be purchased for under $2 in small quantities.

The 2.4-inch, 320×240 TFT color display costs about $7 and uses the SPI interface (see Chapter 4) to communicate with the BP. The volume control on the front panel controls a 3 W amplifier. There is a headphone/speaker jack and key jack in the upper-right back corner of the case. The other control is a rotary encoder with switch that controls the menuing system. We added a dc-dc buck converter that outputs 5.0 V. The BP uses 3.3 V, but has a regulator on it that can accept 5 V. Note that the legend in Figure 7.7 later in this chapter shows that many of the STM32 pins are *not* 5 V tolerant. All other pins are 3.3 V. The 5 V also feeds the amplifier and the display.

**Figure 7.2** shows the MCT parts outside the case. As you can see in Figure 7.2, there are not many components to the MCT. Rather than investing time in a PCB, we just decided to build it on a proto board instead. The simplicity of the circuit, its low parts count, and its low cost makes this a good candidate for a club build. (Again, check for the availability and price of PCBs based on this design.) Indeed, the MCT costs less than a popular code practice oscillator (CPO) that has none of the features of the MCT. If you build one, we hope you're willing to share it with your club members.

Maybe this project will help some of you to fall over the edge and into the CW pool. The water's great...take the plunge!

## Learning Morse Code

First, I learned Morse code the wrong way. I learned by memorizing dits and dahs. The process of learning Morse that way requires a cumbersome translation sequence. First, you listen to the dits and dahs coming in. Second, your brain does the equivalent of a binary search to find the character that matches that sequence of dits and dahs. Finally, you then write the letter on a piece of paper. All not good.

The better way is to forget about dits and dahs and listen to the rhythm of the characters. Very few high-speed CW operators write anything down. Indeed, I find I'm hard-pressed to copy at 15 WPM while writing the message, let alone cruising along at 30 WPM. Enviable CW operators simply close their eyes and read the code as it scrolls across their eyelids ("eyelid speed"). If they're in a contest, they might type the log entry as they listen...it's a thing of beauty to watch.

But, if you're just starting out, how can you sense a code rhythm when the sequence is coming at you at 10 characters per minute?

You don't.

**Figure 7.2 — MCT components.**

## The Koch Method

The Koch Method of learning Morse code, developed by Ludwig Koch, has been around since the 1930s. Essentially, the Koch Method says that you learn code best by sending it at a fairly high speed. The Koch Method starts out sending just two letters at a time, but sends them at relatively high speed. Once you achieve success at copying pairs of letters at that speed, the method adds a new letter to the mix. Given a high initial speed, you won't have time to count dits and dahs. By necessity, you end up listening to the rhythm of the characters rather than engaging in dit/dah counting and its attendant counting/translating/searching. The spacing between letters is often stretched out somewhat over what would normally be used for letter spacing (for example, 4 times a dit length — more on this below). The spacing between "words" is also stretched out.

Clearly, there's an initial hump to get over before the Koch method produces results. At first, the code simply sounds like a bunch of angry bees. Eventually, however, if you practice and stick with it, things start to fall into place.

## The Farnsworth Method

The Koch Method is not the only suggested method for learning code. The Farnsworth Method starts by sending letters/words in code at your ultimate "goal speed." However, the Farnsworth Method inserts relatively exaggerated delays between words. Again, this forces you to listen to patterns, not dits and dahs. For me, that's easier said than done because of the way I learned Morse code.

Today, another successful method is the combination of the two methods, and that's the approach taken here with the Morse Code Tutor. To illustrate, suppose you'd like to be able to ultimately send/receive code at a "goal speed" of 30 words per minute (WPM). There's no way that you can start out at 30 WPM in the morning and hope to be successful in a 30 WPM QSO that evening. So, we throw in a little of the Farnsworth Method and set the inter-word spacing to a much slower speed. Perhaps we set the "Farnsworth Spacing" at 5 WPM. When you first try this approach, it will sound like a burst from a chainsaw (word 1) followed by a slight pause followed a second burst from the chainsaw (word 2). Don't get discouraged...it's worth the effort.

Some hypothetical numbers can illustrate this process. For the old FCC code requirement, a "word" was taken to be 5 characters per word (CPW). (Technically, the word PARIS was the timing benchmark.) So, if you want to copy at 30 words per minute (WPM), that translates to 150 characters per minute (CPM), on average. Therefore:

```
150cpm = 30wpm * 5cpw                   // 150 characters per
                                        // minute (cpm)
.4spc = 60secondsPerMinute / 150cpm     // 0.4 seconds per
                                        // character (spc)
```

Therefore, the characters are coming at you at rate that is faster than one character every half second. That's the bad news. The good news is, throwing in a 5 WPM Farnsworth delay, yields:

```
25cpm = 5wpm * 5cpw
2.4spc = 60secondsPerMinute / 25cpm   // 2.4 seconds per
                                      // character
```

This means, if we send at 30 words per minute and use a Farnsworth delay of 5 WPM, you have 2.4 seconds between chainsaw bursts! If you chew on those numbers for a few seconds, you'll realize the way to get "up to speed" is to keep collapsing the Farnsworth delay to the point where it, too, is at 30 WPM. Simple!

## Which One to Use?

People learn at different rates using different methods. The Morse Code Tutor (MCT) allows you to adjust its parameters to use a method that you like best. You can blend the Koch and Farnsworth methods as you see fit. Keep in mind that there are no shortcuts to learning Morse code — it takes time, perseverance, and patience to master. Most CW teachers say that listening for 15 to 30 minutes a day works best. That said, it is truly a worthwhile investment of your time.

## Morse Code Tutor Feature Set

We're going to explain the MCT feature set within the framework of the menuing system that was described in Chapter 4. **Figure 7.3** shows the entire menu system at once. Figure 7.1 shows the actual MCT starting menu.

The main menu appears at the top of the page and always uses horizontal scrolling. So, turning the encoder clockwise while on the *Receive* option would advance the active menu choice to *Send*. The menu "wraps" in either direction (that means turning CCW on *Receive* goes to *Config*, and turning CW on *Config* would activate *Receive*).

```
Morse Tutor Menus

Receive      Send         Config       (Main Menu)
Letters      By Two's     Speed        (Sub-Menus)
Numbers      Mix          Encoding
Punc         CopyCat      Tone
Words        Flashcard    Dit Pad
Let-Num      Exit         Save
Call Sign                 Exit
QSO
Exit
```

Figure 7.3 — The Morse Tutor Menuing System.

## Receive Submenu

First, unlike the main menu, all submenu options are vertically oriented. Turning the encoder clockwise advances to the next submenu option listed below it. If you are on the last submenu option in the list (*Exit*), the cursor wraps around to the first submenu option. Likewise, if you are on the first submenu option (*Letters*) and rotate the encoder counter-clockwise, you wrap around to the last submenu option (*Exit*). All submenus operate in this manner.

The *Letters*, *Numbers*, *Punc*(tuation), and *Words* options send random sequences of that option at the currently-set words per minute. If Farnsworth encoding is active, the code inserts an appropriate Farnsworth delay into the stream at the appropriate time (such as at the word or letter end). *Let-Num* simply sends a random collection of letters and numbers. *Call Sign* sends only randomly-generated call signs. We did this as call signs seem to be a stumbling block in learning Morse code, probably because a call sign mixes letters and numbers.

The *QSO* option generates what might be called a "typical" sequence of words that you might encounter in a CW contact. If you look near the top of the MCT's .ino source code file, you will see a number of arrays that are pointers to character arrays. These arrays contain subject-related strings, such as:

```
char *rigNumbers[] = {"7851", "7700", "7600", "7410", "7300",
          "7100", "718", "78", "9100",           // ICOM
          "990", "480", "2000",                  // Kenwood
          "991", "891", "1200", "3000", "2000",  // Yaesu
          "1500", "6300", "6400", "6500", "6700" // Flex
          };
```

By using a pseudo-random number generator to generate an array index number, we can make each practice QSO a little different. There are other subject-strings for names, cities, common QSO words, antenna, weather forecast, and a random call sign generator, too. This helps keep the practices varied and will keep you from "memorizing" the content of a practice QSO. Indeed, it's the string space resource demands that made us pick the STM32 for the project.

Again, looking in the .ino source code file, you will see other arrays that hold state abbreviations, ARRL Section abbreviations, and ARRL Sweepstakes contest entry class divisions. We have not used them *per se*, but they are there if you want to modify the code to use them. (If you are modifying my source code and don't want to use these, comment them out, which will save you some memory space.)

Of course, the best way to practice is to listen to CW on the air. Really? Why not just use the MCT for practice? Well, several reasons. First, the MCT sends almost perfect code using the ideal spacing ratios between code atoms (dits and dahs), characters, and words. Few operators send perfect code, so practicing with different "fists" makes that practice just that much more useful. Second, we can't include all of the words that might pop up in a QSO. The subject matter of QSOs ranges all over the spectrum, and we can't possibly include that depth of word variety. Also, because a QSO is a real conversation, it will be more interesting than the random QSO we generate.

Each submenu ends with an *Exit* option. Selecting that option sends program control back to the main menu.

## Send Submenu

You're going to find that sending Morse comes quicker and is much easier to learn than receiving. That's why there are relatively few *Send* submenu options. The *By Two's* option is the starting point for using the Koch method of learning Morse. It sends only letters. We encourage you to activate the Farnsworth delay, too, as it makes for a more successful start than using the normal code spacing. You can activate these options using the *Config* menu, setting the Koch target rate (such as 30 WPM) and implementing a Farnsworth delay (5 WPM, for example). Our experience is that, if you don't experience some initial success, you will get discouraged and give up. Not good.

The *Mix* option is similar to the *ByTwo's*, except digit characters (numerals) are also sent. The speed and spacing depends on how you set your *Config* options.

I really like the *CopyCat* submenu option. As it is currently coded, it generates a random call sign and sends it at your target speed. (We use call signs because beginners have a little more difficulty with a mixture of letters and digit characters.) The code then waits for you to send the call sign back to the MCT. If you do not send it correctly, a question mark appears on the display and the same call sign is sent again and the program awaits your response. The MCT waits in this pattern until you get it correct, at which time it generates a new call sign. You can terminate the sequence by pressing the encoder knob to activate its built-in switch. You could, of course, substitute a separate normally-open SPST switch for the same purpose if you wish.

## Other Learning Tools

Jack was a professor for 40 years and one thing he learned was that one approach to teaching a concept might yield an ah-ha moment for one student and a deer-in-the-headlights for another. So, it rarely hurts to attack a problem from multiple sides. Something we call the Flashcard approach is one such attempt. The *Flash* option was suggested by Joseph Street and we are unaware of an app that is using this approach to learn Morse code. Essentially, the *Flash* approach uses an electronic equivalent of flash cards to learn code. That is, you hear the letter at high speed (again, it's all about rhythm) and, after the Farnsworth delay, it presents an image of that character on the display. The image is much larger than normal text so it does take on a flashcard feeling. The image of the character remains on the display for FLASHCARDVIEWDELAY milliseconds. (You can change this delay easily by editing it in the MorseTutor.h header file.)

I added the flashcard option because everyone learns in slightly different ways, so if this works for you, great. If not...well, don't choose it! Like they say, if the only tool you have is a hammer, don't be surprised if all your problems look like a nail. The *Flash* approach is just another tool to hang on your belt if you choose to do so. As it currently is configured, the MCT only sends letters using the *Flash* option. It would be useful to extend this to include commonly-used words in a QSO. The code includes the top 100 most commonly-used words in normal conversation. Feel free to change these to suit your needs.

## Config Submenu

The Config submenu allows you to set program-wide preferences that can remain in effect even if power is removed from the MCT.

The *Speed* option allows you to set the base words per minute (WPM) that is used for practice sessions. You should change this whenever you feel comfortable at the current WPM. You want to continually push yourself toward higher speeds, with the goal of reaching "eyelid speed." If you plan on using the Koch/Farnsworth methods, we'd like to see you be ambitious and set it fairly high (for example, 25 to 35 WPM).

The *Encoding* option is used to activate/deactivate the Farnsworth encoding scheme. This choice plays a role in how the practice code is sent to you in the *Receive* mode. This option lets you set both the Koch (target) speed (say, a goal speed of 30 WPM) and the Farnsworth "gap" speed (perhaps 5 WPM if you're just starting out). In the program source code, you will see the goal speed stored in a variable named *targetSpeed* while the gap speed is stored in *farnsworthDelay*. Here again, you should continually advance the *farnsworthDelay* toward faster speeds to push yourself. Your goal is to get the *farnsworthDelay* speed to match the target speed so they are both occurring at the same speed, thus making the gap speed zero.

The *Tone* submenu option simply lets you adjust the audio note used as a sidetone when sending Morse. Most CW operators seem to cluster around 700 Hz, but it's a highly personal choice. Rotating the encoder changes the tone. The tone changes in real time when you turn the encoder. Turning counter-clockwise (CCW) lowers the tone and clockwise (CW) raises it. When you've found one that sounds good to you, simply press the encoder shaft switch to store it in EEPROM. (See note on EEPROM later in this chapter.)

The *Dit Pad* submenu option lets you alter the paddle lever that is used to send a stream of Morse code dits. Yes, we do assume you want to use a paddle, as sending good code with a straight key at more than 20 WPM is not easy. When you're two years younger than dirt, it gets even more difficult. You can also change the dit paddle by changing its assignment in MorseTutor.h header file. Look for the symbolic constants:

```
#define DASHPADDLE    PB7    // tip
#define DITPADDLE     PB8    // ring
```

and change as needed. Clearly, it's better to change it through program control so your code remains the same as everyone else.

## STM32F103 EEPROM

The *Save* submenu option is used to write the setup information in the *Config* menu option to EEPROM...well...sort of. For this project, we selected the STM32F103 microcontroller. We did this mainly for the memory resources. The Arduino family of controllers simply do not have the SRAM memory for all of the strings that are used in the program. The Arduino Mega2560 has enough flash memory, but is still pretty skimpy on SRAM. Clock speed wasn't really an issue, since human reaction times are glacial to these microcontrollers.

> ## Storing Data
> Note: You do not want to use EEPROM for data that is frequently changed. Instead, use EEPROM for storing data that does not change that often, like the configuration data we store in EEPROM. The reason is because flash memory (hence, our EEPROM) has a finite write cycle of about 10,000 writes. While that may sound like a lot, it isn't in many applications (such as a data logging program).
>
> If you want to use the STM32F103 for data logging, consider adding an SD read/write module. They are cheap and can be used to store very large data sets. An interesting add-on would be for you to include an SD card to the program, store *Gone With The Wind* on it in ASCII, and then have it play back the book in Morse. The MCT might help pass the time on those long commutes!

However, you can buy an STM32F103-based controller for under $3, compared to over $10 for the Mega. Given the relative price points and the resource bases, the choice was a no-brainer.

One downside of the STM32F103, however, is that it does not have any EEPROM on board. Actually, neither does the Teensy or ESP32. They get around this limitation by emulating EEPROM in the flash memory that is available. The Teensy 3.6 fixes the EEPROM to 4096 bytes and sets the memory pages accordingly. Because the EEPROM memory space is emulated in flash memory, EEPROM writes on the Teensy 3.6 cannot occur at clock speeds in excess of 120 MHz. Because the Teensy's native clock speed is 180 MHz, the processor slows to the lower clock speed during EEPROM writes. While this is normally not an issue, it could be if you wanted to do some form of high speed data logging.

The STM32F103 is a little more flexible in that it allows you to determine how much EEPROM space you need. Flash memory starts at 0x0800 0000 and extends upward for 128 K. The first thing the program must do is define where the EEPROM is to reside in flash memory. We wrote a function named *DefineEEPROMPage()* to do this. Because the EEPROM demands are very small in this application, we define a small 1 K block for use as program EEPROM.

## Powering the MCT

The STM32F103 series is a 3.3 V device, but, as mentioned earlier, most of the I/O pins are 5 V tolerant. In addition, there are 5 V and 3.3 V regulators on the BP board so you can power the board from either voltage source. However, it's probably a good idea to power the small audio amplifier and the TFT display from the 5 V source, not the BP. For that reason, we are adding a buck converter to the circuit so that you can feed it anything from about 6 V to 15 V and the buck converter will output the 5 V needed for the display and also the BP.

You have a bunch of choices for the buck converter. Two basic types are shown in **Figure 7.4**. The BP is at the top of Figure 7.4 and the two buck converters are on the bottom. Both have input voltage ranges from about 3.8 V to 35 V, but the one on the left can handle about 3 A of current while the one on the right is limited to about half that level. The cost of the one on the left is

Figure 7.4 — Comparative sizes of two buck converters relative to the BP.

under $1.50 and the other is about half that price. Both can handle the project load, but see that little Phillips head screw in the lower-left edge of the low priced spread? That screw is used to adjust the output voltage and has a tendency to pop off if you look at it wrong. In this case, go big and spring for the more expensive converter. The trim pot makes adjusting the output voltage a lot less risky.

While the BP has enough power to drive a set of headphones, we added a small audio amp to the project. The audio amp shown on the left in **Figure 7.5** is based on the cockroach of audio amps, the LM386. It draws about 2 mA at idle, and costs $0.92 online. The audio amp in the middle is more expensive ($0.99), but has a potentiometer that makes it easy to adjust. However, the fact that the pot is fixed to the PC board may limit your mounting options. The amp on the right is the one we selected. It also costs about $1 in small quantities and is more than adequate for the task.

Figure 7.5 — Small audio amps.

The board on the right does not include a volume control. We chose to add one to the front panel of the case (the left knob in Figure 7.1). While you could wire a fixed resistor in the audio line, adding a pot to the circuit just makes sense. The amp puts out enough sound to make it useful in a classroom setting if needed. The specific amplifier board and speaker used aren't critical. Fidelity isn't an issue so shop around for what suits your needs.

The TFT color display we are using is a 2.4-inch, 240×320 pixel display and uses the 4-wire SPI interface. It uses the ILI9341 driver chip, for which there are a number of graphics libraries that can be used. We are using the Adafruit ILI9341 library for the graphics routines in this project. However, in this project we only use the text methods on the display. Adafruit makes good products and provides useful software/libraries free. We encourage you to support them as much as you can.

We opted for a 2.4-inch display because we wanted to keep the project fairly small. There's nothing to keep you from using a larger display other that cost, case size, and perhaps a little more power consumption. **Figure 7.6** shows the backside of our display. We have not utilized the SD card reader that is provided on the back of the display. If you elect to use the SD card, there are plenty of resources you can use to guide you. Simply do an internet search on "using Display SD reader." The unused pins for the SD card are on the left side of the display in Figure 7.6. The pins on the left side of the display are used by the SPI interface to the BP.

**Table 7.1** shows the connections between the TFT display pins shown on the left side of the board in Figure 7.6 and the BP pins.

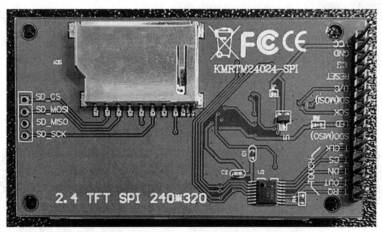

**Figure 7.6 — Backside of the 2.4-inch TFT color display.**

### Table 7.1
**TFT display to BP**

| TFT display | SDO/MISO | LED  | SCK | SDI/MOSI | DC/RS | RESET | CS  | GND | Vcc   |
|-------------|----------|------|-----|----------|-------|-------|-----|-----|-------|
| Blue Pill   | PA6      | N/C* | PA5 | PA7      | PA0   | N/C*  | PA1 | GND | 3.3 V |

*The LED and RESET pins connects to the Vcc pin. The LED connection is through a 10 kΩ resistor. The two pins do not connect directly to the BP. See Figure 7.8.

# Microcontroller Pins versus Their Function

The BP connections refer to the pins as labeled on the BP board. The are many images of the BP online that you can download/print and you may want to do that so you can reference it. However, those images use BP legends instead of the board labels. **Figure 7.7** is a common image used to describe the BP pins.

Note that most of the pins are capable of more than one function. For example, A5 on the BP board appears several ways in an image: physical pin number (15 on the right side of the image); the STM32F103 general input/output (GIO) pin name (PA5); its analog name (ADC5); and its SPI name (SCK1). Sometimes there are other interface names (I2C, CAN Bus, USB, etc.) depending upon which pin you are examining. This means that there are a bunch of ways to refer to (and use) the same pin, usually based on the context in which you wish to use it.

Because pins can be referenced using multiple terms, it can be confusing which pin is actually under discussion. This can be particularly vexing with you are looking at other sources of information. Other references, books, and articles may use other pin names based on the context in which they are being referenced. As a general rule, we will use the pin names that are silk screened

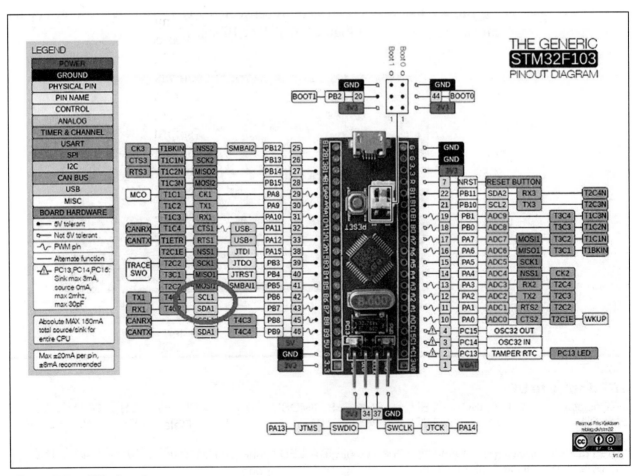

**Figure 7.7 — BP pinouts.**

7-12    Chapter 7

onto the board itself. That way you won't need to reference a printed image of the various pin names. However:

> *When we reference pin names in program code, we have no choice but to use the STM32F103 names.*

This restriction is because the core libraries that must be used within the Arduino IDE for the BP use the STM32F103 pin names, not the physical pin names. Any narrative in this book makes it very clear which pin is under discussion.

## The Bill of Materials (BOM)

There are very few parts needed to build the MCT. The Bill of Materials (BOM) presented in **Table 7.2** details the parts you'll need. When you buy rotary encoders, make sure when you order that the encoder shaft is threaded and that they provide the mounting nut. This is true for any panel-mounted connector (such as stereo jacks, power connectors, etc.). Many vendors do not supply threaded shafts with the mounting hardware. Also, buy encoders 10 at a time, as the discount can be significant and there are lots of projects in this book that use them. In fact, it makes sense to order many components in quantity because of the discount or minimum order quantity. Appendix A has a list of vendors that we often use and some don't have minimum order quantities. If you really feel that you would never use any of the extra components, check with members of your club who may need what you have left over.

Also, our recent experience is that it is now taking much longer to get components from China. A two-week delivery window was common several years ago, but now we are finding many orders taking over 40 days to arrive. While it may cost a little more, we are ordering more components from domestic suppliers. Some overseas vendors have learned that buyers are now aware of the shipping delays so they are using middlemen who advertise as a US company. However, they appear to be little more than a drop-ship storefront where, once your order is received, they place the order overseas. This usually means the same (or worse) long delay even thought they say they are a US vendor. Best

**Table 7.2**
**Bill of Materials for the MCT**

| Item | Description | Cost[1] |
|---|---|---|
| TFT Display | 2.4-inch, 240×320 TFT LCD display, SPI | $6.50 |
| BP | STM32F103 microcontroller | 3.00[2] |
| Audio Amp | LM386 or equivalent small audio board | 1.00 |
| Buck Converter | LM2596S dc-dc 3-A buck adjustable step-down converter | 1.50[2] |
| Rotary Encoder | 20 PPR (pulses/revolution) with built-in switch (e.g., KY-040) | 1.00[2] |
| Stereo socket | 5-pin 3.5-mm audio connector stereo jack socket | 0.50[2] |
| Power socket | 5.5 mm × 2.1 mm with nut | 0.50[2] |
| 10K pot | Audio taper | 0.75 |
| Small speaker | 1 - 2 inch diameter will do | 2.00 |
| | Total: | $16.75 |

**Notes**
1. Cost is approximate and derived from online searches. See Appendix A.
2. Cost is derived from a quantity purchase (e.g., quantity 10) for one unit.

way to check is to ask if they can send it by overnight delivery. If they can't, you may want to look for another vendor.

The total cost shown in Table 7.2 does not include a case or power source. Chances are you probably have a suitable case and wall wart in your junk box. Also, some may want to opt for a larger TFT display, which would likely affect the size of the case. Your call.

## The Circuit

The circuit is very simple with most of the work being done by the software driving the BP. **Figure 7.8** presents the schematic for the MCT. The circuit is sufficiently simple that there's little need for a PCB. Instead, we constructed ours on a proto board. The layout of parts is not critical, but several entry points are

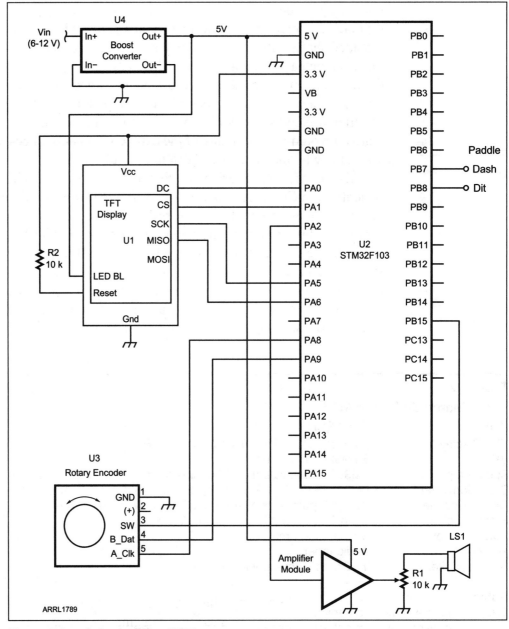

**Figure 7.8 — Schematic for MCT.**

needed in the case: 1) the connector for the power source; 2) the paddle connector; 3) the volume control; 4) headphones/speaker jack; and 5) a cutout for the display. We used a small speaker and mounted it on the top of the case, but that's not required. You could just use headphones and omit both the amplifier and the speaker. In that case, run the BP output through a dropping resistor to the headphone jack. However, if you plan to listen while driving a car, do *not* use headphones. You need to be able to hear what's going on around you while you're driving. And, of course, don't even *think* about sending while you're driving.

If you plan on using this in a classroom environment, you will want to add a speaker. The small speaker we used is rated at 12 W (it was laying around, neglected and unused), which is an H-bomb-to-kill-an-ant in this use, but the amplifier pumps out enough power for the MCT to be heard in a fairly large room.

## The MCT Software

The software for the MCT is available from our website (**groups.io/g/ SoftwareControlledHamRadio**). The code follows the standard conventions discussed earlier in this book. MCT uses the menuing system discussed in Chapter 4, so that aspect of the program shouldn't need any explanation. If some part of the menuing system seems fuzzy, a quick review of Chapter 4 should clear up any questions you may have.

Some of you may be puzzled when you download the source code because the ZIP file contains nine source code files (especially if you didn't read the early "programming" chapters). One file is the usual Arduino IDE .ino file, which always must contain the *setup()* and *loop()* functions. However, there are also two header files: one for the overall project (MorseTutor.h) and one for the menuing system I wrote (Menu.h). That leaves the other six files, which are all .cpp (C++) files. True, you could lump all the code into a single .ino file, but doing so throws away a lot of advantages that multiple files bring to the table.

First, each file has its own tab in the Arduino IDE. This makes it easier to navigate around the project. Using meaningful file names (such as Menu.cpp, ProcessCode.cpp, RotaryCode.cpp) makes it easier to locate/test/debug the code. The first tab in the Arduino IDE is always the .ino file; makes sense, since that's where program execution begins. All of the remaining files are in alphabetical order — again, making it easy to find stuff.

Second, using separate files allows the compiler to perform "incremental compiles." For example, following our discussion in Chapter 2, suppose you're having problems with one file and you keep changing it, but the other files in the project don't get changed between compiles. With incremental compiling, the Arduino compiler "keeps" an image of the unchanged files and merges them with the newly-compiled file you're working on. While this may not mean much on small files, I've been involved in a project with over 11,000 lines of code spread out over 19 source code files. I figure I save about 20 seconds (on a *very* fast machine) per compile by doing incremental compiles. That doesn't sound like much, but when you do 50 compiles per day, it adds up to more than a half hour saved each day. Given that I've been working on the project almost every day for 18 months, that's almost 250 *hours* saved. And that's just the time savings on compiles. Add saved debugging time and...who knows how much time I've saved.

Third, our assessment is that the GNU C++ compiler, which is what the Arduino IDE uses, does a better job of type-checking when the files are split up. The compiler seems to catch more mismatched function parameters than when a single file is used.

Finally, we believe that using multiple source code files actually makes debugging easier. Multiple files make us more aware of the dangers of using global data in a program. Global data is the data that are defined outside of any function or class. As such, it means that every line of code in your program has access to, and the ability to change, that data. When a bug is born, where do you start looking for the father of the bug? By defining all data within the confines of a function or method, at least you know where to start looking for the bug.

## The MCT is Simple Code...Mostly

Most of the code presented in the MCT is much like you've seen in earlier chapters. However, there is one area that is a little different. It's likely that you've seen, or perhaps even written, code for an electronic keyer. It boils down to sensing when the dit or dah paddle has been closed. The code could care less about what the contact closure means. Not so for the MCT.

For example, if you're sending the letter 'A', it's up to you to decide how long to hold the dit and dah contacts closed. However, what if we are using the MCT to practice *listening to*, not sending, Morse code? Is that dit the code just sensed the letter 'E', or is it the prelude to an 'A', 'W', 'P', 'J', or several other possibilities? When a dit or a dah is part of a larger character sequence, we call each dit or dah an *atom*. Therefore, each Morse character is made up of one or more atoms.

How can you determine whether the dit (or dah) is an atom or the letter 'E' (or 'T')? The only way to be certain is to measure the time interval between atoms. When you apply power to the MCT, it reads the current setting for words per minute (WPM) from EEPROM. Knowing the WPM mathematically determines how long a dit is (1 time unit) and how long a dah is (3 time units) since a dah is supposed to be three times as long as a dit. The time interval between atoms is 1 time unit. The time between characters is 4 time units. The time between words is 7 time units. Therefore we can distinguish atoms from characters by measuring the time interval to the next atom. That is, if we measure 4 time units between atoms, what preceded that interval was a completed character. If we measure 7 time units between atoms, we just read a word. Pretty simple.

## Decoding Characters

OK, so what's the best way to store the Morse sequences for each letter, number, and punctuation mark? If you have a mega-munch of memory, you could store the letters as character arrays (i.e., strings):

```
char letters[] = ".-",        // 'A'
                 "-...",      // 'B'
                 "-.-.",      // 'C'
```

and so on. Messages, like the practice QSO that MCT generates are stored in ASCII (American Standard Codes for Information Interchange) which are

**7-16    Chapter 7**

numeric values. If you look at an ASCII table, you'll see that 65 represents the letter 'A', 66 is 'B', and so on. Therefore, if the code reads an 'A', it is easy to figure out what to send:

```
index = letterRead - 'A';
messageString = letters[index];
SendLetter(messageString);
```

which says:

```
index = 'A' - 'A';
index = 0;
messageString = letters[0];
SendLetter(".-");
```

If you're running low on memory and string space is scarce, you can also code each character in binary, where a dit is a 1 and a dah is a 0:

```
char letterTable[] = {   // Morse coding: dit = 0, dah = 1
          0b110,      // First 1 is the sentinel marker
          0b10111,    // B
          0b10101,    // C
          0b1011,     // D
          0b11,       // E
```

The index is calculated as before, but here the most significant binary 1 is a sentinel, or marker, bit and the character code follows. For the letter 'A', stripping away the sentinel bit leaves 10, which is dit dah. Functions named *dit( )* and *dah( )* actually send correct pulses to the I/O pins.

That takes care of sending, but what if the program has to receive and interpret your code? There are a bazillion ways to order a list and then search it. The easiest is a "brute force" search. That is, if MCT receives the letter 'E', we can start at the beginning of the *letterTable[ ]* and just march through the array until we find a match. Truthfully, the BP is so fast that approach would work fine in this application. But, that's sorta like driving a finishing nail with a backhoe. Instead, let's use a binary search.

## Binary Search

When Jack presented an earlier version of the MCT at the Four Days In May conference in 2019, a number of attendees stopped him after the talk with questions about the binary search technique used in the program. This section is a detailed explanation of how it works. If programming makes your eyes glaze over, you can skip this section. However, binary search algorithms can be used in many types of searches and the ideas presented in this section can be used in other programs. Your choice...

The game show *The Price Is Right* had contestants come down to the stage and try to guess the price of an item to the closest dollar. If you guessed a number that was too high, the host said "Too high." If the guess was too low, the host said "Too low." The contestant was supposed to determine the correct price within so many seconds. The problem was that most items cost over a thousand dollars, so a "brute force" approach starting at $1 wasn't going to work.

Viewers quickly learned to make a guess in what they thought was the actual price. For example, if they thought the price was less than $1000, their first guess would be $500. Suppose the price is $672. The host would say "Too low." Immediately after only one guess, the first 500 numbers can be eliminated. Therefore, the next guess would be the midpoint of the remaining numbers

```
NewGuess = (UpperLimit - midPoint) / 2 + midPoint
    $750 = ($1000 - 500 / 2) + $500
    $750 = $250 + $500           // Second guess
```

Now the host says "Too high." Because $750 wasn't right and it's too high, the contestant revises their guess to:

```
NewGuess = ($749 - 500) / 2 + $500
    $624 = $124 + 500            // Third guess
```

Again, the host says "Too low," so the new guess becomes:

```
NewGuess = ($749 - 625) / 2 + $625
    $687 = $62 + 625             // Fourth guess
```

Now the price is "Too high," so the next guess becomes:

```
NewGuess = ($686 - 625) / 2 + $625
    $655 = $30 + $625            // Fifth guess
```

Once again, "Too low," so:

```
NewGuess = ($686 - 656) / 2 + 656
    $671 = $15 + 656             // Sixth guess
```

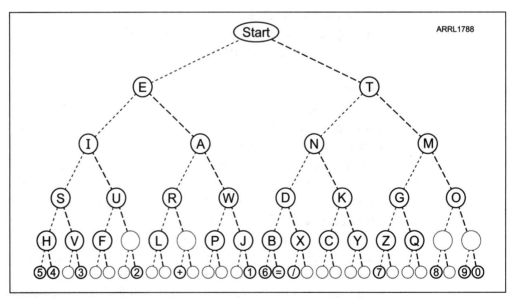

Figure 7.9 — Morse Characters in binary tree.

which is "too low." At this point, the contestant usually just starts racing through the numbers between $672 and $685. Because our price was $672, they nailed it in six guesses. Had the contestant used a brute force search, they would have had to blurt out 671 bad guesses before getting the correct price.

We can use the same binary search algorithm to search our Morse code possibilities. We can illustrate this with **Figure 7.9**. Recall our example array where we used a sentinel value of 1 and then followed it with a 0 if the atom was a dah, or a 1 if the atom was a dit. In Figure 7.9, the word "start" is the sentinel so it is assigned the value 1. If the first atom we hear is a dit, it's a value of 1. Thus, at this point, we would have binary 11. Repeating the Morse table we saw earlier:

```
char letterTable[] = {   // Morse coding: dit = 0, dah = 1
            0b110,       // First 1 is the sentinel marker
            0b10111,     // B
            0b10101,     // C
            0b1011,      // D
            0b11,        // E
```

you can see that binary 11 (0b11) is the letter 'E'. If the time interval after the dit is more than that of a dah, we know we just received the letter 'E'. If you understand the binary numbering system, you know that 11 in binary is 3 in decimal.

We can express the information in Figure 7.9 as a character array:

```
char myBinarySearchSet[] = "##TEMNAIOGKDWRUS##QZYCXBJP#L#F
    VH09#8###7#####/-61########2###3#45";
```

Recall that arrays in C start with element 0, not 1. Look at the string named *myBinarySearchSet[ ]*. In that array, element 0 is a '#', element 1 is also a '#', element 2 is a 'T', and element 3 is an 'E'. Therefore, if we hear a single dit, it equates to binary 0b11, which is 3 and using 3 and an index into the *myBinarySearchSet[ ]* array, we find the letter 'E'.

Let's try another letter...the letter 'C'. Its representation is 0b10101 which, in decimal, is 21. Therefore:

```
'C' = myBinarySearchSet[21]
```

The code to do the binary search involves bit shifting. The code is presented in **Listing 7.1**.

The binary tree gives us a very efficient way to determine the array index for each Morse code character. To illustrate how this works, let's assume we just received an 'N', which would mean *s = '-.' (dah-dit, with the dah represented by '-' and the dit represented by '.'). First, we define a *static int* named *num*. (Quite honestly, there's no good reason to make it a *static*, other than it's allocated on the heap, not the stack, so it might be a few instruction cycles faster.) We set *num* equal to 1 at the outset, as that is our sentinel marker. (That corresponds to "Start" in Figure 7.9.) The *while* loop examines each character in the string and,

**Listing 7.1**
**Source Code for Determining the Array Index for a Morse Character**

```
/*****
 Purpose: To perform a binary search on the pattern strong of dit and dah
          to find the corresponding letter.

 Parameter list:
   char *s    the input pattern string to use for the search

 Return value:
   int        an index into the search set character array.

*****/
int Decode(char *s)
{
 static int num = 0;

 num = 1;              // Set the sentinel
 while (*s) {          // More characters??
   num = num << 1;     // Yep, so shift what we have left one bit
   if (*s++ == '.') {  // Are we looking at a dit?
     Num++;            //    Yep...add 1
   }                   //    Nope, leave it a 0
 }
 return num;           // We're done, shove num in the backpack and go home
}
```

if it's not a NULL character, the code bit-shifts the value of *num* one bit position to the left. So the binary representation of *num* on the first pass through the *while* loop is now 0b00000010.

After the bit-shift, the code examines the character in the string to see if it's a dit. Because the first character is a dah, nothing is done to *num,* so it remains 0b00000010. The post-increment on *s* (i.e., *s++) advances to the next character in the string, which is a dit. The code once again bit-shifts the value in *num* one position to the left, so we now have 0b00000100 in *num*. Because the next character is a dit, we add 1 (increment) *num*, so it now equals 0b00000101. The code then post-increments the pointer, which means were are looking at the third character in '-.'. Because C strings are terminated with a *NULL* character ('\0'), the *while* loop test on *s is logic false and the loop ends. The code then puts *num* (binary 0b00000101) "in the backpack" and returns it back to the caller.

As you know, 0b00000101 is 5 in base 10 (decimal) arithmetic. This means we need to extract *myBinarySearchSet[5]* from the string:

```
char myBinarySearchSet[] = "##TEMNAIOGKDWRUS##QZYCXBJP#L#F
VH09#8###7#####/-61########2###3#45";
                ^
                |
          myBinarySearchSet[5]
```

As you can see, index 5 into the search string yields a ... wait for it ... 'N'!

Could we have used brute force to search string representations of the Morse character set? Sure, but that puts us back into "pounding-a-finishing-nail-with-a-backhoe" mode of coding. While using the binary search here is overkill, it's a simple illustration of how powerful a binary search can be. True, in terms of processor time, the human movements to send a Morse characters is slower than continental drifting. However, hang the binary search method on your tool belt as there may come a time when you need the performance boost a binary search can offer.

## Conclusion

We honestly believe that at least one MCT (or equivalent) belongs in every club. The MCT gives the members a cost-effective tool for learning Morse on their own, but has enough audio power to be used in a classroom setting, too. Even with all of the string space that is used in this program, there is still plenty of both code (flash) and data (SRAM) space to enhance its performance. Adding an SD card reader is inexpensive and easy to do, but can yield up to 8 *gigabytes* of additional storage space for more practice strings. Indeed, many TFT displays, like the one shown in Figure 7.6, include an SD card reader. Finally, because the circuit is so simple, it would be a good project for a single-day club build.

We hope you'll help new hams learn Morse code using the MCT so we can meet them on the air.

# CHAPTER 8

# Programmable Bench Power Supply

Okay, why another power supply? After all, we just built a power supply (PS) in Chapter 5, why do we need another one? Well, there are a number of differences between the power supply built in Chapter 5 and the power supply built here. First of all, the power supply in Chapter 5 runs off the ac mains, whereas this supply is powered by your 13.8 V source (from Chapter 5). Second, this PS provides functionality far and above anything most hams have in their shack. Indeed, if you aren't into designing and testing electronic circuits, the PS in Chapter 5 is probably all you need. However, if you design circuits, you may need power applied to a circuit in the form of a sine, square, ramp, triangle, or other waveform. In other words, this PS has a lot of useful features that can be used by those who are into experimenting with circuits.

Every ham shack needs a good bench power supply. While there are endless numbers of power supplies available, very few are Programmable Bench Power Supplies (PBPS) available at a reasonable price. For several hundred dollars you can get a power supply that hooks up to your computer and allows a dedi-

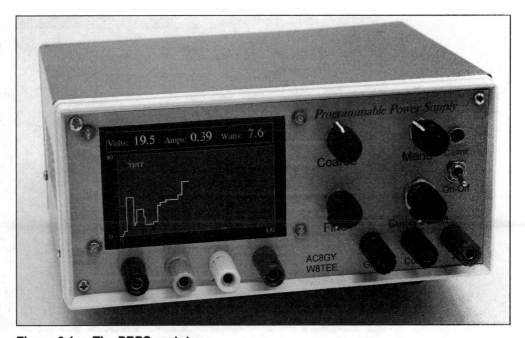

Figure 8.1 — The PBPS prototype.

cated app to control the power supply. Forget that! How about a programmable supply with built-in functions, support for user-defined functions, presets, timed output, current limiting, SD card storage of your favorite functions and data, and with a range of 0 to 25 V with up to 3 A (expandable to 10 A if you wish)? Such a power supply is exactly what you are going to build in this chapter.

The good news: this supply only costs about $40 to $50 (not including enclosure), depending on what you have in your parts bin. The design is both flexible and modular, making it easy to tailor to your specific needs.

A major benefit of the design described here is the ability to use almost any common 12 to 25 V dc supply as the input voltage source. This eliminates the need to add large, expensive, transformers to the system. Every ham shack has at least one 13.8 V supply for transceiver power, which can do double duty as the input to the PBPS. (The PS from Chapter 5 can be used.) **Figure 8.1** shows our prototype.

## Feature Set

As you can see from the following list, this is much more than a "standard" power supply. A partial list of its features include:

- Voltage range — 0 to 25 V dc
- Current — up to 3 A (with minor additions can be increased to 10 A)
- Regulated voltage output — maintains voltage under load variations
- Low noise — less than 5 mV RMS broadband noise
- Variable current limit — front panel control and viewable on the display
- Menu-driven user interface using three rotary encoders (touch screen can be added)
- Voltage control via coarse and fine tuning encoders or built-in keypad

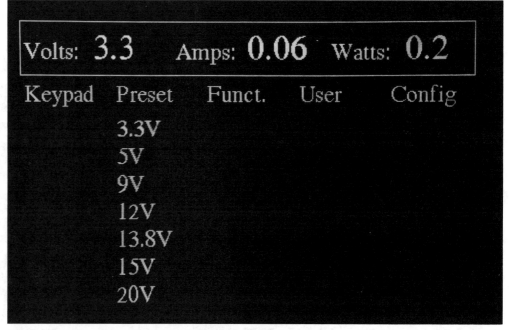

Figure 8.2 — Menu, showing voltage presents.

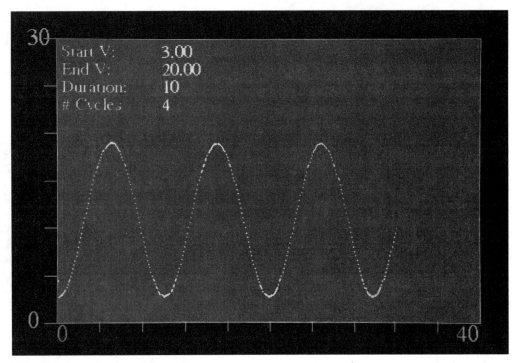

**Figure 8.3 — Sample sine plot**

- Voltage presets for easy selection of frequently-used voltages (See **Figure 8.2**)
- Voltage time-based functions (ramp, step, sine, half-sine, triangle, pulse, timer, trapezoid — see **Figure 8.3**)
- Common configuration parameters stored in EEPROM for quick recall
- User-defined functions
- SD storage for data

The *PBPS Users Manual* provides more details on how each of these features is used. The manual is available from our website (**groups.io/g/SoftwareControlledHamRadio**).

## DC Power Input for the PBPS

The basic design of the PBPS is modular, which gives you considerable flexibility in your choice for the dc source used as the input module for the PBPS. You likely have at least one 12 to 13.8 V dc supply in your shack that is suitable for the dc input source. **Figure 8.4** shows some common examples that could be used as the dc input source. Of course, the PS from Chapter 5 can also be used.

Regardless of the dc voltage input source that you select, it should have certain minimum performance capabilities. These minimum demands for the dc voltage source are that it is:

1) driven by ac mains for its voltage source
2) capable of supplying 12 to 25 V dc at 6 A
3) well filtered.

You can change some of these requirements if your needs are substantially different. You could, for example, increase the current capability of the PBPS

**Programmable Bench Power Supply    8-3**

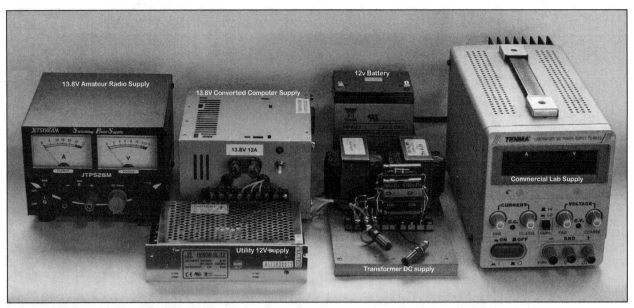

Figure 8.4 — Possible sources of dc voltage input.

to 10 A, but that would also make higher demands upon the dc input voltage source. Some users may not feel the SD storage for test data is necessary. The hardware (and software) make it easy to leave this out of the project should you elect to do so. Likewise, if you change your mind later, it shouldn't be too difficult to retrofit the SD card. You can build the project on perf board if you want to do so.

## Uses for Programmable Power Supply

OK, so a programmable supply is kinda neat, but exactly what sort of uses would I have for the PBPS? (See the Examples section later in this chapter for more details on five specific uses.) Obviously, any project that needs a fixed voltage within the limits of the PBPS would be a common use. Many microcontroller projects need either 5 V (most of the Arduino family) or 3.3 V (almost every other controller). The same is true for projects that use LCD or TFT displays.

If you do circuit design, sometimes the ability to test a circuit with varying voltages is needed. You could use the Ramp function to automate such a design test. Voltage stress or linearity tests are also sometimes required of new designs, and the programmable feature of the PBPS makes such tests much simpler. You can also use it to generate characteristic curves for diodes, transistors, or other devices. Perhaps you want to examine the hysteresis characteristic of non-linear analog devices such as relays or digital Schmidt triggers. Maybe you need a 3 A pulse function for testing. (Try that with your signal generator!) You could use the timer function to place a device under load for some specified period of time and then automatically shut down. What if you want to charge one of those fussy batteries that gets really cranky if you apply too much current?

No doubt other uses will also come to mind (for example, how much difference does it make on output power if the voltage to your µBITX transceiver goes from 12 V to 18 V?). In any case, we think you will find the PBPS a worthwhile addition to your bench.

## Circuit Description

**Figure 8.5** shows the block diagram of the Programmable Bench Power Supply. Everything in the PBPS starts with the ac input, which provides the starting point for the project.

The AC To DC Module supplies the voltage from the ac mains. This (external) module should be capable of 12 to 25 V at 6 A. A good rule of thumb is the current capacity of this module needs to be double the amperage capability of the PBPS. That is, a 12 V, 6 A capability on the AC To DC Module supports 3 A out from the PBPS.

The output dc from the AC To DC Module feeds the Boost Converter. The boost regulator increases the voltage from the 12 to 25 V input to 30 V, which becomes the input voltage to the variable linear regulator, which is the heart of the PBPS system. The circuit uses an LM338, which is capable of handling more than 3 A up to 37 V. However, the actual current handling capacity is determined by the pass transistor, which is a MJ2955 Darlington pair transistor with a current capacity of 10 A. This allows the LM338 to work comfortably within its operating region.

An STM32F103 (BP) microcontroller controls virtually everything the PBPS does. Three encoders serve as data input sources for the BP software. A 480×320 pixel TFT color display based on the IL19488 controller is used to display visual output data to the user. (You could use a display based on an ILI9341 controller,

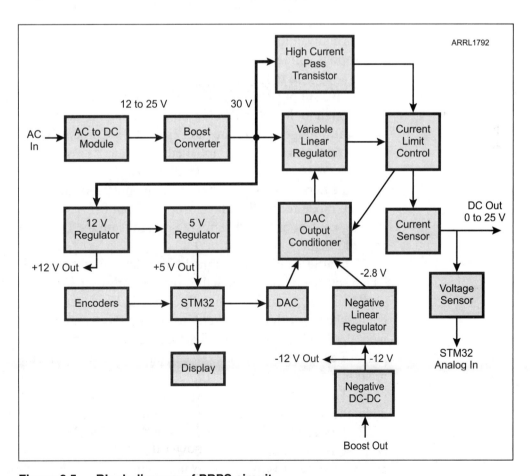

Figure 8.5 — Block diagram of PBPS circuit.

but you would need to modify the display elements of the software to reflect the lower resolution.) The DAC (digital-to-analog converter) controlled by the BP feeds a DAC Output Conditioner, which provides the control currents needed to set the output from the variable regulator. A negative dc-dc converter and linear regulator provide the –2.8 V necessary to allow the LM338 to regulate all the way down to 0 V. (It would be a minimum of +1.2 V without it.) An op amp, some diodes, and a control pot form the Current Limit Control. The current and voltage sensors allow the BP to display the system outputs. Linear 5 V and 12 V regulators are used for the BP and several other functions.

## Full Circuit Description

The full circuit diagram is shown near the end of this chapter in Figure 8.33. The basic control element of the PBPS is a variable linear regulator — LM338T — combined with a high-current pass transistor. The output voltage is determined by the STM32 microprocessor and the current limit circuit. The STM32 digitally controls a DAC (digital-to-analog converter) by providing control voltages to the LM338T circuit. The LM338T output is normally varied by changing the current to its adjustment pin, usually by varying a resistor to ground. In this circuit, however, we emulated a resistor using a J176 JFET whose series source-to-drain resistance is controlled by the JFET gate voltage. The JFET gate voltage is determined by the output of a DAC controlled by the STM32. The JFET yields a voltage controlled resistor, so instead of a pot to control the output voltage, we can use the microcontroller.

Normally, a linear regulator cannot control the output to 0 volts, because of the dropout voltage of about 1.2 V. The solution is to provide a negative voltage of –1.2 V to the Adjust pin circuit to provide control all the way to 0 volts. A negative supply and a dc coupled op amp allow the adjustment voltage to go low enough to drive the LM338 output voltage to zero. Trim pots on the op amp allow the voltage maximum and zero to be set precisely.

Current control is accomplished by means of a series resistor to sense the current, an op amp that serves as a comparator, and a couple of diodes that direct the current flow. When the voltage output of the comparator is greater than the diode forward voltage drop of about 0.7 V, the diode begins to conduct, shifting control of the adjustment voltage from the DAC output to the current control circuit.

The current adjustment threshold is controlled by the current adjustment pot, which sets the point at which the comparator's output causes the diode to conduct.

The LM338 can safely handle 3 to 5 A. However, to afford additional current capacity for those who require it, a PNP pass transistor was added to augment the LM338. The MJ2955 actually is a Darlington pair in a TO3 case. With proper heat-sinking the MJ2955 can pass 10 A and is rated for 100 V. Keep in mind that the output current limitation of the PBPS is affected by several factors:

1) The heatsink must be large enough to handle the heat load of the MJ2955 under the worst-case situation, which would be a low voltage at maximum current.

2) The ability of the external input AC To DC Module to supply enough current. Note that, if the input voltage is 12 V, the supply must be able to provide approximately two times the desired output current at maximum voltage output. Thus, if the output is to be 10 A at 25 V, a base supply of 20 A at 12 V is neces-

sary with a wattage rating of 250 W.

3) The capacity of the boost supply. The limitation here is the current to be supplied, but most importantly the power capacity of the boost supply. It is best to utilize a boost supply that has a considerably larger capacity than required.

4) Maximum current capability of the current sensor. The sensor used in our base design is rated for 3 A. Other sensors may be substituted to raise this maximum. The specified sensor was used for simplicity.

Finally, two linear regulators, one at 12 V and the other at 5 V, are included. The 5 V regulator drops the input voltage to the required 5 V for the STM32. The 12 V regulator is included to reduce the boost supply output and, more importantly, to allow a fixed +12 V out to complement the –12 V from the negative dc-dc for op amp experimentation. (Op amps are often used with positive and negative supplies for simplified biasing.)

The microprocessor portion of the circuit consists of the STM32, a TFT display with an SD card reader, and three encoders. The STM32 is an ARM processor running at 72 MHz. The STM32 has ample on-board storage for the program code and SRAM for program execution. It has lots of GPIO (General Purpose Input/Output) pins for both digital and analog input. Various interface protocols are supported, including SPI and I2C.

The display is a serial interfaced (SPI) 480×320 TFT LCD unit. Touch screen support is available, but not used in version 1 of the PBPS. STM32F103 is programmed using the Arduino IDE, which has been extensively described in the early chapters of this book. The three encoders are interfaced to the STM32, one for menu activation, one for Coarse voltage adjustment, and the final one for Fine adjustment. Each encoder has a built-in switch, used for:

- Menu selection and exit,
- Coarse switch for exiting certain functions and,
- Fine switch for Voltage on-and-off.

**Figure 8.6 — The PBPS splash screen.**

The SD card reader uses the SPI interface and is utilized for storing the various user data files. Both SD and microSD cards are supported. The data are written using standard ASCII formats, which makes it possible to input the data into other applications such as *Excel*.

When the PBPS is powered on, the user is shown a splash screen, similar to that shown in **Figure 8.6**. The user is free to change the splash screen if they wish to do so, but the comments in the source files must remain as distributed by us. (As you know, comments are not part of the executable program and leaving these comments in the source code has no effect on the program execution or resource requirements.)

Once the splash screen is displayed, the display changes to the main program screen, as shown in **Figure 8.7**. Rotating the Menu knob causes the top menu highlight field to move horizontally. When the user has the desired menu choice highlighted, pressing the Menu knob activates the encoder's pushbutton switch and that menu option is activated. What follows is a brief description of what each of those top five menu options does.

## Keypad

There is a software numeric keypad build into the PBPS source code (**Figure 8.8**). Its inclusion is to provide an alternative method for entering a voltage. While there are "preset voltages" that are easily selected, there may be times when you need a weird voltage for some particular reason. The keypad provides a precise way to do this.

To enter a voltage digit, use the Menu encoder to move to the digit of choice and then press the encoder switch to select that digit. Repeat this process for the remaining two digits, which then display the input voltage as XX.X in the Volts field of the display. If you need a voltage that is less than three digits, enter 0 for the first digit (i.e., 0X.X). The "C" key clears the voltage input buffer while the "B" key erases the last digit entered and awaits a replacement digit to be entered. Any voltage input request over 25 V reverts to 25 V.

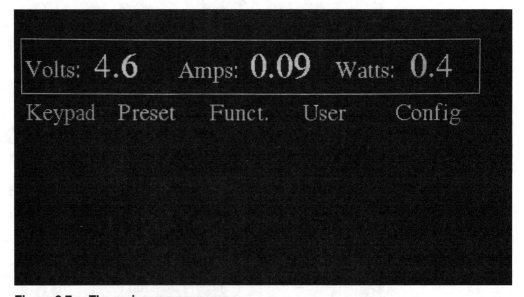

**Figure 8.7 — The main program screen**

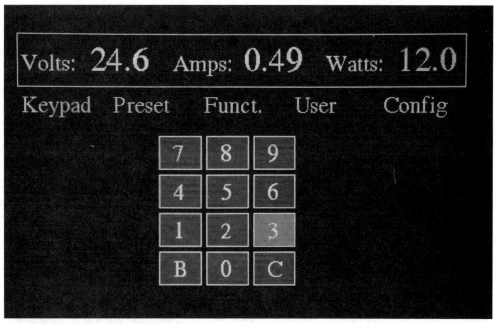

Figure 8.8 — Software keypad.

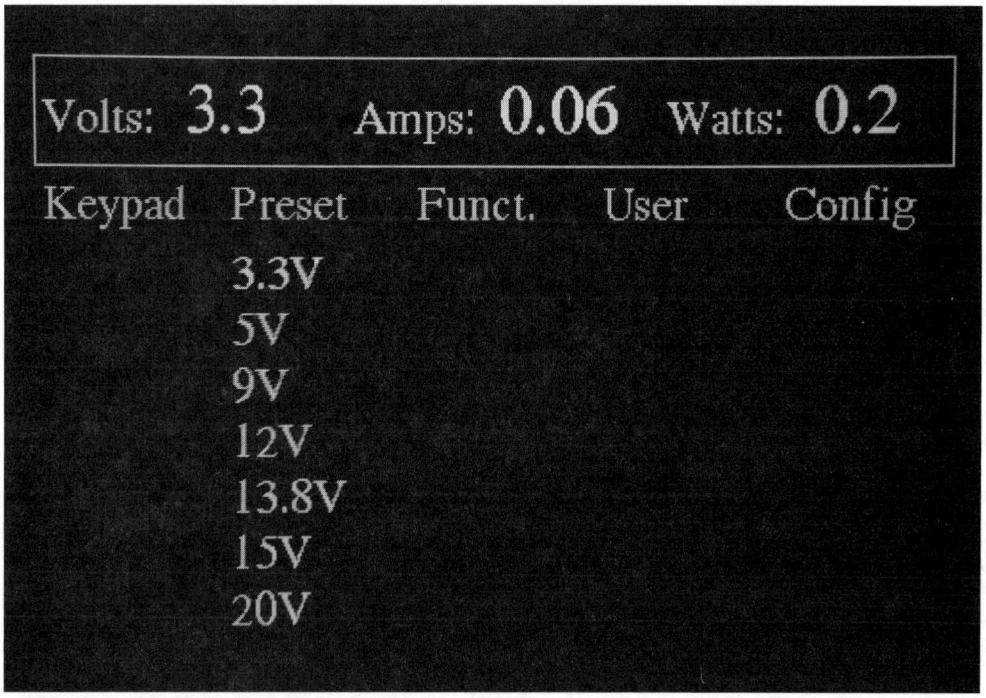

Figure 8.9 — Preset voltage options.

## Preset

There are seven voltage presets available. Simply scroll to the desired voltage and select with the Menu encoder switch. Pressing the Fine encoder switch toggles the voltage on and off. Pressing the Menu switch again exits the Preset. The preset voltages are shown in **Figure 8.9**.

Simply use the Menu encoder to scroll to the correct voltage option and press the Menu encoder. Pressing the Menu encoder a second time reverts to the Main program menu. Users may wish to edit the source files to set different preset voltages to suit their needs.

## Functions

This Main menu option presents you with a list of the predefined functions available for use in the PBPS system. See **Figure 8.10**. Each of the functions has user-defined parameters, which are stored in EEPROM. Uses for each function were touched on earlier in this chapter, but they are certainly not limited to those examples.

All functions are displayed graphically in real time on the screen. The time axis maximum value is set by the product of number of repetitions and duration of each cycle. An example of these built-in functions was shown in Figure 8.3, the sine wave output.

## User

This menu option allows the user to activate user-defined functions as supplied either on SD card or internally-created in the program space. These are likely be similar to the functions described above, but you control specifically the voltage output and time interval.

## SD

This option lets you create a Voltage function file with an external text editor and transfer these files to the PBPS on a SD or microSD card. Files are stored on the SD card in the standard comma-separated value (CSV) format, terminated with a # character.

Each file name must be given a unique name, with 8 or fewer characters for the name. Voltage values are in volts, step duration time is in seconds (integers only, no fractions for the seconds) as shown here:

**Example (Voltage,Time Pairs)**

| | |
|---|---|
| V1,T1 <CR> | 1.5, 3 <CR> |
| V2,T2<CR> | 5.3, 10<CR> |
| V3,T3<CR> | 12, 5<CR> |
| V4,T4<CR> | 20, 60<CR> |
| # | # |
| | Use this format. |

An unlimited number of Voltage-Time pairs may be entered, but the list of pairs must be terminated by the "#" character. These voltages sequences are not saved to EEPROM but are read each time one pair at a time.

**NOTE:** A valid card, either SD or microSD must be inserted for initial startup to proceed. If no card is present, the user must insert one and restart to use the SD Card option. If no card is detected, the startup process will continue, but the User SD card function will not be available. If the SD card is removed during operation and the SD menu item is selected, the system will attempt to continue to access the nonexistent card and slow down significantly.

The SD card must be formatted as FAT32 or FAT16. Other formats may not be recognized. You should do an internet search for a formatting program that is suited to your computer. Use only 16 GB or smaller SD cards. Anything above

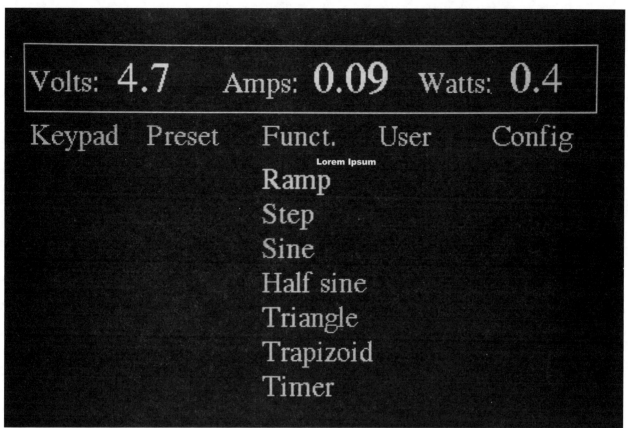
**Figure 8.10** — Pre-defined program functions.

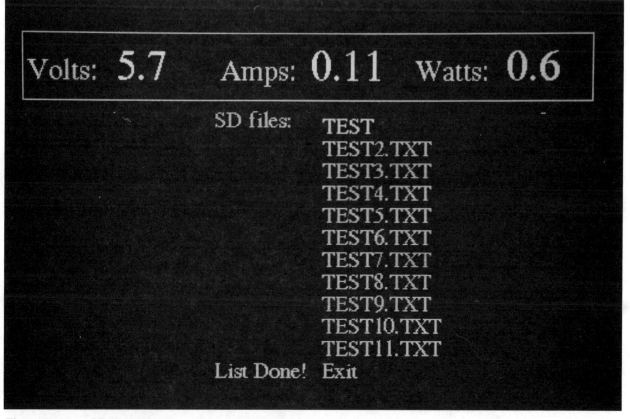
**Figure 8.11** — SD file names.

Programmable Bench Power Supply

16 GB either won't work properly or, if it does, it cannot access the extra storage space because of the way the file allocation table is handled.

When the SD menu item is selected, a list of available files will be shown (see **Figure 8.11**). Scroll to the one desired and press the Menu encoder switch to select the desired file. Voltage pairs are read in sequence from the SD card and plotted on the screen as the voltage level is output.

After the last pair is accessed, the function ceases and reverts to the previously set voltage. To exit without making a selection, choose Exit.

### User 2, User 3

These menu options are like the SD Card menu, except the data are entered internally from the Config Menu and stored in EEPROM. Each User function is limited to 20 Voltage-Time pairs.

### Config

The Config menu option allows the User to enter Function parameters and set up the User Functions. The submenu options are shown in **Figure 8.12**.

### General

This menu option provides the means by which the user modifies the parameters for the various internal functions, except Time. Modified parameter values are stored in EEPROM and are used as defaults upon startup. Again, the Menu encoder is used to scroll through the options.

### Time

The Time menu option allows the user to modify the parameters for the Time Functions, Modified Time parameter values are stored in EEPROM and will be used as defaults upon startup. Select items with the Menu encoder switch.

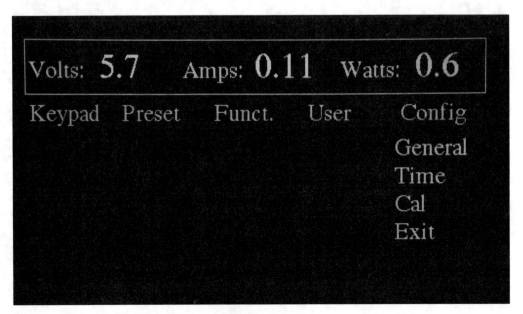

Figure 8.12 — The Config submenu options

## Cal

The Voltage displayed on the screen can be calibrated to match an external reference voltage measurement.

## Building the PBPS

The PBPS may be readily constructed on perf board or on a user-supplied PCB. A PCB is also available from **QRPGuys.com**. The construction steps shown here use the perf board option. All wiring is point-to-point, and the user must use wire sizes appropriate to the function.

Some general construction notes:

- High-current paths should be made with #18 AWG wire or larger, preferably stranded. Low-voltage insulation is adequate. If you are fabricating your own PCB, make sure the high current traces are large enough to carry the expected current.
- High-current paths should be as short as possible.
- Digital signal paths may be connected using smaller gauge wire, #24 to #30 AWG. The examples shown have many of the digital connections made using #30 AWG solid wire-wrap wire.
- Screw terminal blocks are used in a couple of locations to allow easy connection and re-connection during testing and development. The builder may choose to simply hard-wire and solder these connections.
- The display and encoders are connected to the main board using IDC 2×5 and IDC 2×3 connectors for ease of connection. The use of these connectors is encouraged since it makes fabrication easier, especially if you are using a relatively small case.
- Adequate heat-sinking is necessary for the regulators and high-current pass transistor.
- The STM32 and op amps should be mounted using appropriate sockets.
- All other components should be soldered directly to the perf or PC board.
- Select an enclosure that allows adequate space around the heat-generating components. This is not a project to skimp on case size.
- The MJ2955 should be mounted on an external, isolated heatsink. If higher power options are elected, a small fan-cooled heatsink should be considered.
- The enclosure may be isolated electrically from the circuit ground, if desired.

## Construction

The first step is to select the enclosure you wish to use. We elected to use a metal case that measures 7.5 × 6.5 × 3.5 inches, but we think building would have been a little easier had we elected to use a slightly larger case.

Select a perf board that is large enough to hold all the components and fits easily into the enclosure. We used a 3.5 × 4.5-inch board with 0.1-inch hole spacing (standard).

The limiting factor on the PBPS current output is the boost power supply that feeds the PBPS. The boost power supply should have enough current/voltage capability to handle the design output. During development, we used the power supply that drives our transceivers. We later dropped back to an open-frame supply we got from an obsolete piece of cannibalized computer equipment. This power supply was rated at 12 V (adjustable to 13.5 V), 6 A and 75 W.

By the way, many secondhand stores and donation centers have used computers that are still working, just old. The power supplies are often perfect for this type of work. Also, the hard disk ribbon controller cables are very useful. Many of these cables are 40 to 50 conductor cable that can sell for as much as $10/foot. We picked up two 3-foot cables for 25 cents at a church collection center. The disadvantage is that the cables usually are not color coded.

The boost converter must be able to output 30 V or more, for an input of 12 to 25 V. The boost supply should have adequate heat dissipation capability and be mounted with enough space to allow air to circulate freely around it. If you purchase a boost converter online, they are frequently advertised in terms of amperage, but be aware of any total power limitations and select a supply that will deliver enough power at the rated voltage and amperage. It's best to select

**Figure 8.13 — Main parts for perf board construction.**

one that has higher specs than your design requirements. The additional cost will be a few dollars, at most, and it is cheap insurance against letting the magic smoke escape at some point down the road.

## Parts

If you are going to do a perf board build of the PBPS, the main parts you'll need are shown in **Figure 8.13**. We encourage you to use sockets for the STM32 microcontroller and the op amps. To avoid bending the socket pins during construction, we used bolts in the corner holes of the perf board as standoffs to keep the perf board off the table surface during construction. For some

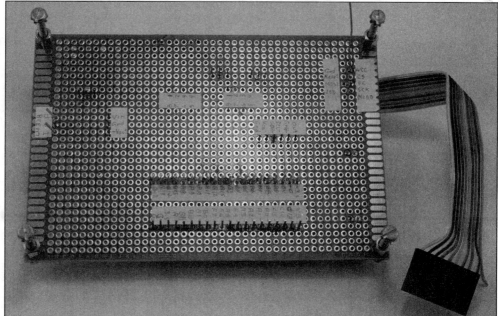

Figure 8.14 — The "dry fit" placement of parts (left-top, right-under).

**Programmable Bench Power Supply  8-15**

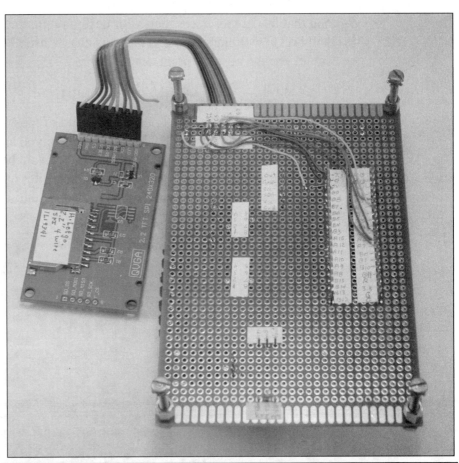

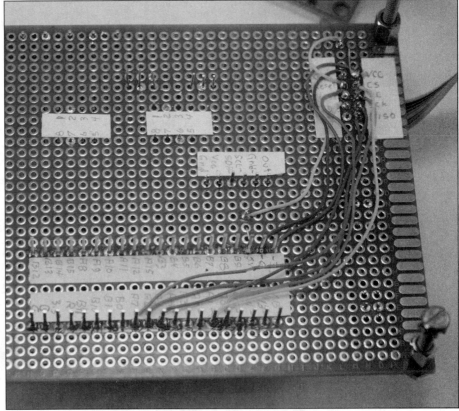

**Figure 8.15 — Underside wiring with labels in place.**

reason, sliding the board over the construction surface and applying power later can produce unwanted surprises.

First do a "dry fit" of the components you are using for the PBPS to get a feel for their placement. Keep in mind the cooling suggestions at this phase, too. Once you're happy with the layout, place and solder IDC male PCB-mount connectors. At this stage it may be a good idea to tack-solder just the end pins of the connectors. That way, if you need to reposition them later, it's fairly easy to do.

You should now mount the four voltage regulators, but do not attach the heat sinks yet. If you elected to use terminal strips for inputs and outputs, mount those at this time, too. IDC connectors are not appropriate for power input or output from the PBPS because of the high current involved.

Now add the DAC module, negative dc-dc converter, and current sensor module. Note that both a negative dc-dc converter and a negative linear regulator are used because variable negative dc-dc units are hard to find. Add the two trim pots next and then wire the connections to the STM32 microcontroller. Finally, wire up the display, the rotary encoders, DAC, and the power and ground connections.

**Figure 8.14A** shows the parts layout that we used for the PBPS. The ICs are placed near the middle of the board with the IDCs on the left edge and the voltage regulators along the right edge. Figure 8.14B is the underside of the board with the display ribbon cable also shown.

**Figure 8.15** shows two views of the underside of the board after point-to-point wiring between the STM32 and the display connector has been finished. The white strips in the photos are adhesive labels that can be purchased at any office supply store. Al had a great idea that is a real time saver because the labels identify the pins for the microcontroller, the op amps, and the display connector. These labels will save you several thousand "board flips" because you can't remember which pin is which as you wire them together. Since Al and I have a hard time remembering what we had for breakfast, the use of labels saved us a huge amount of time.

You can now wire up the op amps and add the appropriate resistors and capacitors as detailed in the schematic presented in Figure 8.33 near the end of this chapter. The wired board is shown in **Figure 8.16**. Again, notice the labels below the IDC connectors and next to the terminal strip.

Once you have finished the wiring of the cables, connect the current-limit pot and current-limit LED.

You are now ready to prepare ribbon cables for the rotary encoders, the display, and the current limit pot and LED (see **Figure 8.17**). If you elect to use the SD card reader capability, mount the SD card reader at this time, too. While one end of the cables houses the IDC connectors, we recommend soldering the other end directly to the component associated with that cable. Use a piece of string to measure the length each cable needs to be to reach the component once it's mounted in place. If you believe that there will be a considerable amount of movement while placing the component in its final position (for example, the display), a healthy blob of hot glue on the soldered connections is probably a good idea for strain relief.

Figure 8.16 — Finished board, top views.

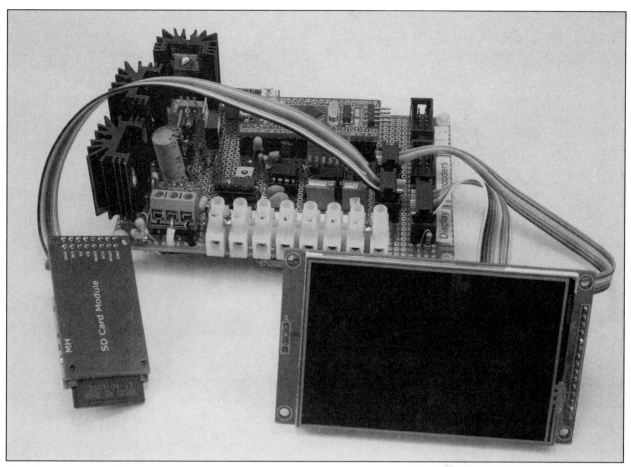

Figure 8.17 — Using ribbon cable for off-board components (display and SD card reader shown).

## Mounting Components in the Case

Now place major components in the case as a trial fitting. Once you're happy with the case layout, mark and drill the mounting holes for the boost converter and the perf board. The trial fitting is shown in **Figure 8.18**.

## Front Panel

It's now time for you to lay out the front panel for the controls and display. While there are many different ways to accomplish the final look of the front panel, what follows is a description of how we did it. (You can find additional details on final construction details for projects in general in Chapter 16.)

The front panel overlay is made from a printed sheet of photo paper with all of the component control labels printed on the sheet. The first step is to cover the metal front panel completely with masking tape. Now draw the placement of all components and the display on the masking tape exactly as you want the panel to appear. Our "masking tape" panel is shown in **Figure 8.19**.

Then mark and drill all of the holes on the front panel. To cut out the display opening, we used the tried-and-true method of drilling corner holes in the display opening and then used a jig saw with a fine-tooth blade to finish the cutout. Two and a half days with a file and a couple of six-packs produced a nice smooth opening for the display. Naw…just kidding. The aluminum panels on our case

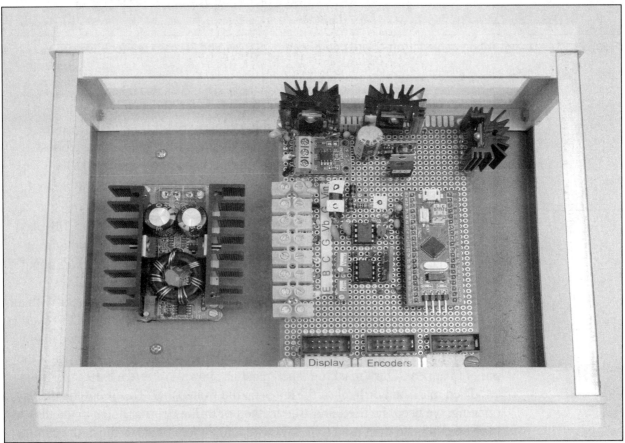

**Figure 8.18 — Board placement in case.**

didn't require much time at all to get a smooth cutout for the display. **Figure 8.20** shows what our panel looked like after drilling the holes, cutting out the display opening, and removing the masking tape. Note that we leave the protective film on the display surface until the very last moment to avoid scratches. **Figure 8.21** shows the various components and the display in place on the front panel.

To create the front panel overlay, scan the front cutout panel with a photo scanner or "all in one" printer and use the scan as a template in a drawing ap-

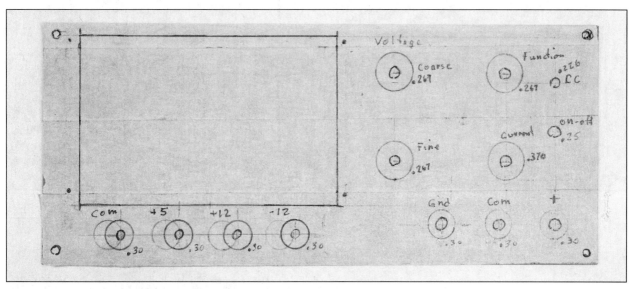

**Figure 8.19** — The "masking tape" layout.

**Figure 8.20** — Front panel with holes and display cutout and masking tape removed.

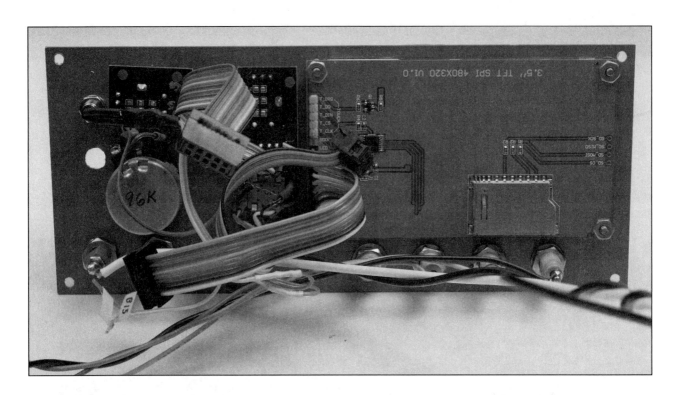

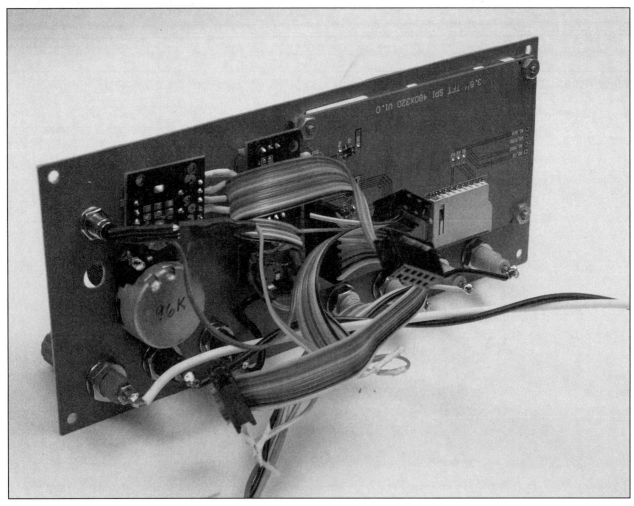

**Figure 8.21** — Two views of front panel wiring.

plication to add labels and text. We used the Open Office *Draw* option to create the overlay. This is shown in **Figure 8.22**.

Once you are satisfied with the front panel overlay, print the panel overlay on high quality photo paper. If you don't have a printer capable of photo-quality printing, most office supply stores can do it for you at nominal cost. Make the overlay slightly larger than the panel and trim later. **Figure 8.23** shows the printed front panel display. A final step is to coat the printed overlay with a lacquer-based fixative spray to protect it. Do this prior to cutting out the display and other component holes.

Now take the printed front panel and carefully align the overlay to the metal panel. While holding or clamping the two together, glue the sheet to the metal front panel, starting at one end and then doing the rest. We used a gel superglue to attach the panel, which gives you about 30 seconds working time. Let the glue set completely. Once the glue is completely set, cut out the holes with a

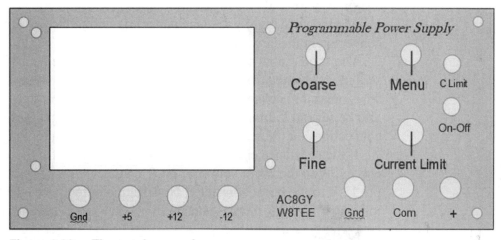

**Figure 8.22 — The overlay panel**

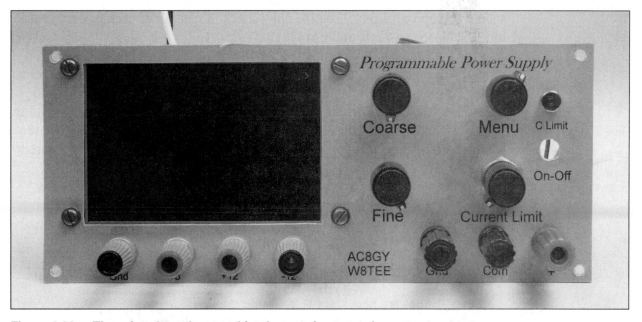

**Figure 8.23 — The printed overlay panel in place on front panel.**

**Programmable Bench Power Supply   8-23**

very sharp Xacto knife. Note that we purposely fit and cut the display opening to hide the edge display wires. If done correctly, you should have no need for a display bezel.

Now mount the components and their knobs on the front panel. The panel should look similar to what is shown in Figure 8.1 at the start of this chapter.

## Wiring Everything Together

Using the appropriately-sized wiring, connect all the components and input/output connectors to the interior components. When completed, it should look similar to **Figure 8.24**. Use cable ties to position and tidy up the internal wiring.

## Initial Testing

Several test steps are necessary before you can actually use the PBPS. First, do NOT install the STM32 microcontroller, display, or op amps yet. *You need to perform a wiring double-check before applying any power to the board.* We find a good way to do this is to make a copy of the schematic and then use a yellow magic marker to highlight each connecting wire after you are sure it is properly connected in the circuit.

You should now perform resistance checks on the board inputs. First, measure the resistance from dc input to ground. It should be several kilohms. Likewise, measure between the input from the boost converter to ground. It should also be several kilohms. If the measurements are suspect, examine all of the connections on the board, being especially vigilant in looking for shorts.

If the resistance checks look good, you can perform the initial power checks. With the boost converter NOT connected to the main circuit board, apply 12 V to the boost connector. Now set the output from the boost converter to 30 V. This adjustment is usually done via a small trimmer pot on the boost converter board.

With the power off, connect the boost converter to the main circuit board. Turn on the dc power to the boost converter and monitor the current draw. It should be approximately 120 mA. Now check the output of each power regulator (+12 V, +5 V, −12 V, and LM338).

Now attach the display and encoders to the main circuit board and insert a properly-formatted SD card if you are using one. Do NOT apply the 12 V power yet. Plug in the USB connector from your PC to the STM32. Compile and upload the software as discussed in earlier chapters. The display should light up and you should see the splash screen, then the "SD OK" message, and then the startup screen, as seen in Figure 8.1. Exercise the menu encoder to make sure it is functioning properly.

You are now ready to do an initial first full power-up. Attach your power source and apply its dc output to the dc input of the PBPS. The current draw should be approximately 400 to 500 mA. The display should repeat its power-on sequence, but now all encoders should be functional.

## Calibration

Several steps are necessary to calibrate the PBPS.

1) First, attach a reliable dc voltmeter to the PBPS output.

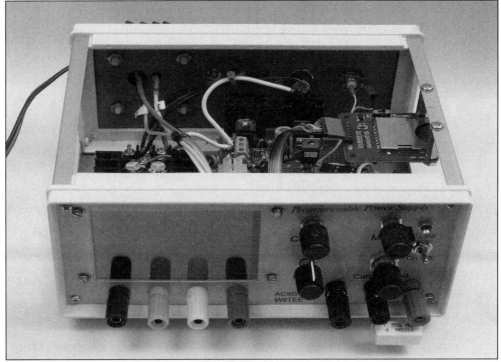

Figure 8.24 — Final assembly view of PBPS.

Programmable Bench Power Supply     8-25

2) Open the Config menu and select "Cal". You should see a screen print of the encoder output in the form of the DAC "set-point" value which goes from 0 to 3999. Rotate the encoders to get a set-point of as close to "zero" as possible.

3) Now adjusted the "Zero" trim pot to yield 0 volts on the voltmeter. You should do this by first adjusting for a very small positive reading and then carefully reduce the output until you read 0 V (or just a few positive mV).

4) Next, using the encoders, increase the on-screen setpoint to approximately 3999. Adjust the "Max voltage" trimpot to yield an output voltage of 25 V.

5) Repeat the Zero and Max adjustments until no change is observed and the voltmeter readings are 0 and 25 V respectively.

6) Select the Menu entry "Screen V Cal".

7) Using the Menu encoder adjust the Cal value until the on-screen display of the voltage is the same as the external voltmeter.

8) Press the Menu encoder switch to save the Cal value.

## The Software

The Programmable Bench Power Supply consists of nearly 3000 lines of source code spread over 11 files, written using the Arduino IDE and its underlying Gnu C++ compiler. You should have the Arduino IDE installed with appropriate extensions for the STM32 processor as described in Chapter 1. A number of "standardized" code features (described in earlier chapters) are used here as well, prominently featuring a menu library Jack wrote specifically for this project, but that can easily be used elsewhere.

Standardized libraries such as the Adafruit GFX graphic library, Rotary library for encoder support, as well as support of the ILI9488 based display and the MCP4725 ADC make using specialized peripherals much easier. Any non-standard libraries (i.e., libraries that are not shipped as part of the Arduino IDE or patch installations) have the download URL as a comment after that library's #include directory in the source code file.

The rest of this section describes some of the more unique features of the code, in no particular order. Also see comments in the body of the code for more detail.

The code is separated into logical files, divided by function. The 11 files are

- *PS_d_3-14-19_ILI9488f.ino* — This is the entry point file and contains the requisite *setup()* and *loop()* functions as required by the Arduino IDE. This file always appears first in the tab sequence of the projects source code files.
- *Config_General.cpp* — Code to allow the user to enter Function parameters.
- *Config_Time.cpp* — Code to set the Time Function.
- *Encoders.cpp* — Code for the three rotary encoders.
- *Graph.cpp* — Graph routines to present the on-screen functions.
- *Keypad.cpp* — Provides routines to input voltage values using an on-screen telephone-format keypad.
- *Menu.cpp* and *Menu.h* — Source code and header files containing the routines for the PBPS menu library system.

- *ProgPS.h* — The primary header file for the project. Contains the variable declarations, and global variable definitions that appear in the project's .ino file, as well as function prototypes and #include statements, plus the #include statements for libraries and all user-defined function prototypes.
- *SD_Card.cpp* — Routines to support the SD card reader.
- *Volt_Funct.cpp* — The various time-based predefined functions.

The keypad source file has some interesting and reusable functions in it. It creates a display that looks like a standard keypad but is built from the base menu system. When used in conjunction with a rotary encoder, it allows the user to "rotate" to the desired digit and press the encoder switch to select that value. Repeating this sequence allows the user to input any number they wish by pressing the Menu encoder switch when the input is complete. The input value is fairly flexible and can be interpreted as either floating point or integer value. (See the *DSCConvert()* function as an example.)

If you do not want the SD card feature, you could modify the code to eliminate it. However, we discourage you from doing this as you may change your mind later. If you do keep the SD card feature, you need to have a properly-formatted SD card in its holder which is on the back of the TFT display. (If you have an old TFT display that does not have an SD card reader, you can buy one and easily add it to the circuitry.) It is fairly common for factory formatted SD cards to *not* work with most SD libraries. If that happens to you, read: **learn.adafruit.com/adafruit-micro-sd-breakout-board-card-tutorial/formatting-notes**.

The code in the Graph.cpp file produces the graphs seen in the PBPS and can be used as a model for other graphic displays. These, when combined with the voltage functions, illustrate how these can work together to create new displays, too.

## Menuing Library

In order to better illustrate some of the concepts described in Chapter 3, we have included our own library for menuing, using the rotary encoders and on-screen display. The two files that make up the Menu library are *Menu.cpp* and *Menu.h*. *Menu.h* is the header file that contains all of the symbolic constant definitions and other overhead information necessary for the menu code to work. *Menu.cpp* contains the actual code that gets executed in the main program. Let's look at the header file first.

At the top of the header file are several *#define* preprocesssor directives that define parameters for the number of menus and the menu depth. If you write your own menuing system using this library, this is where you define the parameters for its appearance.

Next is the creation of the library class named *Menuing,* by using the C++ statement *class Menuing*. As usual, the header file does not contain data definitions, but rather data declarations so the class can be instantiated later in your code. The class declaration begins with the opening curly brace ("{") and ends with its matching closing curly brace ("}"). Everything in between the curly braces are the class members (i.e., the people sitting in the windowless black house from Chapter 3) and the class methods (i.e., the doorways in the same windowless house from Chapter 3). The class declaration begins with the class member declarations in the *public* section, such as:

```
int itemCount;              // options in a menu
int defaultOption;          // which is highlighted when shown.
                               Taken from defaultsList[] array
int width;                  // width of display   (480)
int height;                 // height of display (320)
int spacing;                // pixel count available for each main menu option
int activeColumn;           // Which column is active
int activeRow;              // Which row is active
int fontSize;               // Default text size
int foregroundColor;        // Color of normal text
int backgroundColor;        // Background color
int selectForegroundColor;  // Color for active menu option
int selectBackgroundColor;  // Background color for all menus
```

Note that *"public"* indicates that these are variables and methods that are available to the user.

Also included within the class declaration are some method prototypes, which are declared here so they can be used throughout the program code. Note that class methods are like regular C functions, but they are called "methods" inside a class and they can only be called through a class object using the dot operator. Examples of the class declarations are:

```
void deselectMenuItem(ILI9488 myDisplay, char *menu[]);
void eraseDisplay(ILI9488 myDisplay);
void eraseMenus(ILI9488 myDisplay);
```

but can only be called after you have instantiated the class, such as:

```
Menuing myMenu;             // Instantiate a class object
                            // probably some code...
myMenu.eraseMenus(tft);     // We assume the display object is named tft
```

The actual code for the methods is located in the *Menu.cpp* file. (Header files rarely contain executable code.)

After the *public* portion of the header, there is a *private* section, which is not used in this library. The *private* section of a class declaration is used for the variables and methods you might want to manipulate in the class, but don't want exposed to "the outside world." That is, the scope of the *private* elements of a class are limited to the internal class code itself.

The file *Menu.cpp* is a bit more extensive than its header file because it contains all of the actual code for the library methods. The library methods behave like regular C functions, with a couple of important distinctions:

First, methods are set up differently from regular C functions: The code declaring a method has a different syntax:

```
void Menuing::showMenu(ILI9488 myDisplay, char *whichMenu[])
{
```

As explained in Chapter 3, *void* indicates that the class method does not return a value to the caller. The code fragment:

```
Menuing::showMenu(
```

begins the declaration of the *showMenu()* method of the Menuing class. The "::" is called the *scope resolution operator* and can often be translated as "belongs to." You can verbalize the fragment above as: "The *showMenu()* method belongs to the Menuing class." The remainder of the statement contains the parameters that this function needs to operate properly.

Inside the method declaration is the code, including regular C statements as well as library calls to other libraries, as needed, for example the statement body for *showMenu()* is:

```
int i;
myDisplay.setTextColor(foregroundColor, backgroundColor);
activeRow = FONTROWOFFSET;
for (i = 0; i < itemCount; i++) {
  myDisplay.setCursor(i * spacing, activeRow);
  myDisplay.print(whichMenu[i]);
}
activeColumn = activeMenuIndex * spacing;
myDisplay.setFont(&FreeSerif12pt7b);
myDisplay.setCursor(activeColumn, activeRow);
myDisplay.setTextColor(selectForegroundColor, selectBackgroundColor);
myDisplay.print(whichMenu[activeMenuIndex]);
```

Note how this menuing method uses the *myDisplay* object to perform its tasks. There are nine total methods in the Menuing class library, including the special one called "menuing", which is actually the **constructor** for the library, containing any special parameters the library needs to operate. A class constructor cannot return a value.

A couple of the class methods, however, do return values when used in a regular C statement that assigns a value, for instance in the statement from the loop function in .ino file we see:

```
selectedMenuOption = myMenu.selectFromMenu(myDisplay, menuPreset) + SECONDOFFSET;
```

The variable *selectedMenuOption* from the main program is assigned the menu option value returned by the *myMenu* library function call. As you can see, you must use the dot operator to access class methods.

Ok, you ask "Why go through all of this trouble? I could just as well written all of this as straight C functions." The answer is the same as for any library, which is that you can more easily reuse the code because it is generalized and not specific to any one program or code routine. By the way, we could have included this library in the Arduino Library folder. We elected to keep it with the Prog PS code because we have not put it in the public domain as a regular Arduino library, but we might at some time in the future — who knows.

What happens if you forget to instantiate the *Menuing* class object? Well, first, you'll get a lot of compiler errors telling you the object is not in scope. After all, if you declare it, but don't define it, it doesn't exist...the ultimate "not in scope" error. In fact, the compiler is smart enough to just ignore all of the *Menuing* class code if a *Menuing* class object is not instantiated.

What happens if I define two *Menuing* class objects? This question is a little more interesting. If the two definitions appear at the same scope level (for example, global) and the library contains no *static* declarations, new memory space is allocated for each member variable, but the methods are not duplicated. Why not? Because there is no reason to. If those functions do not contain any *static* data, you can use them with either *Menuing* object. Think about it...

Now that you have seen how easy it is to create a class library, try writing one yourself for one of your projects.

### Examples of PBPS Use

No doubt you're going to think of examples that haven't even crossed our minds, but here are a few to get those creative juices flowing.

### Hysteresis Curves

You can use the PBPS to plot the hysteresis characteristics of devices such as relays or digital Schmidt triggers. **Figure 8.25** shows a possible setup for testing a 74LS14N Schmidt trigger. In this example, use the Triangle function to

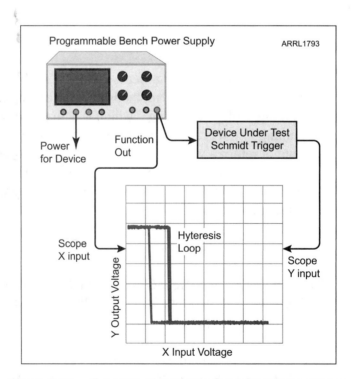

Figure 8.25 — Example of testing a Schmidt Trigger device.

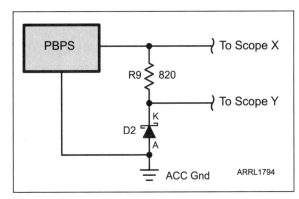

Figure 8.26 — Testing Zener diode.

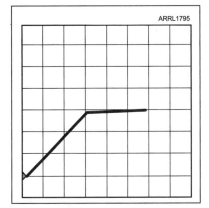

Figure 8.27 — Zener diode characteristic.

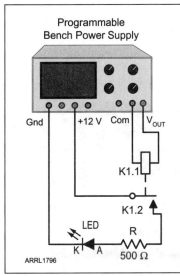

Figure 8.28 — Relay timing test.

compare the ramp up voltage vs. the ramp-down voltage response with the +5 V output from PBPS to power the Device Under Test (DUT). Plot the output vs. input of the DUT on an X-Y storage scope.

## Zener Diode Tests

If you want to determine the characteristic curve of a 15 V Zener diode, you could use a setup similar to that shown in **Figure 8.26** with the linear ramp function or triangle, and then plot the output versus the input as you did in the first example (see **Figure 8.27**).

## Periodic Relay Activation

Suppose a project requires you to turn on a 12 V relay every 10 seconds for a defined time period of 1 second. Using the layout depicted in **Figure 8.28**, you could perform such a test. In this case, use the Pulse function to activate a 12 V relay for 1 second every 10 seconds. Repeat the cycle N times as needed.

## Charge a Battery

Use the Timer Function along with the current limiter to charge a lead-acid storage battery (**Figure 8.29**). Set the Time

**Programmable Bench Power Supply    8-31**

parameters for the charge time, such as 12 hours, 30 minutes. Set the Voltage control to the battery voltage plus 3 to 5 V (for example, 12 + 3 = 15 V). Turn the Current limit to the current value. Attach the battery and turn the current limit until the desired charge current is achieved (for example, 200 mA). Start the timer, which turns off the charging process after the charge time is finished.

Note this process does not replace specialized battery chargers that monitor the state of charge and optimize battery life but will serve in a pinch.

## RF VGA

Radio receivers usually have an automatic gain control (AGC), often implemented using a variable gain amplifier (VGA) stage. A VGA is a gain circuit whose gain is controlled by means of a dc voltage. The gain vs. input control voltage curve is clearly of interest to the designer. In this example we show how the PBPS can be used to automate plotting that curve.

The VGA under test is from another of the authors' projects — the JackAl board for the µBITX transceiver. The active circuit in the VGA is a JFET which is operated in its non-linear operating region.

In this setup the RF amplifier is fed an input signal of 100 kHz. The output is measured with a Fluke 8920A True RMS Voltmeter, the output of which is a dc signal proportional to the $\log_{10}$ of the input. This dc signal is fed to the Y axis

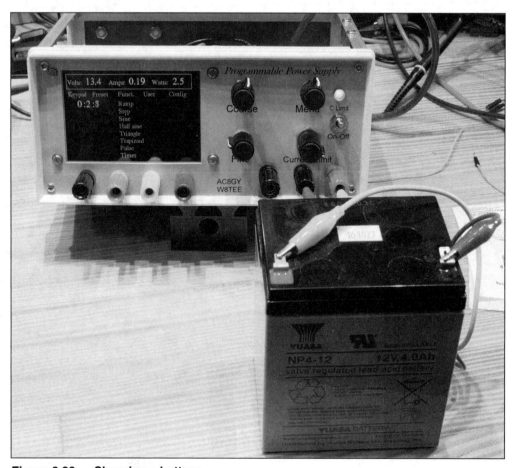

**Figure 8.29 — Charging a battery.**

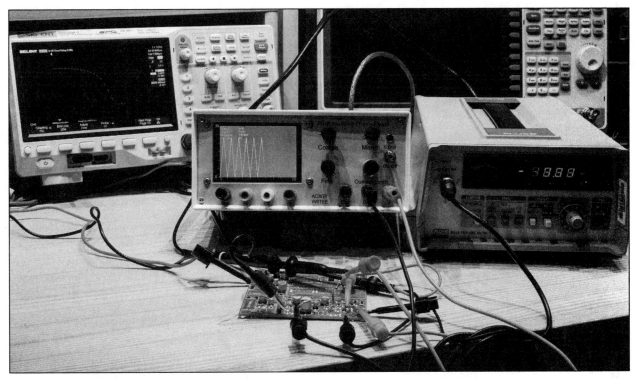

Figure 8.30 — Setup for VGA measurement test.

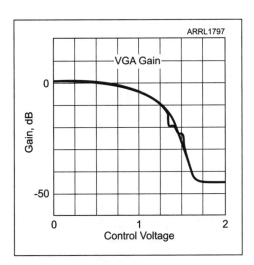

Figure 8.31 — VGA gain curve.

of the storage scope, while the X axis reads the voltage input from the PBPS, also connected to the VGA control input through a 10:1 voltage divider.

The PBPS Trapezoid and Triangle functions are used to vary the input control voltage. The resulting curve on the scope is gain vs. input control voltage — exactly what the designer requires. **Figure 8.30** is the setup for the VGA measurement and **Figure 8.31** is a screen printout of the gain curve.

## Conclusion

This chapter has shown the builder how to add a versatile power supply to their bench. We have also presented some examples of things this supply can do that ordinary bench supplies cannot. To make building the PBPS as easy as possible, there is a PCB available, as shown in **Figure 8.32** and the schematic for the PBPS is shown in **Figure 8.33** at the end of this chapter. Larger diagrams may be downloaded from our web site.

While the PBPS is quite complete as presented, there are some additional features that the builder might wish to add, such as:

- More current capacity — with a different current sensor, larger heatsinks, and a capable input 12 to 15 V supply the unit can handle up to 8 or 10A.

- Additional functions which are tailored to the user's specific needs, such as irregularly spaced steps.
- A touch-screen interface — the display used for the prototype is capable of resistive touch interface. The available PCB has provisions for touch screen implementation.
- Increase the maximum output voltage from 25 to 30 V, if needed. Much higher than 30 V gets quite close to design limits on some components, such as the op amps.

You may have specialized needs we didn't cover, but there are plenty of idle resources in the BP that should let you implement them if you choose to do so.

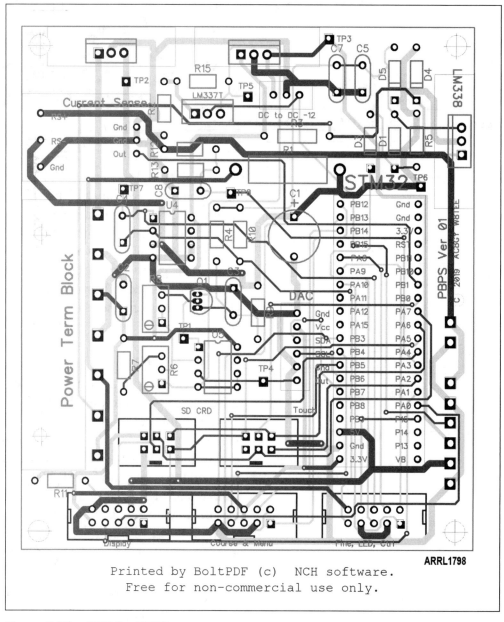

**Figure 8.32 — PCB for PBPS.**

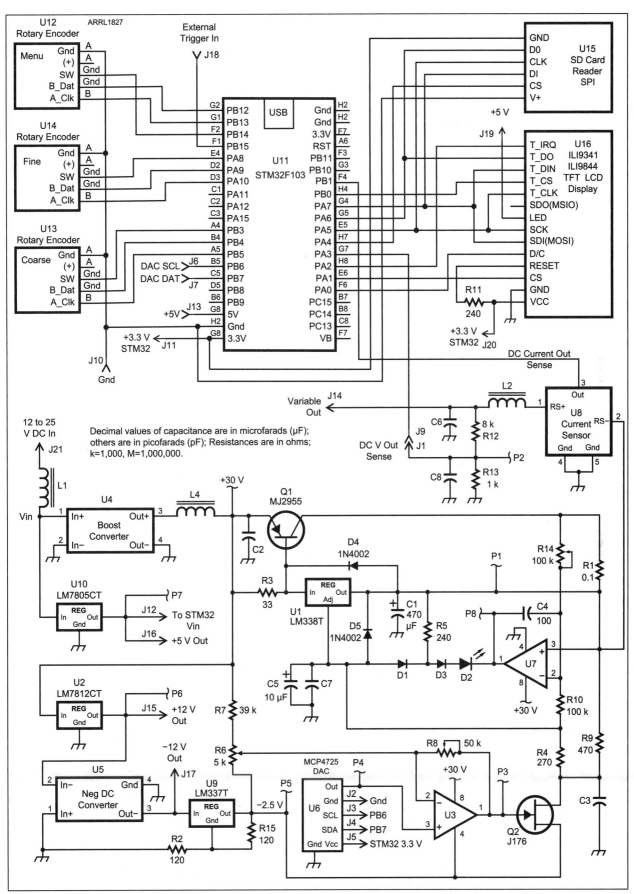

Figure 8.33 — Schematic for PBPS.

# CHAPTER 9

# 100 W Antenna Tuner with Graphical SWR Analyzer

Many of the projects in this book are directed toward the QRP operator (for example, the Mini Dummy Load in Chapter 6). This chapter's project is aimed at the most popular rig found in most ham shacks: the 100 W transceiver. As such, this antenna tuner/analyzer is not designed to be used as a portable (or QRP) antenna tuner. (We'll call it the ATA from now on.) It can be whatever size you want, within the constraints of the hardware of the ATA. One of our units is shown in **Figure 9.1**. We also incorporated the ATA into a portable HF "Go-Box" as shown in **Figure 9.2**.

Some of you are saying: "Another antenna tuner! There are lots of those available." Yes there are, but this one is different. We have added an antenna analyzer with a real-time graphical display that allows you to see the SWR across the entire band in one image and view the effect of changes as you make them. This is particularly useful if your antenna has a narrow bandwidth. Now you can see if you need to re-tune when changing frequencies. The other advantage of this tuner is that it can be very inexpensive, depending on your junk box and skill at finding bargains. Commercial manual antenna tuners run well over $200. This tuner can be built for as little as $40 or $50.

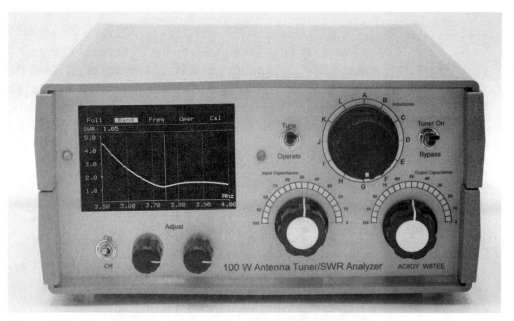

Figure 9.1 — Completed Antenna Tuner with Graphical SWR.

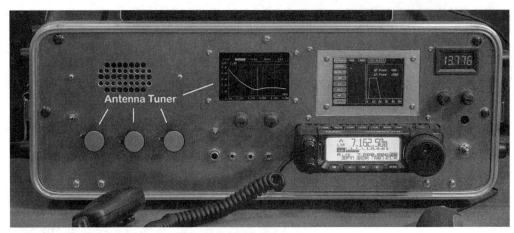

Figure 9.2 — Antenna Analyzer/Tuner Go-Box.

Figure 9.3 — Main screen.

The three large knobs seen on the front panel in Figure 9.1 adjust the two tuning capacitors and the tapped-toroid inductor. A Tune switch sends a small RF signal to the antenna for calculating the SWR for the present tuner settings. The other toggle switch allows you to bypass the antenna tuner completely. The On/Off switch controls the power to the tuner and the TFT display. The two knobs below the center TFT screen are used to navigate the menus and change parameters such as Band selected. The main menu screen is shown in **Figure 9.3**. There are five menu selections:

- Full — graphically shows the SWR for all ham bands from 80 to 10 meters.
- Band — allows the user to select a specific band for an SWR scan.
- Freq (Single Frequency) — allows the user to select a single frequency and shows the SWR for that selection during tuning.
- Oper (Operate) — shows SWR during transmit.
- Cal (Calibrate) — guides the initial calibration of the unit.

## Antenna Tuner Circuit

There are four main parts to the ATA circuit:

1) The Manual Antenna Tuner (Tune) circuit elements — capacitors, inductor, switches.

2) The Tune Impedance/SWR Bridge/Measure (Tune SWR Bridge) section with the impedance bridge and compensation circuit for improved accuracy.

3) The Control and Display function portion, consisting of the microprocessor, graphic display and encoders.

4) The Operate SWR Bridge, which can handle higher powers and is less sensitive than the Tune SWR Bridge.

The block diagram of the ATA is shown in **Figure 9.4**. The signal flow begins at the antenna input, which is coupled to the Tune circuit through the Bypass and Operate switches. The Tune circuit is a classic T configuration with two variable capacitors and a tapped inductor. For a 100 W transceiver, the capacitors must have a voltage rating of at least 600 V and a capacitance range from about 20 pF to about 300 pF. These capacitance values are not critical, but the wider the range, the better.

The output of the Tuner circuit goes to the Tune SWR Bridge, which has a resistive impedance bridge. Three rectifying Schottky diodes sample the voltages across a reference 50 Ω resistor, across the antenna, and between the legs of the bridge. The bridge uses 50 Ω resistors in three of the legs and the antenna

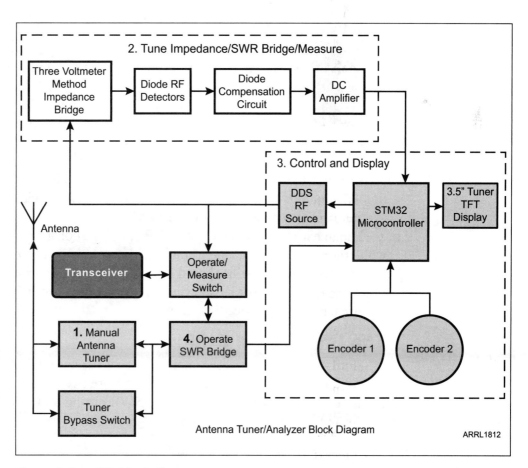

Figure 9.4 — ATA block diagram.

in the fourth. When the bridge is in balance, it gives a null reading when the antenna impedance is *resistive* 50 Ω. When the antenna impedance is 50 Ω resistive, the SWR is 1:1, the lowest possible value. Other antenna impedance values yield higher SWR readings.

The outputs from the diodes are input to JFET op amps, which have very high input impedance so as not to load the bridge circuit. The first op amp in each leg has a matched Schottky diode as the negative feedback element. This feedback circuit compensates for the nonlinear behavior of the diodes at low signal levels. A second op amp provides dc gain to improve signal-to-noise ratio. The linearizing effect of this circuit is shown later in the chart in Figure 9.8. Note that the linear region of the curve is improved by nearly 20 dB compared to the diode rectifier alone.

To perform an SWR measurement, the antenna must be excited at the measurement frequency. This signal is provided by an AD9850 DDS (direct digital synthesis) module that produces sine wave outputs up to about 30 MHz. The DDS is controlled by the microprocessor, an STM32F103. The output from the DDS is at a fairly low level, so an RF amplifier consisting of several transistors is used to boost the signal to approximately 0.5 V and yields an output at 50 Ω impedance. It is critical that sine wave excitation be used to avoid exciting higher harmonics of the antenna system, yielding false readings.

The DDS amplifier input stage is an emitter follower, which presents the DDS with a high impedance. This stage is direct-coupled to the common-emitter configured second gain stage. This transistor boosts the voltage from 400 mV to over 1 V. The next stage is another emitter follower, which provides power gain and drives the final stage, which is a push-pull configuration to provide sufficient drive to excite the antenna. The circuit has a flat frequency response to well beyond 30 MHz.

The microprocessor we used is an STM32F103, which is attached to a 3.5-inch TFT LCD, used in the serial data transfer mode. Two rotary encoders are used for menu and parameter selection.

The final major circuit element is an Operate SWR Bridge, using two toroid transformers to sample the forward and reverse signals during transmit operation. The outputs from the bridge are rectified by two diodes and sampled by another pair of analog inputs on the STM32.

The transformers in the Operate SWR Bridge consist of two T50-43 toroids with 10-turn secondaries and a single wire through the center of the toroid as the primary. The primaries should be at least #18 AWG wire and the secondaries can be #24 AWG (or larger gauge) magnet wire. Just tightly wind the secondaries and evenly space their turns.

## The Impedance Bridge

Accuracy of SWR measurements depends on several factors. First there is the method of obtaining the raw signals. We use two different approaches, one for tuning and the other for monitoring SWR during transmit. These approaches are discussed separately in the following sections.

## Operate SWR Bridge

The most straightforward way to calculate VSWR is to measure the forward and reverse components of the standing waves in the transmission line. The Operate SWR Bridge consists of two directional couplers that yield readings of the forward and reverse voltages in the transmission line. Once these two values are known, the Reflection_Coefficient can be calculated as:

$$\text{Reflection\_Coefficient} = \frac{\text{Voltage\_Reflected}}{\text{Voltage\_Forward}}$$

The VSWR is:

$$\text{VSWR} = \frac{(1 + \text{Reflection\_Coefficient})}{(1 - \text{Reflection\_Coefficient})}$$

Our directional couplers give us the forward and reflected voltage values, from which we calculate the VSWR. The Operate SWR Bridge circuit is shown in **Figure 9.5**.

This approach is quite straightforward and easy to implement but has the limitation of requiring fairly large RF excitation signals, such as are found during transmit. We could, in principle, just require the operator to key the transmitter with a CW power level sufficient to allow tuning to take place. This approach, however, has two major drawbacks:

1) Requires a sufficiently large power level that would be disruptive on-air. That would be impolite of the operator at best.

2) For a sufficiently high SWR presented by the antenna, above 1:5 or so, most transceivers do not allow transmission at all or transmit at reduced power, so the tuning operation could not be completed.

To avoid both of these limitations, we have added a separate, more sensitive, VSWR measurement that operates at very low power levels and does not require the transmitter to provide the RF excitation. An added advantage of this approach is the ability for a microcontroller to program the RF excitation source to sweep the input frequency over a range, allowing real-time graphical presentation of the VSWR results.

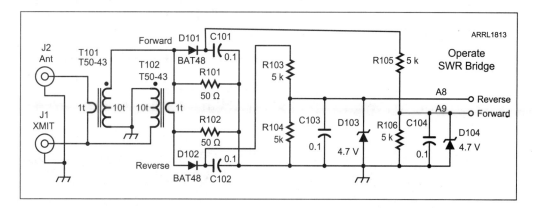

Figure 9.5 — Operate SWR Bridge.

## Tuning Impedance/VSWR Bridge

The simplest way to measure voltages that lead to a pseudo VSWR determination is to use a voltage divider consisting of a known resistor and the antenna. The primary drawback of this scheme is that it measures *only* the resistive part of the antenna impedance. What we really need is the complex impedance, comprised of the resistive part and the reactive part. (If math doesn't interest you, just skip on to the next section.)

Impedance is defined as:

$$Z = R + jX$$

where Z is the complex impedance, R is the resistive component and X is the reactive component. So why all of the fuss? The short answer is that unless one measures the complex impedance, false indications of minimum VSWR may be obtained. This is what happens:

The ideal 50 Ω transmitter match of VSWR = 1:1 occurs when the impedance is purely resistive and equal to 50 Ω; that is, when the X component of Z is zero. Now suppose you are measuring the resistive component only, but your antenna system impedance is actually Z = 50 (resistive) + $j$20 (reactive) Ω. Your resistive-only meter would tell you the VSWR = 1:1, but actually the VSWR using the complex impedance would be VSWR = 1:1.49, a substantial error!

There is a better way to go about these measurements that actually yields a pretty good indication of the true VSWR, using just one additional measurement. The approach is known as the "Three Voltmeter Method" or TVM (see the References at the end of this chapter).

To implement the TVM method, only a simple resistive bridge with three voltage measurements is required. To this bridge we add diode sampling of the RF, converting the RF ac voltages to proportional dc voltages. To minimize low RF voltage level errors, Schottky diodes are used, with forward voltage values in the 200 to 400 millivolt range. This circuit is shown in **Figure 9.6**.

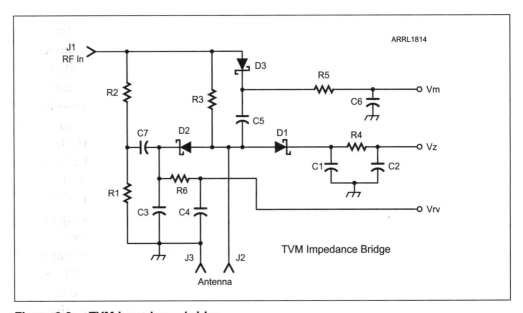

Figure 9.6 — TVM impedance bridge.

From the three voltages, Vm, Vz and Vrv, the complex impedance Z can be determined. First, calculate the magnitude of Z:

$$|Z| = \frac{Vz \times Rm}{Vm}$$

and the complex angle between the R and X components is given by

$$\cos\theta = \frac{Vm^2 + Vz^2 - 4Vrv^2}{2\,Vm \times Vz}$$

Knowing the magnitude of Z and the complex angle θ allows computation of the Reflection_Coefficient, ρ.

$$|\rho| = \sqrt{\frac{(R-Rm)^2 + X^2}{(R+Rm)^2 + X^2}}$$

where Rm = 50 Ω and X = reactive component.

$$X = \frac{|Z|\tan\theta}{\sqrt{1+\tan\theta^2}}$$

where |Z| = magnitude of impedance and θ = complex angle between R and X. Finally, as above, the value of VSWR is

$$VSWR = \frac{1+|\rho|}{1-|\rho|}$$

If all of that math is not your cup of tea, don't worry. We have programmed the microcontroller to do the computations on the fly.

## Diode Compensation

The final part of this puzzle is the reduction of the non-linearities because of the low-level signals presented to the Schottky diodes. When forward biased (that is, with voltage polarity that allows current flow), diodes have two major regions of conductance — above the forward voltage drop and below that value. **Figure 9.7** shows the characteristic curve for the BAT48 Schottky diode. Note the extreme non-linearity below about 0.5 V. Since our excitation voltage is around 0.5 V RF, the diodes are actually operating in the non-linear region most of the time.

The easiest way to compensate for this non-linearity is to use an op amp with the same diode type in the negative feedback loop. Below the linear region, the diode ef-

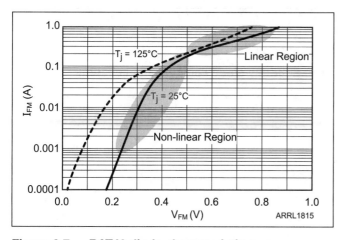

Figure 9.7 — BAT48 diode characteristic curve.

fective resistance increases. In a non-inverting op amp circuit, the circuit gain is giving by

$$G = \frac{1 + R_{feedback}}{R_{Ref}}$$

When $R_{feedback} = 0$, the gain =1. As $R_{feedback}$ increases, the gain goes up as well, in inverse proportion to the change in the diode rectifier's output, thus compensating for the non-linearity.

The compensating op amp is followed by a dc gain stage, to improve signal levels. **Figure 9.8** shows the computed response of a Schottky diode compensation circuit. The circuit shows about 20 dB of additional linear region and a total linear RF detector response range of about 50 dB. The compensation and amplification op amp circuit is shown in **Figure 9.9**.

The complete Antenna Analyzer/Tuner circuit is shown in **Figure 9.10**. The additional components are the microcontroller, display, and encoders.

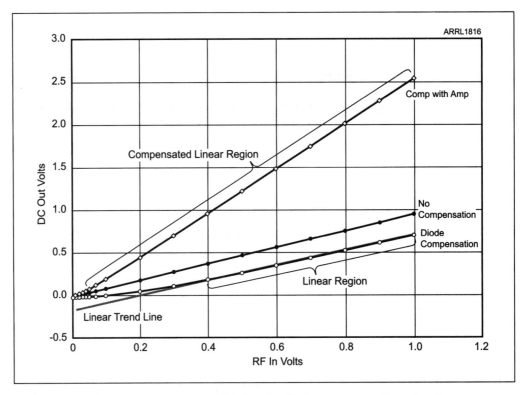

Figure 9.8. — Computed response of Schottky diode compensation circuit.

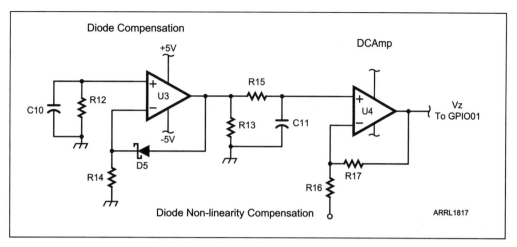

Figure 9.9 — Diode compensation circuit.

## Building the ATA

Generally, the construction is fairly straightforward. However, there are several things to keep in mind:

1) Operation at 100 W generates up to several hundred volts across the capacitors, especially when there is a large impedance mismatch, so use properly rated capacitors and be careful to separate wires carrying the RF signals.

2) Use a shielded case. We elected to use a plastic case, but lined it with copper foil we had available. A metal case is probably the preferred solution, which is what we used for the Go-Box.

3) Because the capacitors are not connected to ground, your body capacitance affects the tuning. We elected to use non-conductive shaft extensions and to locate the capacitors toward the rear of the case, away from our hands. We used old ballpoint pen bodies glued to the shafts. This was a cheap and effective solution. The capacitors and shaft extensions are shown in **Figure 9.11**. The capacitors themselves were found at a hamfest for less than $10 each. All three sections of the stators are wired together in parallel.

4) The inductor is a T157-2 (red) toroid core with 51 turns of #20 AWG stranded wire that has 300 V insulation. The inductor is tapped in 11 places at 2, 3, 4, 5, 7, 9, 12, 17, 23, 32, and 41 turns. We wrapped the toroid with electrical tape first and attached wires at the tap points as we wound the inductor. This arrangement makes a very compact unit that can be attached directly to the switch or mounted separately, as shown in **Figure 9.12**. **Figure 9.13** shows the compact switch/inductor assembly. An alternative to using a toroid is to wind stiff magnet wire around a form to create an air inductor. However, an air inductor must be larger and needs some sort of support structure. For this reason we elected to use the toroid instead.

The remainder of the circuitry is built on three boards, one for the STM32 and AD9850 and the others containing the two SWR bridges. Using multiple circuit boards allows for separation of the RF from the digital circuits, minimizing cross-talk issues. Otherwise, layout is not critical. Just follow the suggestions previously mentioned.

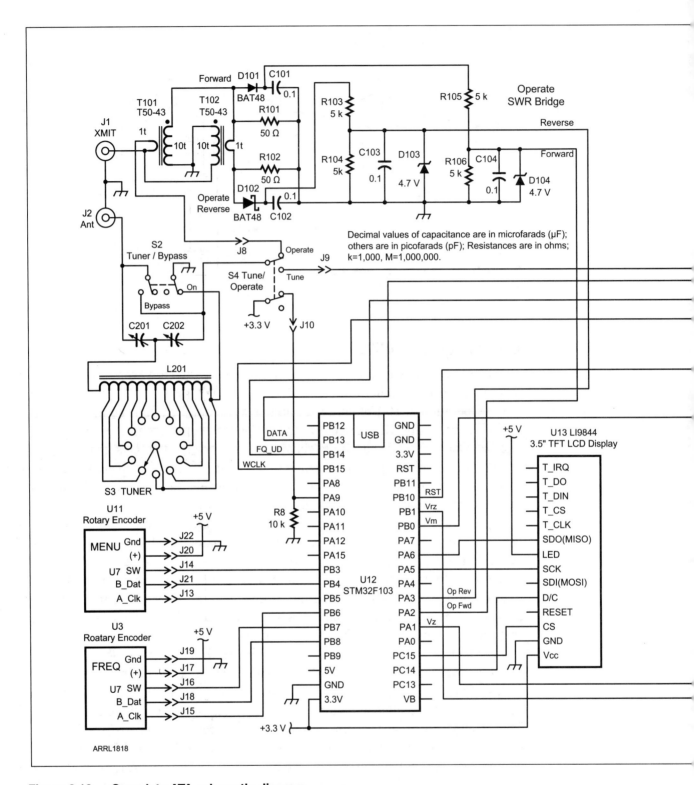

Figure 9.10 — Complete ATA schematic diagram.

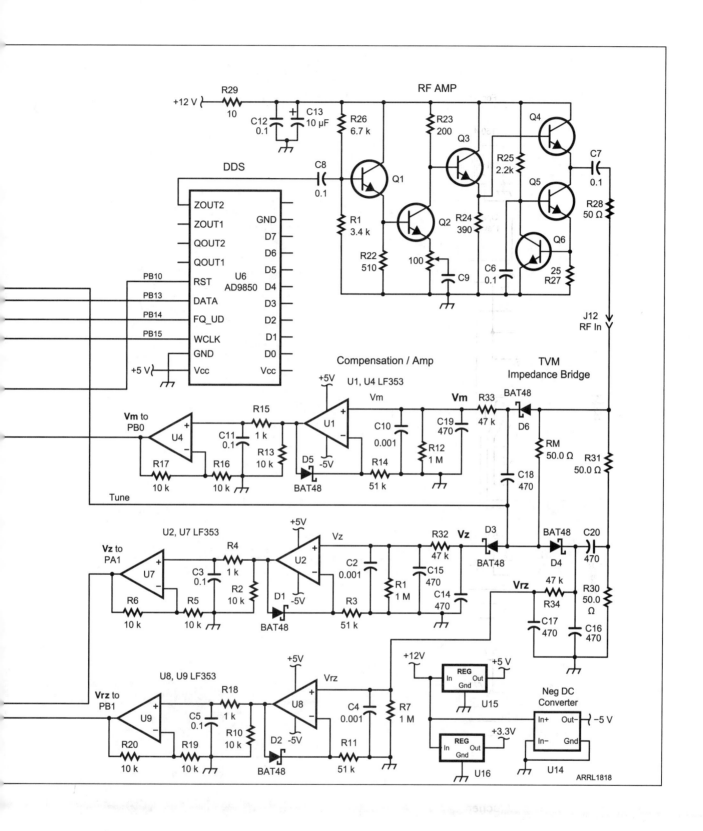

**100 W Antenna Tuner with Graphical SWR Analyzer**

**Figure 9.11** — Variable capacitors used in the tuner.

**Figure 9.12** — Toroid core (left) and the completed inductor wound with taps.

**Figure 9.13** — Inductor switch assembly.

As mentioned, for a case, we would recommend a metal enclosure. We used an existing plastic case lined with copper foil, but this is time-consuming and the copper foil is not inexpensive. Go for a metal case!

We created a front panel overlay with index markings for the capacitors and switch indications as well. This allows you to record settings for later reference. Chapter 16 details the process of creating nice front panels.

For the Go-Box, we divided the ATA into the tuner/bridge circuit and a separate control unit. **Figure 9.14** shows the tuner unit with the impedance bridge circuits, which are housed in a metal box for shielding purposes. Connections to the control unit are by a 10-conductor ribbon cable. The antenna and transmitter connections are on the back panel of the tuner box. **Figure 9.15** contains a view of the control circuits. The display and encoders are mounted on the front panel, connected by ribbon cable. All signal cables to the control unit contain only dc and are heavily bypassed using ceramic 0.1 µF capacitors on each end to reduce noise. The dc input from the 12 V supply is also bypassed going into the control box to reduce noise.

Because the transceiver in the Go-Box has a nice SWR readout, we elected to simplify this unit and not include the Operate SWR Bridge.

**Figure 9.14 — Tuner unit in the Go-Box with the impedance bridge.**

Figure 9.15 — Go-Box ATA control unit.

## Antenna Tuner Operation

After showing the usual startup splash screen, the user is presented with the main menu options: Full, Band, Freq, Oper, and Cal. These options were shown at the top of Figure 9.3. The following sections discuss the main menu options.

Tuning is generally a two-step process the first time you use a specific antenna. First, highlight the Full menu option and press the Menu encoder switch. This takes you to the screen shown in **Figure 9.16**. (The band limits are highlighted by the blue vertical lines on the display.) To start the tuning process at a new frequency or antenna, first set both capacitors to mid-range. Then cycle through the inductor switch positions until the SWR begins to decrease near the band you wish to use. If more than one inductor setting gives a reduced SWR reading, select the one giving the lowest SWR reading. Now adjust the capacitors in turn to get the minimum SWR in the desired band range. This is the starting place for the fine-tuning step. When you are satisfied, press the Menu encoder switch to return to the main menu.

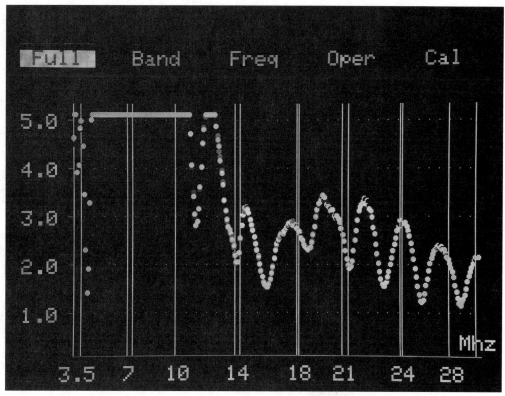

Figure 9.16 — Full frequency scan display showing the 80 to 10-meter amateur bands.

## Band Function

The next step is to refine the tuning and center the SWR plot on your Band using the Band function. First, use the Menu control to highlight the Band main menu option as seen in **Figure 9.17**. As always, turning the control clockwise (CW) moves the highlighted field to the right and counter-clockwise (CCW) moves the highlighted option to the left. If the currently highlighted option is at the end of the list (Cal), another CW movement causes the highlight to "wrap" around to the first menu option (Full). Likewise, given the highlight position shown in Figure 9.16, a CCW rotation would highlight the Cal option.

When you have highlighted the menu choice you want to activate, press the Menu knob. Because the encoder has a built-in switch, this activates the Band-Select submenu. The submenu is shown in Figure 9.17 with the 80-meter band currently selected.

The second step in the Tune process is to highlight the band you wish to use. In this case, because the menu is oriented vertically, a CW rotation of the Menu control moves the highlighted option toward the bottom of the list. A CCW rotation moves the highlighted option upward. Again, the highlighted option "wraps" when the end of the menu list is reached. When you have your desired band highlighted (for example, 80M in Figure 9.17), press the Menu control switch. To exit this menu and go back to the main menu, select Exit and press the switch.

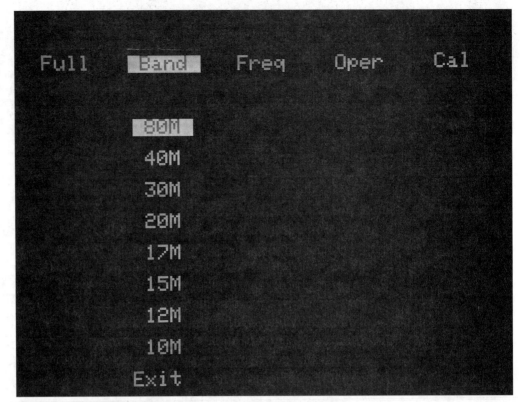

Figure 9.17 — Band select screen.

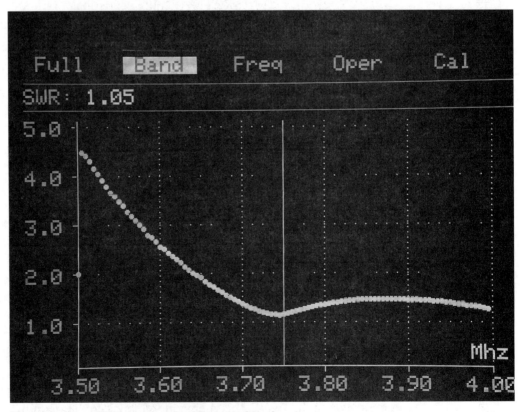

Figure 9.18 — SWR plot for the 80-meter band.

The content of the TFT display now changes to something similar to that shown in **Figure 9.18**. The software displays the center frequency for the band that was selected and calculates the SWR for the current capacitors, and inductor settings for that frequency. If you wish to change the default transmit frequency, this can be done in the .ino file by editing the statement at line 57 *labeled bandCenter[ ]*. The selected frequency is shown on the SWR plot as a green vertical line at the selected frequency.

In addition, the SWR for the full band is shown on the plot. The SWR calculation is refreshed every 50 ms or so and appears in near real-time. To change the tuning, the capacitor and inductor settings may now be adjusted to optimize the SWR across the selected band. Since you previously roughly adjusted the settings in the Full scan mode, only a small amount of "tweaking" should be necessary. With a little experience, you quickly get a feel for the inductor setting for each band, and then adjust the capacitors to minimize the SWR for the selected frequency.

You may want to make a note of the capacitor/inductor settings for future use (for example, checking into a net you regularly participate in). While subsequent tuning may not be exact (things like rain and snow may affect the settings), past settings should get you pretty close to minimizing the SWR. Note that the scan refreshes at a quick rate, so as you change the settings on the inductor and capacitors, the scan updates immediately.

It is easy to optimize the setting to cover an entire band in many cases. Another example is shown in **Figures 9.19** and **9.20**. Figure 9.19 shows a full scan of the HF bands with minimum SWR at 40 meters, and Figure 9.20 shows

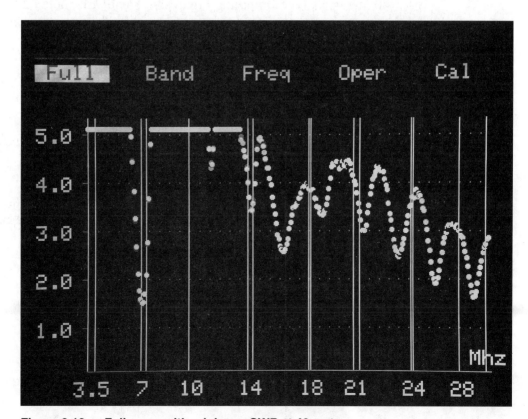

Figure 9.19 — Full scan, with minimum SWR at 40 meters.

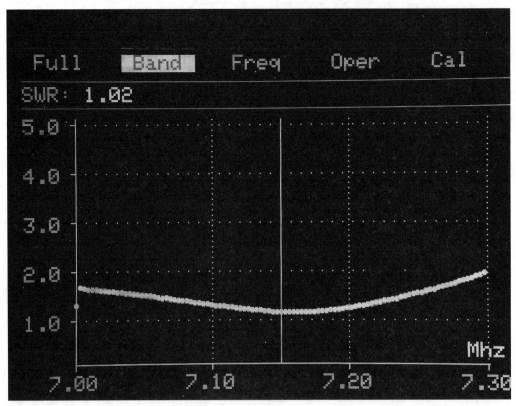

**Figure 9.20 — 40-meter band scan.**

40 meters alone. Comparing Figures 9.18 and 9.20, note that the bandwidth of the antenna at 80 meters is not as wide as for 40 meters, making precise tuning more important.

Once you have found the lowest SWR possible, you are ready to operate the transmitter. *Always make sure you flip the front panel switch from Tune to Operate before you key the transmitter.* Otherwise, nothing good is going to happen.

## Band Edges

The ATA software comes pre-configured to use the US band frequencies as set by the FCC. If you are operating under some other licensing authority, you may need to change these band edges. The band edges are stored in an *unsigned int* array with the clever name of *bandEdges[]*. These band limits are entered as pairs. For US bands 80 meters through 10 meters, these edge pairs are arranged from low to high band edges, as shown here:

```
unsigned int bandEdges[] = {3500, 4000, 5330, 5403, 7000, 7300, 10100,
10150, 14000, 14350, 18068, 18168, 21000, 21450, 24890, 24990, 28000,
29700};
```

If you are subject to different frequency allocations, you should edit the definition of the array as it currently appears near line 40 in the .ino project file.

## Single Frequency Function

High-Q antennas have narrower bandwidths, which may not cover the entire band. This is where the Single Frequency menu setting comes in handy. The operator can pick a target frequency and precisely adjust the SWR minimum to that frequency. To activate this option, select the Freq menu setting from the main menu. Pressing the Menu encoder switch takes you to the screen shown in **Figure 9.21**.

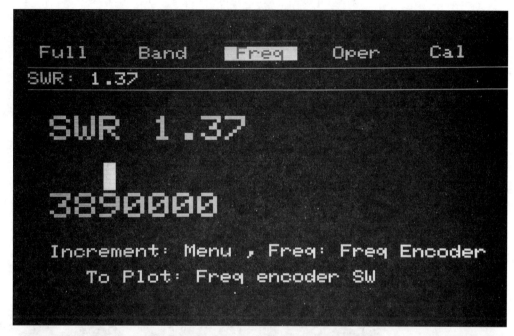

Figure 9.21 — Single frequency screen (SWR is 1.37:1 at 3890 kHz).

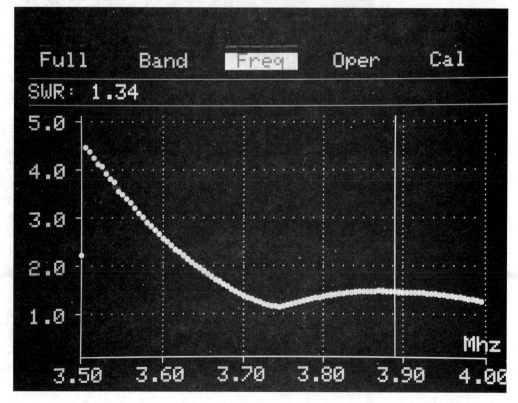

Figure 9.22 — Real time plot with selected frequency (3890 kHz) designated.

The left encoder moves the cursor to the digit to be adjusted and the right encoder changes the value. The corresponding SWR value is shown above the frequency. Minimize the SWR with the capacitor and inductor controls. When the optimum SWR has been achieved, you may enter the scan plot by pressing the right encoder switch and the real-time plot such as shown in **Figure 9.22** is displayed.

## Operate Function

The Oper (Operate) menu option is what you think it is: The option selected when you are using the ATA to operate in conjunction with transmitting. Set the ATA's mode switch to Operate, which disengages the tuning bridge from the transmit circuit — the tuner is still in the line. However, if you have included the Operate SWR Bridge, you can monitor the SWR during transmit by selecting the Operate option in the Main menu. The ATA then continues to monitor the SWR while the transmitter is active. By default, the transmission line is routed through the ATA's SWR sensing unit so it can monitor the SWR in real time. If you wish to bypass the tuner altogether, use the bypass switch on the front panel. The Operate SWR screen is shown in **Figure 9.23**.

Using the same components we used to build the ATA, you should be able to operate the ATA safely with power levels of up to 150 W. Higher power levels would require changes to the variable capacitors and the inductor.

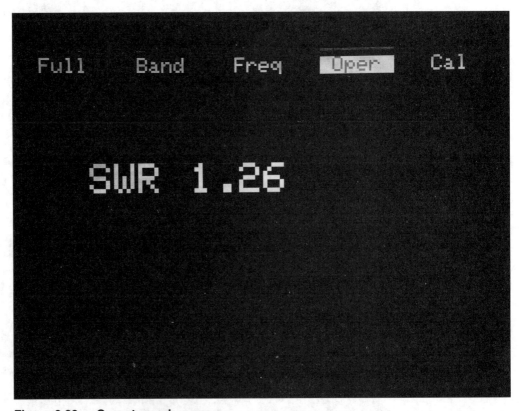

Figure 9.23 — Operate mode screen.

## Calibration

Because of variability in parts values, especially in the op amp circuits, it is necessary to do an initial calibration so the SWR readings are correct. The good news is that you only have to do this once. We debated about whether to include this as a menu function, and decided to provide a guided on-screen calibration procedure.

Calibration is in two parts. First the op amp outputs must be zeroed with a zero-voltage RF input and a known load attached. Then as a second step, the gains of the three op amp circuits need to be matched. This is all very easy and does not require any external instrumentation. All that is necessary is to make up a set of calibration resistors to be inserted in place of the antenna and to use the on-screen readouts to adjust a couple of values. Calibration takes just a few minutes and you are all set. The steps are outlined below.

1) Create a set of calibration resistors for the following values: 25.0 Ω, 50.0 Ω, 100.0 Ω, and 200.0 Ω. If you have an accurate multimeter, it is easy to start with a common nominal value of resistance, say 51 Ω. First measure this resistance, then by adding resistors of higher value in parallel, this value can be trimmed down to the needed 50.0 Ω. To ease the process, you can use an online calculator, such as the one found at **www.1728.org/resistrs.htm** to determine parallel values to be used. Simply connect the resistors together and measure the new value. We suggest tack-soldering the parallel resistors and then waiting for everything to cool down prior to measuring. It may take a few minutes, but you can end up with a nice set of accurate reference resistance values. If you can't or don't want to go through this procedure, just order some 1% or 0.5% resistors to use as references.

2) To make the calibration process easier, we hooked our resistors to a multi-position rotary switch with an appropriate connector to plug right into the tuner. With this setup it is easy to switch between the various values as you calibrate.

3) On the Impedance Bridge board, temporarily remove the RF input cable and short the connection to ground at the Impedance Bridge board. This provides the conditions to "zero" the readings. We added a permanent shorting switch in the line to make things easier. Now attach or select the 50.0 Ω reference resistor in place of the antenna.

4) Next, select the Cal option on the main menu. A list of several variables is shown, along with their current values. The screen is similar to that shown in **Figure 9.24**.

5) With the short in place, select one of the three top variables with the Menu encoder and adjust each value in turn to get as close to zero as possible, using the right encoder. When you are finished, press the left encoder switch and adjust the gains as follows.

6) First remove the short and select the 50 Ω calibrated reference resistor connected in place of the antenna.

7) Select either the linVm or linVz variable and use the left encoder to adjust until the value so that linVm = linVz. The SWR readout should be very close to 1:1.

8) Next select the 100 Ω resistor and adjust the linVrz gain to yield a SWR

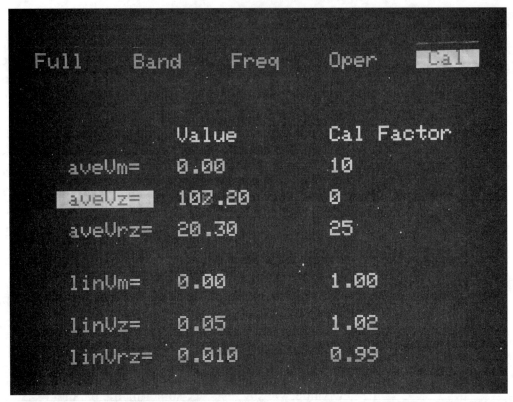

**Figure 9.24 — Calibration screen.**

value of 1:2 or as close as possible. Do the same with the 200 Ω resistor, except this time the target SWR is 1:4.

9) These steps are somewhat trial and error, and the adjustments interact, so be patient. It may be necessary to repeat the sequence a couple of times, refining the values each time.

10) Now check the SWR values with 25 Ω through 200 Ω. The SWR values should be as follows: 25 Ω gives SWR = 1:2; 50 Ω gives SWR = 1:1; 100 Ω gives SWR = 1:2; and 200 Ω gives SWR = 1:4.

11) If the values are substantially different, repeat the zero settings with the new gain values and go through the sequence again. With a couple of rounds of adjustment the results should be satisfactory.

The values are stored in EEPROM as you make the adjustment, but it is a good idea to note values for future reference. This all takes longer to explain that to actually do, so stick with it and you will be rewarded with an accurate instrument.

## Software

The unusual aspect of this ATA project is having the ability to graphically display the SWR in real time for a given band. As such, this is a good place for a discussion about using various graphic elements in a program. The most convenient way to collect the graphic data in one place is by using a C structure. The syntax for using C structures is very similar to that for objects in a C++ class. Indeed, structures and classes have a lot of similarities, except structures cannot have methods embedded within them. The graphics structure we use is shown here:

```
struct grafix {             // Graph structure declaration
  int x;                    // upper left coordinate horizontal
  int y;                    // upper left coordinate vertical
  int w;                    // width of graph
  int h;                    // height of graph
  float minX;               // minimum X graph value, can be negative
  float maxX;               // maximum X graph value
  float minY;               // minimum Y
  float maxY;               // maximum Y
  float xInc;               // scale division between lo and hi
  float yInc;               // y increment
  float currentValue;       // Current value
  int digitTotal;           // total digits displayed, not counting decimal point
  int decimals;             // digits after decimal point
  int barColor;             // Color for bar
  int voidColor;            // Background color in bar chart
  int backBar;              // Background bar color
  int border;               // Border color
  int textColor;            // Color for text
  int backFill;             // Background color for entire graph
  char label[];             // Label text
};
```

First, note that this is a structure *declaration*, not a definition. That is, drawing upon the discussion from Chapter 4, we are presenting a set of blueprints for the structure. The structure declaration does not give us a structure that we can manipulate in our program. To be able to use a structure, we need to define it, using the following syntax:

```
grafix myGraph;
```

The variable *myGraph* now has an lvalue (a place in memory) and we can use it in our program. If you are using a 240×320, 2.8-inch display and want to set the width and height of the display, the syntax is:

```
myGraph.w = 320;
myGraph.h = 240;
```

Note how the dot operator is used to access the members of the structure in the same manner we did in Chapter 4 when accessing an object's class member.

As a general rule, we initialize the structure with a function call that is placed in the *setup()* function, using code similar to the following function:

```
/*****
  Purpose: This sets the default values for the graphics structure

  Paramter list:
    void

  Return value:
    void
*****/
void SetGraphixDefaults()
{
  myGraph.x           = 20;
  myGraph.y           = 100;
  myGraph.w           = 300;
  myGraph.h           = 30;
  myGraph.minX        = 1.0;
  myGraph.maxX        = 3.0;
  myGraph.yInc        = 0.25;
  myGraph.xInc        = 0.25;
  myGraph.minY        = 1.0;
  myGraph.maxY        = 3.0;
  myGraph.digitTotal  = 3;
  myGraph.decimals    = 2;
  myGraph.barColor    = GREEN;
  myGraph.backBar     = DKGREEN;
  myGraph.border      = GREEN;
  myGraph.textColor   = WHITE;
  myGraph.backFill    = BLACK;
}
```

Note in the function code that we did not use the height and width limits of the display when we set those parameters. The reason is because it is fairly common to draw a graph on the display, but you want to leave room for other information on the same display page (such as title, axis labels, and so on). There are, of course, lots of other members that could be added to the structure. For example, the axis lines drawn in Figure 9.22 use solid lines. Perhaps you would like to use a less obtrusive grid, in which case you might not draw a line, but rather turn on every tenth pixel in the line, giving the grid a "dot" effect like the grid lines in Figure 9.22. Therefore, you might add new members:

```
byte gridType;      // 0 = solid line, 1 = every tenth pixel
byte dotDensity;    // the number of pixels skipped when doing a dot grid
```

Because uninitialized global data in C are set to zero or null by default, the default graph grid is a solid line. If I were going to use this feature, I would write a new function named *DrawGraphGrid()* that would look something like **Listing 9.1**.

If you read through Listing 9.1, the *if* statement says that, if we are looking at the "dotDensity-ith" pixel (that is, every 10th pixel), draw the pixel; otherwise just continue on to the next one. The "%" C operator is the modulo operator, which returns the *remainder* after division. So, if we are looking at i = 9, i divided by 10 leaves a remainder of 9. Since 9 is not equal to 0, we do not turn on that pixel. On the next iteration of the loop, i = 10. Because i divided by 10 is 1 with no remainder (the modulo is 0), we do display that pixel. The *dotDensity* and *textColor* members of the graphics structure determines the density of the grid line and its color.

## The Programmer's Dilemma

The *DrawGraphGrid()* function illustrates a common programmer's dilemma. As you know, a dilemma is when you have two or more choices, all of which are bad. The *DrawGraphGrid()* function is bad because it only draws one line on a grid, and only in the horizontal direction. The function also bad because it assumes that the *myG* graph structure has been properly initialized before the function is called. One aspect of writing functions we want to minimize as much as possible is coupling. *Coupling* occurs when a function cannot

---

**Listing 9.1**
**A Dotted Line Draw Function**

```
/*****
 Purpose: To draw a solid or dotted grid line on the current graph

 Argument list:
     struct grafix myG    the graph structure being used
     MCUFRIEND_kbv tft    the display object

 Return value:
    void

 CAUTION: Assumes that graph structure x and y are set for current line
*****/
void DrawGraphGrid(struct grafix myG, MCUFRIEND_kbv tft)
{
  int i;
  for (i = myG.x; i < myG.maxX; i++) {
    if (myG.gridType && i % myG.dotDensity != 0) {// for a 10:1 grid if
                                                  // dotDensity = 10
      continue;
    } else {
      tft.writePixel(i, myG.y, myG.textColor);
    }
  }
}
```

perform its task properly without the aid of some other function or external variable(s). In the case of our *DrawGraphGrid( )* function, we violate the no-coupling principle horribly by relying on the current values of *myG.gridType*, *myG.dotDensity*, *myG.y*, and *myG.textColor*. If those values are not initialized correctly before we call the *DrawGraphGrid( )* function, the function cannot perform as we want it to.

My students would tackle this problem by saying: "No problem, we'll just pass the desired value for each of those variables to the function as a parameter!" But such a solution flies in the face of a second principle of writing good functions: Cohesion. Simply stated, a function is *cohesive* if you can tell me precisely what the function does in two sentences or less. Each additional parameter that you pass to a function makes it just that much more complex to write, test, and debug. This almost always means the function is becoming less cohesive as you add parameters to the function call. Students often want to write functions that are Swiss Army knives: They do lots of things, but none of them well.

It is very difficult to write functions that are totally devoid of coupling. Indeed, some functions require a certain degree of coupling to perform their tasks properly. Functions that manipulate a database must first open that database. Same for files, printer connections, sensor initialization, and similar activities. However, the more devoid a function is of coupling, the greater are its chances that it can be re-used in some other application. Cohesion also involves tradeoffs. We can reduce the number of parameters that must be passed to a function by making all of those parameters global data. However, that tears down one of the pillars of object-oriented programming discussed in Chapter 4: encapsulation.

So, what's the solution? Often the solution becomes more clear if you stand back and look at the forest instead of the trees. That is, ask yourself: Will we ever need a function that can draw a horizontal line of evenly-spaced dots? Perhaps. Will we ever need a function that can draw a vertical line of evenly-spaced dots? Perhaps. Will we ever need a function that can draw a solid horizontal or vertical line? Pretty sure we will. In the end, I would probably write two functions for these tasks:

```
void DrawHorizontalLineFlex(graphObject, startX, endX, Y, color,
   lineType)
void DrawVerticalLineFlex(graphObject, X, startY, endY, color,
   lineType)
```

The two functions pass in the graphic's object we want to use (after all, there could be more than one), the starting and ending pixel coordinates for the line, its color, and finally a line type. I used the word Flex in the function names because they are different than the standard graphics primitives. We can make the functions more flexible if the last parameter does double-duty. If the value of *lineType* is 1, the code draws a solid line. Any value greater than 1 becomes the number of pixels we skip before printing the next pixel. In other words, it becomes the modulo value for the function. If you look elsewhere in this book, you still won't find the code for the two functions mentioned above. We want you to try your hand at writing these two functions. Give it a try...

## Multiple Files

Once again, we break up the project's source code files into three file types:

- **INO file**. This file *must* contain the *setup()* and *loop()* functions. There should be only one .ino file and it appears at the extreme left position of the files that appear above the source window of the IDE. The primary file name must match the directory in which the file appears.
- **CPP file**. These are the C++ files that contain the rest of the project's code. We tend to name these so they reflect their purpose (such as Graphics.cpp, Menuing.cpp, and so on.).
- **H file**. These are the header files that contain any overhead information used in the program.

Our main reason for using smaller .cpp files is twofold. First, it allows you to easily search for various functions. That is, if something is amiss in the program's graphics output, it's easier to look through the Graphics.cpp file than searching through (a much longer) single .ino project file. Second, we discussed in Chapter 2 that the Arduino IDE supports incremental compiles. Simply stated, this usually shortens development time for the reasons explained earlier.

## Header File Content

Let's look at part of the AntennaTuner.h header file:

```
#ifndef BEENHERE
#define BEENHERE

#include "Adafruit_GFX.h"
#include <ILI9488.h>            // https://github.com/jaretburkett/ILI9488
#include <EEPROM.h>
#include <Rotary.h>             // https://github.com/brianlow/Rotary
#include "SPI.h"
#include <AD9850SPI.h>          // https://github.com/F4GOJ/AD9850SPI
#include <AccelStepper.h>       // http://www.airspayce.com/mikem/arduino/
                                // AccelStepper/AccelStepper-1.59.zip
#include <Wire.h>

//========================= Symbolic Constants =========================


//#define DEBUG                        // Comment out when not debugging
#define VERSION        001           // Software version
#define SPLASHDELAY    2000L         // Normally, 4000L
```

The header file begins with two preprocessor directives:

```
#ifndef BEENHERE
#define BEENHERE
```

The very last line of the header file contains:

```
#endif
```

Think about what this means. If the symbolic constant named *BEENHERE* is not defined, we define it with the very next preprocessor statement line. The compile then spins through eight #include directives to read their content into the program. (Note that non-standard header files supply a comment that tells you where you can go to download their associated library files.) Once those files are read into the program, the *#define DEBUG* directive is processed, and so forth for all of the lines in the header file.

So, what happens if the symbolic constant *BEENHERE* has already been defined? In that case, every line within the header file is skipped until we read the *#endif* directive at the end of the header file. Keep this in mind while we show you some addition statements in the header file.

```
//========================== Globals ============================

extern char mySDFiles[][NAMELENGTH];

extern uint8_t latest_interrupted_pin;

extern int FwdOffSet;
extern int RevOffSet;
extern int displaySize;
extern int encoderPassCount;
extern int eepromMinIndex;
extern int filesFound;
extern int menuIndex;
extern int menuDepth;
                                     // Some lines left out...
//=========================== Function Prototypes =============
void AlterMenuDepth(int whichWay, const char *menu[], int len);
void AlterMenuOption(int whichWay);
void DoScan();
void DoXmitSWR();
void DrawBarChartHAxes(int flag);
void DrawBarChartH(int flag);
char *Format(float val, int dec, int dig, char sbuf[]);
```

The Globals section of the header file holds all of the global variables that are used in the program. Why the word *extern* before each data type specifier? Recall from Chapter 2 that the word *extern* means that this variable in not defined in this (header) file, but it is defined elsewhere in the program. However, any statement that uses *extern* does give us a full attribute list for the variable so we can use it in this (or other) source code files. Now look in the AntennaTuner.ino file. Near the top of that .ino file are these statements:

```
unsigned char result1;
unsigned char result2;
char *menuOptions[]      = {" Full ", " Band ", " Freq "," Oper ", " Cal "};
  // Menu text
char *menuOptions2[3][3] = { {"Change Band", "Change Freq"},
  {"Change Band", "Activate", "Set"},
  {"Band Edges"}
};
long bandCenter[] = {3750000, 7150000, 10125000, 14150000, 18118000,
  21225000,24940000, 28850000};
long presetFrequencies[8][PRESETSPERBAND];
char prompt[22] = {"Set frequency for "};

int currentBand;              // Should be 40, 30, or 20
int currentPage;
int currentCapPosition;
int encoderPos = 0;           // counter
int menuIndex;
                              // More statements follow...
```

Therefore, all of the *extern* **declarations** in the header file are **defined** in the .ino file. We still have a problem, however. Suppose the Menuing.cpp file needs to use the *menuIndex* variable. Because *menuIndex* is defined in AntennaTuner.ino, the scoping rules covered in Chapter 2 say that *menuIndex* dies when the end of that file is reached. Not good.

However, suppose we put these directives at the top of Menuing.cpp (and all other .cpp files):

```
#ifndef BEENHERE
#include "AntennaTuner.h"
#endif
```

If *BEENHERE* is **not** defined, the compiler reads the AntennaTuner.h header file into the program before reading the Menuing.cpp code. Because the AntennaTuner.h header file contains all of the *extern* data declarations for global variables, including the one for *menuIndex*, the compiler knows enough about that variable to let you use it in the Menuing.cpp file. If *BEENHERE* **is** defined, all of those *extern* data declarations are already in the symbol table so we don't need to read the header file again. This means we can have our cake and eat it, too. We get all the benefits of multiple source code files (such as shorter search times and shorter compile times because of incremental compiles) and the compiler has no complaints. How cool is that?

The next block is for function prototypes. These are no more than the function signatures we talked about in Chapter 2 with a semicolon at the end. The purpose of a function prototype is fourfold. Consider the first function prototype above:

```
void AlterMenuDepth(int whichWay, const char *menu[], int len);
```

First, it tells the compiler the type of data that the function is designed to return to the caller; *void* in this example. Second, it gives the name (ID) of the function. Programmers misspell things all the time and the compiler gets a little cranky when that happens. Function prototypes make it easier for the compiler to flag these errors early in the compile. Third, the function prototype tells the compiler how many arguments the function has. A function can have zero function arguments (a *void* argument list) or it can have as many as it needs to accomplish its task — three in this case. Finally, the function prototype also provides the details about the data type for any function arguments. This fourth check is incredibly useful because, if you place the wrong data in a function call without function prototypes being in effect, you can blow away the stack pointer and the system could crash. Even worse, the program *doesn't* crash but you don't know it!

Sometimes you'll get an error message that says the compiler doesn't know about such-and-such a function, but the source code is right in the file. However, if you call the function before the compiler sees that function's code, it has no clue what it's looking at. (This is why we put *setup( )* and *loop( )* at the bottom of the .ino file.) To fix the problem, either move the function code to a place in the program code above where the function is called, or put a function prototype at the top of the file (or in its header file).

The actual program code for the ATA is very similar to the code you've seen in earlier projects, so there's no need to kill any more trees repeating those discussions. Still, at least look through the code and mentally trace what the program does. Ideally, it would be most instructive if you did this before you run the program. That way you can assess how well you understand the program. If the program results are wildly different than what you expected, you need to spend some additional time explaining to yourself where you went wrong.

## Conclusion

Most of you likely already have an antenna tuner, but chances are that it doesn't support a real-time graphic plot of the entire band being used, much less the whole HF spectrum from 80 to 10 meters. We think this project is a very useful addition to your station's equipment lineup, especially if you like to experiment with different types of antennas. If your transmitter supports higher power levels than what our design supports, only a few components such as the variable capacitors and the inductor need to be changed for higher power levels.

## Enhancements

Another enhancement that could be useful is to save the resonant tuning quads (cap1, cap2, frequency, inductor) to EEPROM. Then, when the frequency for the band is set, you could scan the EEPROM list and display the quad values. This would give you a starting point for setting each of the caps and the inductor.

Finally, our tuner adjustment range probably does not extend to either 160 or 6 meters, but with suitable modification of the capacitor and inductor values and replacing the DDS with one that covers 50 MHz, the range could be extended. (We don't have access to a 160-meter antenna to try that option.) The impedance bridge circuit as configured certainly covers this wider range, if you attempt to extend the range.

If you were to need a simple impedance bridge, some code modifications would allow readout out of the actual resistive and reactive impedance components, which could be used to create Smith charts and calculate other useful antenna parameters.

As the ATA code currently stands, it uses less than 44% of the available flash memory and about 54% of the SRAM resources. The current source has about 2400 lines of code, so another couple of thousand source lines might be added — plenty of programming resources available for experimenting. If you do enhance the ATA code, don't forget to share it with the rest of us!

## References

Duffy, Owen, "Measuring RF Impedance by the Three Voltmeter Method," **owenduffy.net/calc/z3vm.htm**

Knight, David W., "Impedance and admittance measurement using discrete scalar voltage samples," **www.g3ynh.info/zdocs/bridges/appendix/YZ_scalar.pdf**

Knight, David W., "Radio Frequency Bridges: Part 3," **www.g3ynh.info/zdocs/bridges/part_3.html**

Kuhn, Kenneth, "Impedance Measurement," **www.kennethkuhn.com/electronics/impedance_measurement.pdf**

Steber, George, "A Low Cost Automatic Impedance Bridge," *QST*, Oct. 2005, pp. 36 – 39.

Steber, George, "A Low Cost RF Impedance Analyzer," *Nuts & Volts*, Feb. 2008, **www.nutsvolts.com/magazine/article/a_low_cost_rf_impedance_analyzer**

Sutorius, Gustaaf, "Challenges and Solutions for Impedance Measurements," **www.keysight.com/upload/cmc_upload/All/ChallengesandsolutionsforImpedance.pdf**

# CHAPTER 10

# A CW Messenger

A CW what? This project involves building a device that is capable of fulfilling the role of a memory keyer by sending "canned" messages at the press of a button. Some keyers, such as the K1EL WinKeyer, have buttons on the top of the unit that, when pressed, send a stored CW message. The project constructed in this chapter also has ability to send stored messages and function as a keyer. So far, nothing new. The final project is shown in **Figure 10.1**.

However, we add a small touch screen to the CW Messenger (CWM) that allows you to change the message in the field without using error-prone keying commands to the keyer. Instead, we use a small, resistive, touch screen keyboard for entering/editing the messages. The keyboard is shown in **Figure 10.2**.

The keyboard is similar to a standard QWERTY keyboard, minus some punctuation keys. We use a pen to select the key (yes, the end with the cap on it!) and the keyboard is quite sensitive to the touch, making it very easy to use. The display is a small 2.4-inch TFT SPI display using the ILI9341 driver and costs under $7.00.

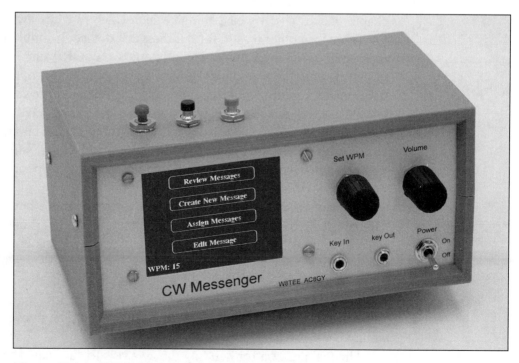

Figure 10.1 — CW Messenger in its project case.

Figure 10.2 — The touch screen keyboard for the CWM.

Why would you need to change a message in the field? Most people change the message array using their home PC, compile and upload the modified message array, and they are done. However, a lot of things can go wrong between your home and the field. Have you ever saved an incorrect exchange message only to discover the mistake in the field? Have you ever done a contest where you needed to use your Maidenhead grid locator, only to find you typed it in wrong? Or perhaps you planned on doing a Summits on the Air (SOTA) activation and, on the way home, you determine that you could activate another one... if only you could change the summit designator. The device presented in this chapter makes it pretty easy to make such changes while away from your PC.

There is another reason for building this device. It should be pretty evident that we like QRP operation and really enjoy CW QRP since it's a low-cost way of getting onto the HF bands. However, since the FCC dropped the code requirement for getting a license, it seems that fewer and fewer new hams know CW. Related to this is the fact that we hams have been relatively unsuccessful in attracting young people to our hobby. After all, a cell phone makes it a snap to talk with almost anyone anywhere in the world. If that's the message young people get for what ham radio is about, we have done a poor job of portraying our hobby to the public.

A recent *RadCom* article discussed the success some hams had demonstrating ham radio to high school students in the UK, and the show-stopper for them was Morse code! So, how 'bout letting a high school student enter their name using the keyboard shown in Figure 10.2, and then press a button and hear their name in Morse code? The CWM allows you to do this, displaying their name on the display as it is heard using the onboard buzzer. Hopefully, some of you might try demonstrating Morse code to students. It might even attract a few young people to our hobby.

The main reason for building the CWM is to automate sending specific CW messages, most likely to be used during a contest. However, it could also

Figure 10.3 — Labeling display pins.

be used to just send the "repetitive" parts of a common QSO, such as the CQ, QTH, and name messages.

One more thing... In **Figure 10.3**, during breadboarding, Al placed a label strip on the front of the display, identifying each of the display pins. This is a simple idea, but a great one. Prior to placing that label on the front side of the display, I flipped the display front-to-back ... oh, I don't know ... probably a bazillion times while I was working with the circuit. It's worth the effort to add the label.

## Hardware

We selected the ESP32-WROOM-32 for the CWM project. The schematic diagram for the project is shown in **Figure 10.4**. The CWM is powered by a 9-12 V dc source. Because of the TFT display, a 9 V battery isn't going to last very long. A Li-ion battery pack would be a better off-grid power source. You could also use a 9 V wall wart if you have access to the power grid. U3 is a simple LM7805 5 V, 1 A voltage regulator that supplies the working voltage for the system. Power demands are small enough that a heat sink is not required. The 5 V is supplied to Vin (physical pin 19, Figure 10.3) to power the ESP32-WROOM-32.

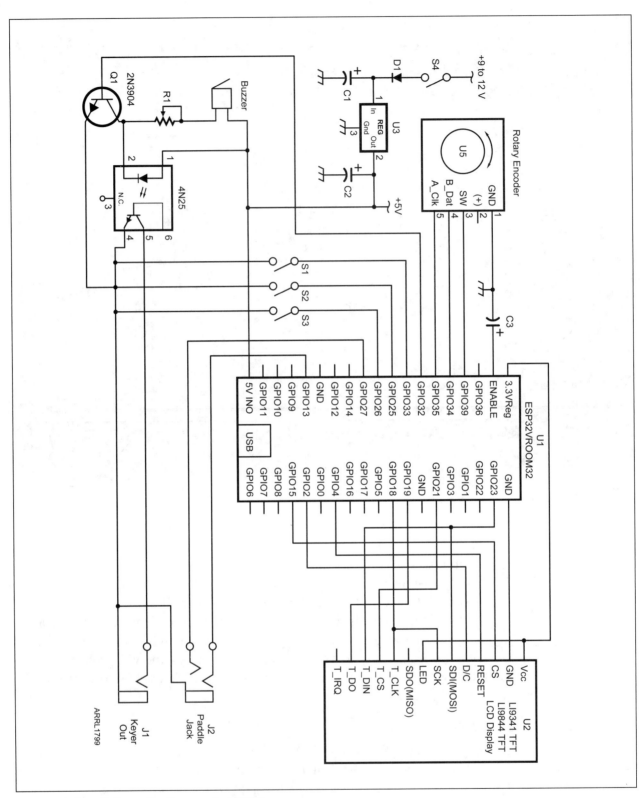

**Figure 10.4 — Schematic diagram for CWM.**

## Display

The display we chose to use is a small 2.4-inch SPI resistive touch screen TFT display that uses the ILI9341 display driver. The displays are fairly inexpensive and sell online for under $10 at the present time. (Even larger 3-inch displays are under $20.) The important things to look for are the driver chip, that it supports a touch screen, and that the display uses the SPI interface. The back of the display should look similar to **Figure 10.5**. The "top" five pins control the touch screen parameters while the remaining pins are used for displaying data. We are not taking advantage of the SD card reader. The pins on the right side of the display in Figure 10.5 control the SD card reader. You can use a larger display if you wish, but keep in mind that larger displays draw larger amounts of current.

You can also use a display that has the Arduino Uno form factor. You'll have to wire it directly to the ESP32 pins rather than just plugging it in. Some of these displays use a parallel, rather than SPI interface, so be sure you pick an SPI device that also includes touch control for this project.

Obviously, it's important that you make the proper pin connections between the display and the ESP32. **Table 10.1** shows these connections. The pins that begin with "T_" (see Figure 10.5) are associated with the touch screen display. Because the microcontroller is relatively fast, we did not feel it was necessary to have the touch screen interrupt-driven. Human reaction time is such that a polling method is more than fast enough to process the touch screen. Therefore, the T_IRS pin is not used. The remaining touch pins (data out, data in, command, and clock) are tied to the general purpose I/O pins shown in Table 10.1.

Likewise, because the TFT display is the only slave device, the SDO pin is unused. The remaining entries in Table 10.1 function to control the display fields. Note that most displays expect 3.3 V to work properly. However, some displays may work with 5 V. If you opt for a different display, make sure you power it with the proper voltage. The small display we are using borrows its 3.3 V demands from the ESP32 board. Again, if you select a larger display, you may need to add a 3.3 V regulator for display for power.

**Table 10.1**
**Connections Between TFT Display and ESP32**

| Display Pin | ESP32 Pin |
| --- | --- |
| T_IRQ | N/C |
| T_DO (MISO) | GPIO19 |
| T_DIN | GPIO23 |
| T_CS | GPIO21 |
| T_CLK | GPIO18 |
| SDO (MISO) | N/C |
| LED | 3.3V |
| SCK | GPIO18 |
| SDI (MOSI) | GPIO23 |
| D/C | GPIO2 |
| RESET | GPIO4 |
| CS | GPIO15 |
| GND | GND |
| Vcc | 3.3V |

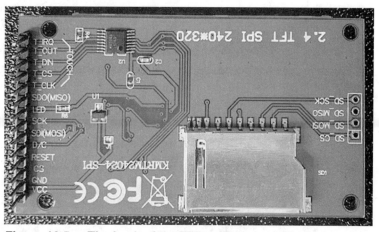

Figure 10.5 — The back of the TFT touch screen display.

## Encoder

The rotary encoder only has one function and that's to provide a fast way to change the words per minute (WPM) sending speed. Figure 10.1 shows the current WPM in the lower-left corner of the display. Turning the encoder clockwise increases the WPM, while a counter-clockwise movement decreases the speed. The built-in switch on the encoder is pressed when you have finished changing the WPM. If you are not a hardcore contest person, you could use a set of Up/Down arrow buttons to change the WPM and do away with the encoder. However, most CW operators appreciate it when you attempt to match their sending speed.

There are a lot of different encoder types that you can use for the circuit. Anything that has a granularity of 16 to 24 pulses per revolution (PPR) will do. For this circuit, an optical encoder is overkill. We have had good luck with Bourns encoders. They are reliable and can be purchased for under $2 each. (Arrow Electronics sells quantity 1, but discounts on as few as quantity 10. If you like this book so far, buy 10!) One thing to pay attention to is whether the shaft is threaded and includes a mounting washer and nut. We ordered some that were pictured with a threaded shaft and nut, but that was not what was delivered when they finally showed up. If you're buying online and the description does not explicitly say the shaft is threaded and nut included, ask the vendor to confirm before ordering, even if the encoder is pictured in the ad with the threaded shaft and nut.

## Switches

We bought a bunch of inexpensive NO pushbutton (non-latching) switches, and those are perfect for this project. (Truthfully, we've had two of the switches fail in the course of a few months use, but the others are working perfectly.) You do need, however, to consider the type of case you're going to use and where the switches are mounted on that case. Our experience with this device is that you have a tendency to slap the switches in the heat of a contest rather than use the dainty push you did while building and testing it. This means you need to consider the size and weight of the case and the placement of the switches.

First, are you left- or right-handed? Chances are your dominant hand will be used for sending CW and logging, and the other for slapping switches. That dominance affects whether you place the switches toward the left or right side of the case. Second, the switches probably should be mounted as far away from the other controls and TFT display as possible. That way, a non-dainty switch press won't change something by mistake. Indeed, for some operators it might be desirable to have a smaller case that just holds the switches, running a four-conductor cable back to the display case. That would give you some flexibility in the placement of the switches in your operating position. Finally, consider the weight of the case. If you're a slapper, a metal case might be a good idea so the case doesn't slide around in use. In any case, a non-skid rubber mat under the CWM might be a good idea. Metal cases are also more durable. Your choice.

## Buzzer Circuit

We chose to use an inexpensive buzzer that we had in our junk box. We

## Using Fast Interrupts

In some applications we've written, we use interrupts to monitor the encoders. As we mentioned earlier, human reaction times are such that you don't have to use interrupts. On the other hand, there may be other projects you build where quick response time is needed. We've seen some encoders that are attached to motors, and the interrupts need to be serviced very rapidly.

The ESP32 has a unique feature to help service the demands of very fast interrupts. The feature is the IRAM_ATTR attribute. The IRAM_ATTR attribute tells the compiler to place the interrupt service routine (ISR) in the ESP32's internal RAM (IRAM) instead of flash memory. So, if your ISR routine is named ISR, you can use:

```
void IRAM_ATTR ISR()
{
        // Code for the ISR
}
```

and the compiler places the ISR code (if it can) in the IRAM memory instead of flash memory. Flash memory on the ESP32 is considerably slower than IRAM memory, so the ISR should be serviced a little faster. The Arduino compiler seems to recognize the IRAM_ATTR compile attribute and does not throw an error on it.

---

placed a 75 kΩ potentiometer in the line to control the volume, as the buzzer we used *really* likes to hear itself. You need to experiment here based on the buzzer you select. The impedance of the buzzer you use may affect the value of the pot used.

Obviously, you need to be able to key your rig directly, too. The paddle jack allows you to use the keyer that is part of the CWM. This allows you to send "non-stored message" types of exchanges when needed.

### 4N25 Optoisolator

It's a good idea to isolate the messenger circuit from your keying circuit. Doing this means we don't have to be as concerned about the transmitter that's being keyed. Some older rigs could have some fairly high voltages on the keyed circuit. We added the 4N25 optocoupler to isolate the CWM from the rig's keying circuitry. The 2N3904 transistor was needed because the internal LED in the 4N25 needed a little extra oomph to function properly. If you are using an old tube-type rig, you may need to add an amplifier stage to the output of the 4N25 to key the rig.

The 4N25 is a little odd because it's a 6-pin IC, while it is more common to have 8 or more pins. While 6-pin sockets are not expensive, most of the online sources force you to buy 10 or 20 pieces for a couple of dollars or, with shipping and handling, the cost is about $1.50 each. However, we've got 8-pin sockets all over the place, so we just use an 8-pin socket. It's not a good idea to leave the 8 pins intact if you're only going to use 6 pins. You could just bend them under the socket or clip them off, but it's just as easy to remove them. Take a pair of needle nose pliers and, from the bottom of the socket, push on the two pins on one end. The two end pins should come out quite easily and your socket should look something like that shown on the right side of **Figure 10.6**.

The rest of the hardware is pretty much standard stuff you've seen hundreds of times before. The final hardware layout is shown in **Figure 10.7**.

## The Software

The μC that we chose for this project is the ESP32-WROOM-32. The board costs under $10, but recall that it has 1 MB of flash memory and 350 KB of SRAM and is clocked at 240 MHz. Also keep in mind the wide variety of ESP32 boards out there, many with differing pin counts. Any one of them has enough pins to use the SPI interface.

Quite truthfully, the ESP32 is an H-bomb-to-kill-an-ant with respect to its flash memory size and clock speed and the resource demands of the project. Both resources far exceed what this project requires. However, the data demands associated with the Morse table, string space for the messages, and runtime data are fairly large. Also, there needs to be room for both the support TFT and touch screen library code, too. The program requirements use just over 302 KB (23%) of flash memory (including the bootloader, which is much bigger than an Arduino bootloader) and 17.2 KB (5%) of SRAM, so there's still plenty of idle memory for enhancing the project.

## Before You Start

There are a few display-dependent issues you need to resolve before you can run the CWM code. First, there is a library named TFT_eSPI that must be included in the program. (The URL for downloading is in the header file code, as usual.) The TFT_eSPI library handles all of the TFT display attributes for the program. If you look in the TFT_eSPI library, it's a little unusual in that it has a user setup file using the clever name of *User_Setup.h*. While the exact line numbers may change, around line 19 you will find lines similar to these:

```
// Only define one driver, the other ones must be commented out
#define ILI9341_DRIVER
//#define ST7735_DRIVER      // Define additional parameters below for this display
//#define ILI9163_DRIVER     // Define additional parameters below for this display
//#define S6D02A1_DRIVER
//#define RPI_ILI9486_DRIVER // 20MHz maximum SPI
//#define HX8357D_DRIVER
//#define ILI9481_DRIVER
//#define ILI9486_DRIVER
//#define ILI9488_DRIVER     // WARNING: Do not connect ILI9488 display SDO to MISO
                             // if other devices share the SPI bus (TFT SDO does
                             // NOT tristate when CS is high)
//#define ST7789_DRIVER      // Define additional parameters below for this display
//#define R61581_DRIVER
```

**Figure 10.6** — Using an 8-pin socket for a 6-pin 4N25 IC.

**Figure 10.7** — Hardware layout.

As you can see, the library is capable of working with a variety of display drivers, of which we are using the ILI9341. In the code fragment presented above, we uncommented the symbolic constant for the ILI9341 as shown by the grayed-out statement.

A little further into the header file are these lines:

```
// For ESP32 Dev board (only tested with ILI9341 display)
// The hardware SPI can be mapped to any pins


#define TFT_MISO    19   // From ESP32 TFT User_Setup.h
#define TFT_MOSI    23
#define TFT_SCLK    18
#define TFT_CS      15   // Chip select control pin
#define TFT_DC      2    // Data Command control pin
#define TFT_RST     4    // Reset pin (could connect to RST pin)
```

Once again, the grayed-out lines show the symbolic constants that must be active for the display to work properly. If, for some reason, you cannot use the specific pins defined in the header file, you need to either change them in this file, or redefine them in the CWM.h header file so your new symbolic constants can override the pin numbers reserved in the header file. Keep in mind that the ESP32 is a little weird in that not all I/O pins are capable of output and some do odd things during boot-up. If at all possible, stick with the defaults shown in the code fragment above.

## Main Menu

The main menu screen is shown in **Figure 10.8**. There are four primary menu functions, as can be seen in the figure. We also display the current speed in WPM being used by the keyer in the lower-left corner of the screen. Although the CWM automates the sending of messages, there is also a jack for your paddle so you can augment whatever messages you are using with the CWM. The rotary encoder allows you to quickly change the code speed should you need to do so. We also added a volume control for the built-in buzzer. (The buzzer actually has a fairly nice tone to it.) Although your rig probably has its own sidetone, we added the buzzer in case you want to demo the CWM (such as the high school demo mentioned earlier) without the need for your rig. You could also use it as a code practice oscillator.

## Review Messages

We have limited the messages to 10 as stored in the Messages.h header file, the contents of which is shown in **Listing 10.1**. The reasons for limiting the number of messages were 1) we couldn't think of more than 10 contest messages that we needed, and 2) to do away with the need to scroll the message display when adding/editing messages. Also, because you can edit the messages "in the field" without a PC, you can always "re-purpose" an unused message. If you still feel the need for more messages, you can always add scroll management code — there's plenty of memory resources for it.

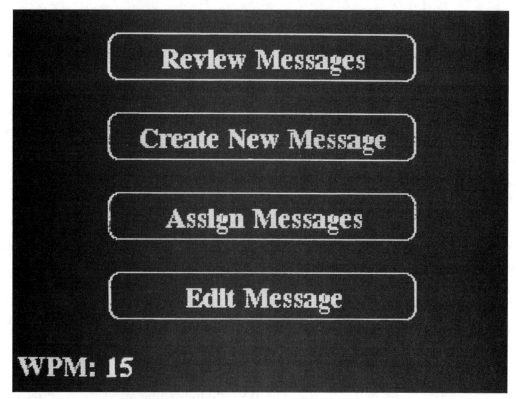

Figure 10.8 — The Main Menu screen.

**Listing 10.1**
**The Message.h Header File**

```
// Message for Contest/common info:
char contestExchanges[MAXMESSAGES][MAXMESSAGELENGTH + 1] = {

    "                              ",  // DO NOT CHANGE. 30 spaces for 1-based array
                                       //    and erasing msgs

    "CQ CQ CQ DE W8TEE",               // General CQ. Change as needed
    "599 OH",                          // DX CW contest exchange
    "1D OH",                           // Field Day See: http://www.arrl.org/files/file/
                                       //    Field-Day/2016/2016%20Rules.pdf
    " A W8TEE 54 OH",                  // SSCW This message is preceded by an incrementing
                                       //    QSO number:http://www.arrl.org/sweepstakes
    "SKN 12345",                       // Straight Key number. Send before RST.
                                       //    Might add member number
    "CINCINNATI, OH",                  // QTH
    "NAME JACK JACK",
    "RIG HB 5W QRP",
    "ANT EFHV SLOPE to 60 FT",
    "WX HR IS "                        // NOTE: No comma after last entry
};
```

Obviously, the easiest way to add or edit a message is to use the Arduino IDE to edit any given message in the Messages.h header file. The Messages.h header file we distribute has our call, name, and QTH in it. We're pretty sure you'll want to edit those messages.

The messages shown in Listing 10.1 are just plain, ordinary, text so you can use virtually any text editor (or the Arduino IDE) to edit it. (We use Notepad++ from **notepad-plus-plus.org/download/** as a general purpose editor.)

Before you start mucking around in the *contestExchanges[ ]* array, there are a few rules to follow. First, the maximum number of characters in any one message is 30 (MAXMESSAGELENGTH). (The extra space is for the null termination character.) The maximum number of messages is 10. Note that, actually, we have set MAXMESSAGES to 15, but all of the processing code uses 10. If this bothers you, you can reduce its rank to 11 (10 text messages plus the first "spaces" message).

## Create New Message

Actually, this option is a little bogus in terms of its usefulness. We added this option so you could create messages from within the CWM project itself. Truth be told, that's a dumb way to do it. Instead, load the Messages.h header file into your text editor and change the sample messages to whatever you feel you need. You'll need to edit the sample messages currently in the code to suit your personal information, making sure those messages fall within the limitations mentioned above. Once you're done editing the messages, compile/upload the code and you're done.

If you do wish to use the Create New Message menu option, the display looks similar to that shown in **Figure 10.9**. The top line of the display says that we are adding message number 7. Note that this really *is* message number 7 because the zeroth element is a blank message, so our list of messages is from *contestExchanges[1]* through *contestExchanges[10]*. The "T" you see in Figure 10.9 is the first letter of the new message. When you finish entering the message via the keyboard, press the Store button, which writes the new message to the *contest Exchanges[ ]* array. The Clear button clears the new message array buffer, which functions as an erase-and-start-over button. Pressing the Cancel button returns you to the main menu without making any changes to the message array.

Note: Any changes made using any of the CWM editing features are temporary. That is, the message change remains in effect until you lose power to the CWM device. The next time you apply power, the

**Figure 10.9 — The Add New Message display.**

message array appears the same as it was before you made any edit changes. The reason is because the contents of the array are stored in the heap space when you compile the project on your PC. When you use the Edit feature, the change appears only in the stack space. When you run the program, global data (like the message array) are move into the heap space, which is actually part of the SRAM resource base. So, when you edit that array, you are changing the SRAM "copy" of the array, not the "real" compiled array that is stored in the heap space.

You could get around this limitation in several ways, including the addition of an SD card to hold the array, but we didn't think it was worth the effort. (Many TFT displays have an SD card reader built into the back of the display.) You could also use the emulated EEPROM memory of the ESP32 to permanently store the new message, but again, we didn't feel it was necessary. Most of the time, any changes needed to the array would be made before the CWM is needed, using the PC to make those changes to the Messages.h file itself. The real purpose of the edit facilities is to fix minor screw-ups that you discover after you are set up in the field (for example, you discover your Maidenhead locator is off by one digit in a message). We hope that anyone who expands this to include SD storage shares those changes with the rest of us.

## Assign Messages

There are three normally-open pushbutton switches (SPST, NO) on the project's case. When you press one of these buttons, the message from the *contestExchanges[]* array associated with that switch is then sent. Near the top of the CWMessengerVer007.ino file (approximately line 12) is the following line:

```
byte switchIndexes[] = {1, 7, 6};
```

and it defines the relationship between the buttons and the messages. For example, if I press Button 1, *contestExchanges[1]* would be sent by the CWM. Looking at the message array in Listing 10.1, you can see the result is that CWM sends:

```
"CQ CQ CQ DE W8TEE"
```

Button 2 sends my name (message #7) and Button 3 sends my QTH (message #6). The Assign Messages menu option allows you to change the message associated with each button. When you select the Assign Message menu option, the display changes to that shown in **Figure 10.10**. The three pushbutton switches are indicated by the SW1, SW2, and SW3 buttons at the bottom of the display. The number immediately above each switch is the message number. It shows that Button 1 is associated with message #2, Button 2 with message #6, and Button 3 with message #1. You use the Up ("^") button to move the highlight message selection upward and the Down ("v") button to move the selected message downward. When the message you wish to assign is highlighted, press the appropriate switch button (SW1-SW3) to assign that message to that button. Press Done when you've finished assigning the messages to the buttons.

When you press the Done button, the *switchIndexes[]* array is updated with your new choices and then written to the ESP32's EEPROM. That way, the next

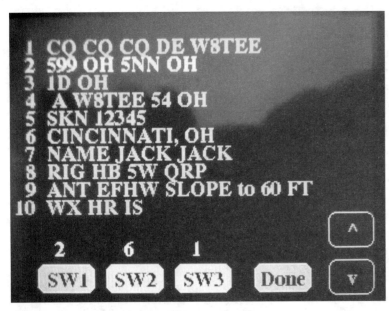

**Figure 10.10** — The Assign Messages menu screen.

time you power up the CWM, the EEPROM is read and assigns the values into the *switchIndexes[ ]* array.

Note that, like all but the Arduino controllers, the ESP32 doesn't have "real" EEPROM. Instead, it emulates EEPROM by setting aside a chunk of NVS (Non Volatile Storage, or flash memory) as defined in its partition table. Without getting too detailed, there's enough memory partitioned by default to service our meager needs. (The only other variable we store in this partition is the words per minute used by the keyer.) We have more to say about using EEPROM later in this chapter.

## Edit Messages

If you need to edit a message in the field, use the Edit Message menu option shown in **Figure 10.11**. The Edit Message menu option works similar to the way you assigned a message to a switch. Use the Up/Down arrow buttons to highlight the message you wish to edit. Once the correct message is highlighted, press the Edit Highlighted button. At that point, the display changes and looks very similar to Figure 10.10, except it displays the message being edited at the top of the display. You then enter the new message that will replace the message that is shown at the top. Note that you must retype the entire message, as there are no real editing features with this menu option. You are actually *replacing*

**Figure 10.11** — Edit Message menu option.

the existing message with whatever it is you wish to enter. When you are done, press the Store button to change the message array. Again, this is only a temporary change because we are storing the array in SRAM, which is volatile storage. You can only change one message at a time, but you can use this option as many times as you wish.

## Some More Software Details

Most of the software used in the CWM project is standard boilerplate stuff you've seen before. However, there are some differences in the way the ESP32 handles things.

### ESP32 EEPROM Processing

Reading and writing the ESP32 EEPROM is a little different because the ESP32 doesn't have any conventional EEPROM. The EEPROM library that is used is from the Arduino path starting with *hardware\espressif* and is *not* the EEPROM library presented in the Arduino's *libraries* directory. The two libraries are very different.

**Listing 10.2** shows the code used to read and write the *wordsPerMinute* variable. The first 3 bytes of EEPROM memory space (0 – 2) hold the *switch Indexes[ ]* array discussed earlier with respect to the three buttons responsible for sending the messages. Byte 4 (memory location 3) holds the *wordsPerMinute* variable. The two *writeByte()* method statements:

```
EEPROM.writeByte(3, wordsPerMinute);
EEPROM.commit();
```

show the way any data byte is written to the EEPROM memory space. The arguments to the *writeByte()* method are: 1) the EEPROM address followed by 2) the variable being written. The second statement is unique to the ESP32 EEPROM processing. The *commit()* method call is required to permanently change an (emulated) EEPROM memory byte. If you look in the ProcessEEPROM.cpp source file, you can see how this same general approach is used to read/write the *switchIndexes[]* array using a *for* loop.

The *ReadWordsPerMinute()* function is used to retrieve the current EEPROM value for *wordsPerMinute*. The statement:

```
wordsPerMinute = (int) EEPROM.readByte(3);
```

does all the work. Note that, because we store *wordsPerMinute* as an *int* in the code, we cast the byte read from EEPROM to an *int* before assigning the value into *wordsPerMinute*. Because it's unlikely that any human is going to exceed a WPM of 255, we probably could have used a *byte* data type for *wordsPerMinute*. However, the calculations for sending the code all use *int* values, so changing *wordsPerMinute* to a *byte* data type probably doesn't make much sense.

Note what happens, however, if EEPROM byte 3 holds any value less than MINWPM (5) or greater than MAXWPM (60). In that case, we write a default value of MESSAGEWPM (15) to that EEPROM memory location. We do this because the first time you run the program, some random value is stored in that

**Listing 10.2**
**EEPROM Read and Write Functions for *wordsPerMinute***

```
/*****
  Purpose: Write the words per minute to EEPROM

  Parameter list:
    void

  Return value:
    void

  CAUTION:
*****/

void WriteWordsPerMinute()
{
  EEPROM.writeByte(3, wordsPerMinute);
  EEPROM.commit();
}



/*****
  Purpose: Read words per minute from EEPROM

  Parameter list:
    void

  Return value:
    void

  CAUTION:
*****/

int ReadWordsPerMinute()
{
  wordsPerMinute = (int) EEPROM.readByte(3);
  if (wordsPerMinute > MINWPM && wordsPerMinute < MAXWPM) {
    return wordsPerMinute;
  } else {
    wordsPerMinute = MESSAGEWPM;    // If no value written yet...
    EEPROM.writeByte(3, wordsPerMinute);
  }
  return wordsPerMinute;
}
```

EEPROM address and we want to make sure it's a reasonable value. If you aren't happy with the limits imposed by those symbolic constants, you can edit them in the CWM.h header file.

Clearly, you can add the messages themselves to the EEPROM, as there is plenty of space to do so. We simply didn't see the need.

## Processing CW Characters

Obviously, the CWM must have a way to convert the ASCII text held in the *contestExchanges[ ][ ]* array into the Morse dit and dah atoms. Because we are not decoding Morse code, but rather just converting from ASCII to Morse atoms, we can use a simple array to accomplish this. Near the top of the CWMessengerVer007.ino file is a table that defines the Morse character set (the 'b' in the constants means it is the binary representation of a data item) for letters, numbers, and commonly-used punctuation. Consider the letters table:

```
char letterTable[] = {    // Morse coding: dit = 0, dah = 1
  0b101,            // A     first 1 is the sentinel marker
  0b11000,          // B
  0b11010,          // C
```

Let's say that the message being sent contains 'CQ' as part of the message. The ASCII code for 'C' is the number 67. Suppose I write:

```
char letter = 'C';
int index;
// some code...
index = letter - 'A';   // Very common syntax to
                        // index into an alpha array
```

As you might guess, the ASCII code for "A" is 65. Therefore, the last statement resolves to:

```
index = letter - 'A';
index = 'C' - 'A';
index = 67 - 65;
index = 2;
```

Now look at *letterTable[index]*:

```
letterTable[2] = 0b11010;
```

Suppose we interpret a binary '1' as a dah, and a binary '0' as a dit. Now look at the least significant 4 bits of the binary number (the shaded bits below):

```
0b11010
```

which is interpreted as dah-dit-dah-dit. Hmmm...sounds like the letter 'C'.

Okay, but what's the 0b1 that is in front of the letter 'C'? First, the 0b is a C data prefix that tells the Arduino compiler we are using a binary (base 2) representation of the data. The '1' that immediately follows the "0b" is called a sentinel. *Sentinels are used to mark the start or end of a piece of data.* In this case, the computer finds the first '1' bit in the number and knows that everything *after* this sentinel bit is the actual data we want to use. Now look back at the *letterTable[ ]* array and look at the second entry:

```
0b11000,           // B
```

The program reads to the sentinel bit:

```
0b1
```

and tosses that information away. It then uses what's left:

```
1000
```

to construct the data, which becomes:

```
dah-dit-dit-dit
```

or the letter 'B' in Morse. If you look at the *SendCode()* function in the ProcessCode.cpp source file, you can see how this approach is coded in C. It's a fairly efficient way to convert from ASCII to Morse code. Once we have the "atom representation" of the ASCII character, we simply call the *dit()* and *dah()* functions to send the code.

As a way of checking your understanding, what does this do?

```
int number = 7;
int index;
// some more code...

index = number - '0'    // that's a zero at the end of the
                        // expression, not an "Oh"
SendCode(numberTable[index]);
```

Spend some time figuring this out, as this is another common way of indexing into an array.

We could have written the bit-coding scheme presented here using a pointer array and a binary tree search, but what's the point? Even the simple code used is fast enough to find any letter a thousand times before you can blink an eye. This leads to another programming corollary: simple is better than complex if simple gets the job done in an acceptable way.

## Interrupts versus Polling

We've discussed the difference between using interrupts and polling to respond to some external event in Chapter 2. We thought this might be a good place to expand that discussion because this project uses both interrupts and polling.

## Interrupts

The CWM needs a way to alter the CW sending speed, or words per minute (WPM). If you've ever worked a CW contest, you're aware that the WPM speed of the contestants can vary widely. Since most contests require some form of information exchange, which needs to be accurate for logging purposes, someone may ask you to slow your sending speed (QRS). Some keyers require you to use a rather cryptic menuing process to change the WPM. The CWM uses a rotary encoder.

We want that encoder to respond as quickly as possible, so we use interrupts to read the encoder. We have defined the encoder pins A and B as:

```
#define ENCODER1PINA          25
#define ENCODER1PINB          26
```

We use an encoder library to decipher the pin data from pins 25 and 26 to determine whether the encoder is moving clockwise or counter-clockwise. However, to be able to do that, we need to inform the program that we want to use interrupts for reading the encoder data. These statements tell the program which pins to monitor for an interrupt:

```
attachInterrupt(digitalPinToInterrupt(ENCODER1PINA), rotate, CHANGE);
                                                              // WPM encoder
attachInterrupt(digitalPinToInterrupt(ENCODER1PINB), rotate, CHANGE);
```

We are using the *attachInterrupt()* library call to tell the program: 1) which pins to monitor for an interrupt (for example, E*NCODER1PINA, ENCODER-1PINB*); 2) the name of the interrupt service routine (ISR) to call when an interrupt occurs (for example, the *rotate()* function); and 3) the state of the signal for the interrupt (CHANGE). We want our program to respond to any change on the encoder pins, so we are using the CHANGE symbolic constant for the last function argument. (The interrupt can respond to RISING (for example, moving from 0 V to 3.3 V), FALLING (for example, 3.3 V to 0 V), or CHANGE (any state change). There are other ways of using interrupts on the ESP32, but they are not portable to other processors, so we are sticking to the *attachInterrupt()* method here.

When a state change occurs on either *ENCODER1PINA* (pin 25) or E*NCODER1PINB* (pin 26), the code does an immediate jump to the *rotate()* function, as shown in the code fragment below.

```
void rotate()
{
  unsigned char result = rotary.process();

  encoderDirection = 0;
  switch (result) {
    case 0:              // No interrupt
      return;
      break;

    case DIR_CW:         // clockwise
      encoderDirection = 1;
      break;

    case DIR_CCW:        // clockwise
      encoderDirection = -1;
      break;

    default:
      encoderDirection = 0;
      break;
  }
  UpdateWPM();
}
```

Simply stated, the *rotate()* function calls the *process()* method of the rotary object. The *process()* method can return 0, which means the encoder is doing nothing, or DIR_CW, a symbolic constant that means the encoder is turning clockwise, or DIR_CCW which means the encoder is being turned counter-clockwise. You can see the variable name *encoderDirection* hold a direction value. If the value is 0, the function ends via a *return* statement. Otherwise, *UpdateWPM()* is called because the user is either trying to increase the wpm (*encoderDirection* = 1) or decrease it (*encoderDirection* = -1).

The key to notice in the *UpdateWPM()* function is this statement:

```
wordsPerMinute += encoderDirection;
```

The rest of the function does little more that display the new wpm value on the display screen. Because *encoderDirection* is either 1 or -1 from the interrupt service routine, the *wordPerMinute* either increases (1) or decreases (-1). Think about it.

One more thing: we define the *encoderDirection* as:

```
volatile int encoderDirection;
```

The *volatile* keyword should be used with any variable used in the ISR, but defined outside of it. The reason is because optimizing compilers, like the GCC compiler, often store frequently-used variables in a register. Doing so means the compiler doesn't need to generate the code necessary to look up the variable's lvalue, go to memory, and fetch the rvalue back across the data bus and back into a register. Instead of the fetch-and-store process, keeping the value in a register is called *data caching*. However, caching optimization also means it's possible for an interrupt to use an out-of-date value for a variable. By using the word *volatile* in the data definition, you're telling the compiler to always fetch a "fresh" copy of the variable.

Because we use interrupts to set the code speed, you can change the WPM "on the fly" during a QSO. The encoder is much quicker to use than a menu-driven system.

## Polling

The CWM needs to provide a way to send Morse code directly from a set of paddles. During a contest, for example, you may need to send a call sign or other information that is not in the message list. The CWM needs to provide a means of directly keying the transmitter using the CWM keyer. To do that, we need to know which I/O pins on the ESP32 have been assigned to the dit and dah contacts of the keyer paddles. The way to let this pin assignment to be known in the program is to use symbolic constants. The CWM.h header file defines these as:

```
#define DITPADDLE     13    // I/O pins for paddle
#define DAHPADDLE     33
```

which simply says ESP32 I/O pin 13 has been assigned to the dit paddle and pin 33 to the dah paddle. So why not just hardcode these numbers as needed throughout the source code?

Well, you could, but suppose you plug your paddle in and you find out it's wired "backward." That is, when you press what you expect to be the dah paddle, the CWM sends out a string of dits. Not good. Of course, you could simply rewire the paddle plug to set things straight. Or you could just reverse the symbolic constants (make dit = 33, dah = 13). Also, some left-handed operators may need to make the same change. By the way, pay attention to which I/O pins you choose. Use Table 1.3 in Chapter 1 as a guide. If you can, stick to pin 13, or pins 16-33 for general purpose I/O on the ESP32.

Unlike interrupts, which fire more-or-less automatically once they are defined in the source code, polling requires the code to test them continually as the program executes. This continual monitoring demands that the pins are polled in the *loop()* function. Consider the following code snippet:

```
while (digitalRead(DAHPADDLE) == LOW) { // Same here
    Dah();
}
while (digitalRead(DITPADDLE) == LOW) { // Send atom until released...
    Dit();
}
```

There are other ways to write this, but this is about the easiest way to do it and it works fairly well. Note that we read each of the paddle key pins using a *while* loop. As long as one of the paddles is closed, we should continue to send out that dit or dah stream. When the key is released, we stop the stream and reset the paddle I/O pin to high. The paddle keys are read on every pass through the *loop()* code. Because there is a fair amount of code in the *loop()* function, the keyer speed seems to be a tad slower (perhaps 1 or 2 WPM) than the speed indicated on the main menu screen (Figure 10.8). If that deviation in code speed bothers you, write an equation to better reflect the "real" speed and change it! After all, you have the source code!

## Conclusion

The CWM project is a useful device both for contesting and for casual CW operating. Boring stuff like calling CQ is reduced to pressing a button. Indeed, most contest exchanges can be completed just by button presses. Even casual QSOs could automate the first exchange, which is usually an RST report, name, and QTH. From that point on, use your paddle as always. Being able to edit the messages without being tethered to a PC is a very useful feature. We hope this will encourage some of you to build a CWM and make arrangements to demo it to some students. Who knows, they may catch the bug, too.

There are lots of ways to extend the project. It would be fairly easy to expand the message list beyond the 10 that we are using. All you need to do is manage scrolling the message list. You could also convert the polling of the paddle keys to use interrupts. It wouldn't be all that difficult to tie in an ASCII keyboard that would allow you to send Morse without actually using a paddle. Another, much more advanced feature, would be to have the code sent by the CWM stored on an SD card. After all, there is a card slot on the back of the display. Add a Real Time Clock (RTC) such as the DS3231 and you could log the date and time, plus the call that you sent. You would likely need to figure out an easy way to save their exchange report. It's not a trivial task, but it might be doable.

If you have other ideas for enhancements, we hope you'll share them with us.

# CHAPTER 11

# DSP Post Processor

While most modern transceivers have extensive digital signal processing (DSP) functions for post-processing, older rigs do not. Also, there are many inexpensive transceivers that have no DSP capability. Regardless of their shortcomings, many amateurs still like to use their older (vintage) or less-expensive units on a regular basis. Indeed, sitting on a park bench operating *al fresco* with a $10,000 transceiver probably isn't going to happen.

Al had an older Yaesu that was not up to modern standards of signal processing, so he created an all-analog post processor unit with "tone controls" and low-pass filters. It works fine, but is still not what he wanted. He then discovered the Teensy microcontroller and its Audio Library, which has extensive digital signal processing functions in the audio frequency range. Now it is possible to readily put together digital filters, digital compressors, and lots of other useful functionality in one reasonably-priced package. It is the bundle of features presented in this chapter that has formed the basis of the post-processor Al really wanted.

So, what can a post processor do for you? Some of the possibilities are:

- Variable low-pass, high-pass, and band-pass filters, with up to 160 dB/decade filter skirts
- Notch filter with variable Q
- Automatic notch filter
- Automatic level controls and compressors
- Multi-band equalizers
- FFT display
- DSP noise reduction
- Any combination of the above
- Power final amplifier to drive less efficient speakers
- Multiple inputs
- Dual channel

There are several commercial units with some of the functions mentioned above, but typically at relatively high prices. The unit we are describing does all of the functions listed above and costs less than $100 if you build it yourself. In addition, the software, which is the heart of the unit, is open-source, so you can extend and tailor the processor to your needs. We also present some ideas of ways to augment the features of our DSP Post Processor (DPP), if you choose to do so.

## User Interface Overview

The stuff under the hood of our DPP is quite interesting, but the every-day interaction with the user interface (UI) is what makes the high-power DSP technology accessible. We have spent a significant amount of time and countless prototypes streamlining the way our DPP interfaces with you and the functionality it brings to the table.

A UI has several parts, most important of which are: 1) the control/data input mechanism, 2) feedback to the user, and 3) the ease with which the user can make changes. We have elected to present data about the state of the processor using a high-resolution TFT color LCD screen. Data input/control uses the touch capability of the TFT screen as well as rotary encoders, push button switches, and even analog level controls and input switches. Our experience has shown that this combination of physical controls and TFT touch screen input is quite efficient and minimizes the number of steps necessary to make changes and set the state of the unit.

**Figure 11.1** shows our final front panel iteration with the TFT display, two

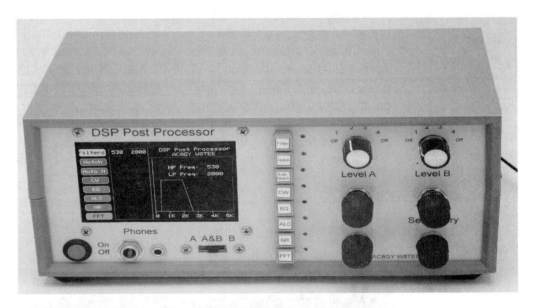

Figure 11.1 — DSP Post

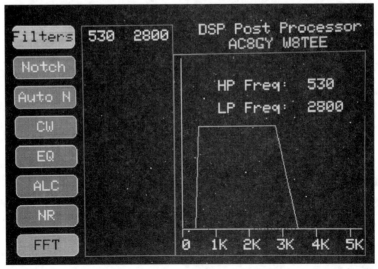

Figure 11.2 — User interface.

level controls, two rotary encoders, eight pushbutton switches to turn functions on and off, and a pair of six-position rotary switches for input selection. With the TFT display, the user gets graphical feedback as to the state of the unit's functions and can select functions to set parameters. A real-time FFT display shows the spectral content of the various signals, after DSP. Because of graphics involved and the use of a touch screen, the display necessarily has higher resolution (480×320) than most projects in this book. Still, a high-resolution touch screen TFT display costs less than $10.

**Figure 11.2** shows a close-up of the TFT screen with the graphical display of the low-pass (LP) and high-pass (HP) filter setting function. Here the user can select the LP and HP cutoff frequencies while listening to the results through the speaker or headphones. Real-time interaction makes fine tuning the unit a breeze. The continuously variable filters mean fine control of the signal quality is very easy. The UI is explained in detail later in the chapter.

## Circuit Description

This project is about 25% electronic hardware and 75% software, so let's get the circuit out of the way first and then dive into the software.

**Figure 11.3** shows a block diagram of the circuit. The heart of the unit is the Teensy microcontroller and the Teensy Audio Adapter. We are using the Teensy

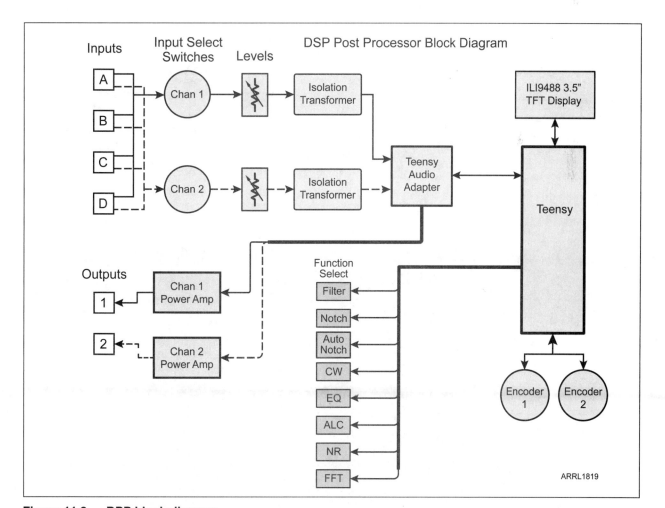

Figure 11.3 — DPP block diagram.

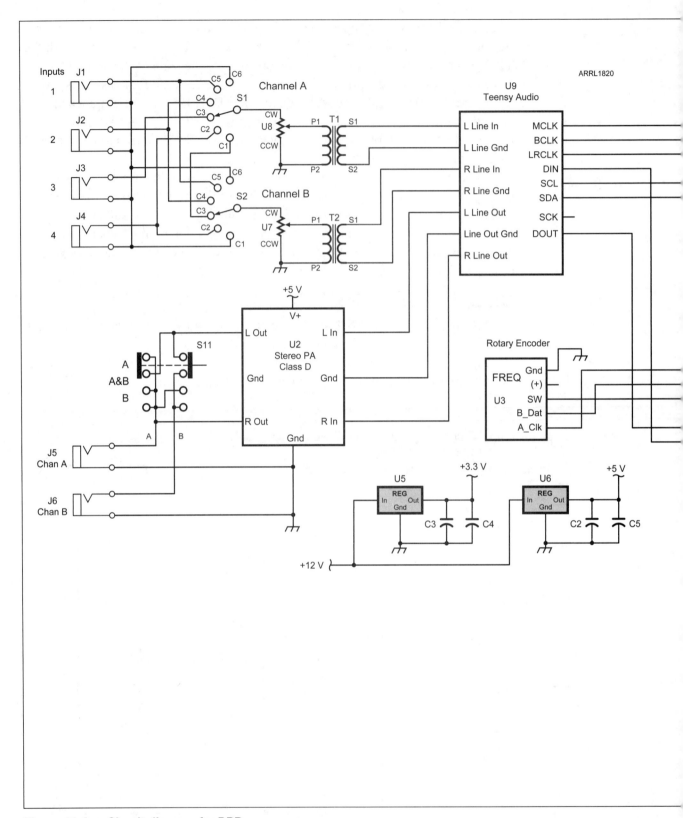

**Figure 11.4 — Circuit diagram for DPP.**

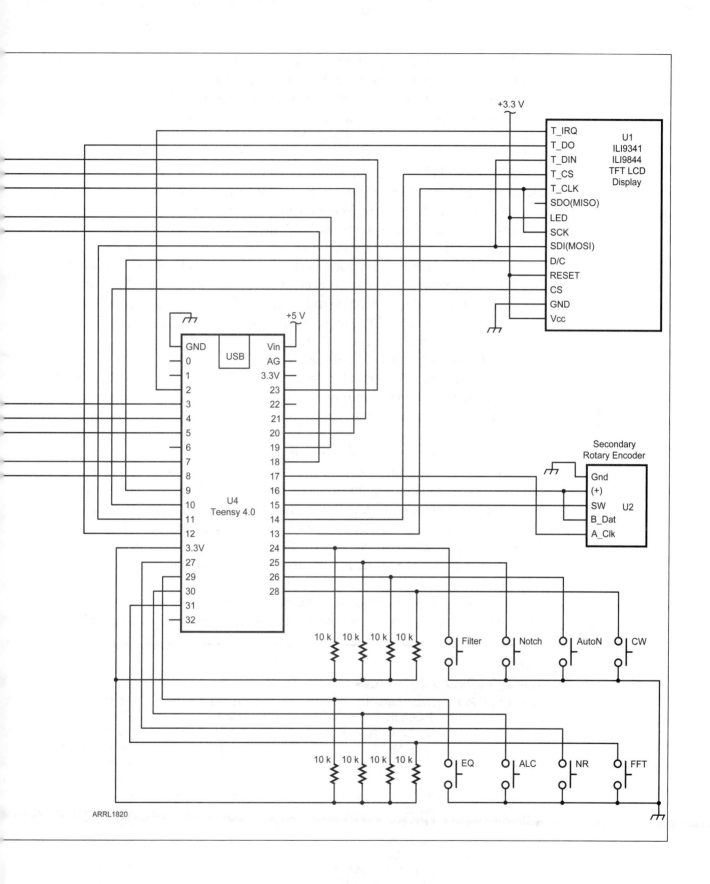

4.0 (T4 from now on), a powerhouse clocked at 600 MHz with more than enough memory and enough GPIO pins to do everything we want to accomplish.

The second most important component is the Teensy Audio Adapter (TAA), which contains all of the audio I/O, such as ADCs and DACs necessary to capture the audio signals, send the digital data to the T4, and convert and output the processed signals back to analog audio again. The TAA has 16-bit conversion and operates at a 96-kHz conversion rate, enough for full 16-bit, 44-kHz CD-quality audio. The TAA can directly piggyback onto the T4, too, significantly simplifying the circuit wiring. Finally, the T4 has enough horsepower to run a real-time FFT process at the same time as doing the DSP filtering on the audio channels.

Starting at the input, we see that provision is made for four inputs that can be routed to either of the two computational channels. At this time, only Channel A has DSP implemented. However, it would be very easy to add the DSP functions to Channel B. We did not need two independent DSP channels, so this is left as an exercise for the reader. Channel B is a straight pass-through, with no DSP. However, each channel has its own Level control (Level A and Level B), so two separate inputs with different levels can be accommodated. Note that each switch has an off position at each end of its range, so either channel can be easily muted, without changing any other setting. Outputs from the Level controls go to isolation transformers and then to the Line Inputs of the Audio Adapter. The isolation transformers eliminate the possibility of ground loop noise and isolate the grounds of the transceiver and Post Processor.

Line Outputs from the Audio Adapter go to dual 10 W power amplifiers, which boost the signal to be able to drive a modest-sized speaker. For best efficiency, we have employed inexpensive Class D amps that are 90%+ efficient and run cool, even at full power. Class D amps have a reputation of being quite noisy because of the switching technology employed. To minimize any interference with receiver performance, the power supplies have been bypassed, with extra filtering throughout.

The remainder of the circuit is comprised of the ILI9488 SPI interfaced display and UI inputs and controls. Two rotary encoders allow easy manipulation of

**Table 11.1**
**Teensy T4 Pin Assignments**

| Teensy 4.0 Pin | Attachment | Teensy 4.0 Pin | Attachment |
|---|---|---|---|
| 2 | Touch IRQ | 17 | Encoder2 A |
| 3 | Encoder1 A | 18 | Audio Adapter |
| 4 | Encoder1 B | 19 | Audio Adapter |
| 5 | Encoder1 SW | 20 | Audio |
| 6 | N/C | 21 | Audio |
| 7 | Audio | 22 | N/C |
| 8 | Audio | 23 | Audio Adapter |
| 9 | LCD D/C | 24 | Filter PB |
| 10 | LCD CS | 25 | Notch PB |
| 11 | SPI MOSI | 26 | Auto N PB |
| 12 | LCD, Touch MISO | 27 | CW PB |
| 13 | LCD, Touch SCK | 28 | EQ PB |
| 14 | TOUCH CS | 29 | ALC PB |
| 15 | Encoder2 SW | 30 | NR PB |
| 16 | Encoder2 B | 31 | FFT PB |

parameters such as the corner frequency of the filters, CW filter width, and other characteristics. The eight latching, on-off switches are used to turn on any of the eight primary functions.

The detailed circuit diagram, including pin assignments, is shown in **Figure 11.4**. Note that some Teensy pins, such as USB, are not available because of system use. Two voltage regulators provide power for the Teensy, display, and LED indicators. The power amplifier is operated straight from the 12 V input. A 5 V fixed regulator drops the 12 V to the T4 Vin pin. A 3.3 V regulator is also used to power the display and indicator LEDs. This regulator needs to have a heat sink attached. Note the enhanced power supply bypass, using ceramic capacitors at several locations. **Table 11.1** presents all of the interconnections between the Teensy and other devices.

## Building the DPP

One advantage of the audio frequency range is the relaxed requirements on layout. The DPP can be constructed using any of the popular methods. We built our prototype on a modified breakout board. The Audio Adapter plugs directly onto the T4, so control wiring is minimal. We always use sockets for the microprocessors for ease of insertion and removal, if necessary. The other components are connected using IDC connectors and ribbon cable for convenience. No special precautions are required. We like to use #30 AWG solid hookup wire because of the close 0.1-inch hole spacing in the proto-board (same for most perf boards.) The main power connections use #24 AWG wire (or larger) to minimize resistive losses. When mounting the board to the chassis, keep the T4 USB connector as clear as possible in case you want to modify the software.

Eight of the T4 connections, which go to the pushbutton switches, are located

**Figure 11.5 — DPP circuit board (top).**

on the bottom of the T4. We connected to these by soldering a small wire to each and attaching the wires to an 8-pin male header, which can be plugged into the circuit board sockets as shown in **Figure 11.5**. Alternatively the user could use spring-loaded connector pins on the circuit board, but we felt that soldered connections would be more reliable.

Figure 11.5 shows the top of the main circuit board, with the Teensy and Audio Adapter installed. Note the use of multi-pin IDC connectors for display, front panel connections, and encoders. **Figure 11.6** shows a view of the bottom of the circuit board. **Figure 11.7** shows the front panel wiring, with the ribbon cables and female IDC connectors. The back panel is shown in **Figure 11.8**.

Finally, a top view of the assembled unit is shown in **Figure 11.9**. Shielded cables are used to carry the low-level audio signals from the front panel to the Audio Adapter and from the inputs to the switch assemblies.

For most of the projects in this book, we assume the user has a 12 V or 13.8 V power supply available, similar to the one discussed in Chapter 5. The current project doesn't require much power, except for driving the speakers at high levels. Even then, the power demands are modest.

If an adequate supply is not available, a small switchmode 12 V supply capable of a couple of amps works just fine. Be sure to add adequate filtering of the incoming 12 V supply to minimize noise from getting into the audio path.

We made a simple and inexpensive 3D printed case with a custom faceplate as shown in **Figure 11.10**. Once again, Chapter 16 has ideas on how to finish your projects and give them a professional look.

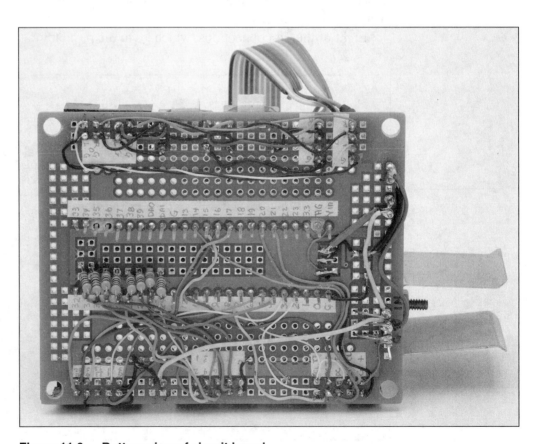

**Figure 11.6 — Bottom view of circuit board.**

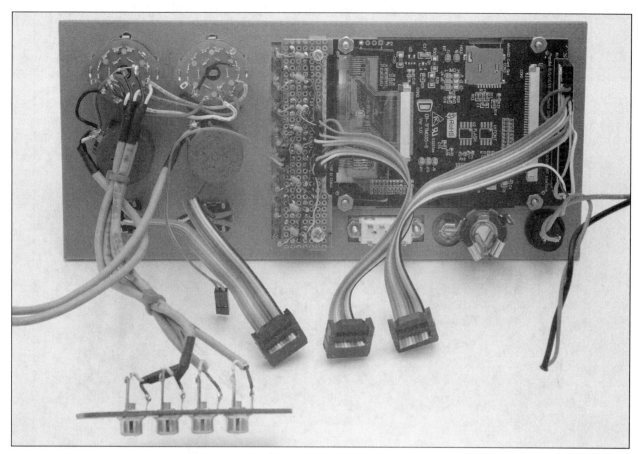

**Figure 11.7** — Front panel wiring.

**Figure 11.8** — Back panel wiring.

**Figure 11.9** — Top view of complete unit.

**Figure 11.10** — The DDP at home in its case.

## Operating the Post Processor

Most of the difficulty of using the processor is in the setup and, as we indicated previously, we spent a lot of time trying to make the user interaction as easy and intuitive as possible. The operational steps are a simple 1, 2, 3 step process:

1) After hooking up the DPP to your rig and some speakers, you must set the parameters for the functions you want to use. To enter the "Set" mode for any function, touch the on-screen function button (such as Filters or Notch — see Figure 11.2) to bring up the setup routine for that function. The current status is displayed next to the screen function button on the left side (see **Figure 11.11** for an example). Using the two encoders, set the parameters for that function. A detailed description of each function's setup is given below.

2) Turn each function on or off using the pushbuttons on the right side of the display. The status of the function is then displayed by the color of the button for that function and by means of the indicator LED next to the respective pushbutton. The pushbutton is depressed when the function is on and extended when it is off.

    a. Note that some of the functions are mutually exclusive, particularly the CW Filter, which cannot be used at the same time as the LP or Notch functions.

    b. All of the others functions are available with LP, Notch, CW, or each other.

3) Select the inputs for each channel and set the levels. You are now ready to use the DPP!

## Function Detail

Each function is set up in a slightly different way, but the general process is similar for all options. In each case, the desired function is selected by enabling its "Set" mode button as seen on the left side of Figure 11.11.

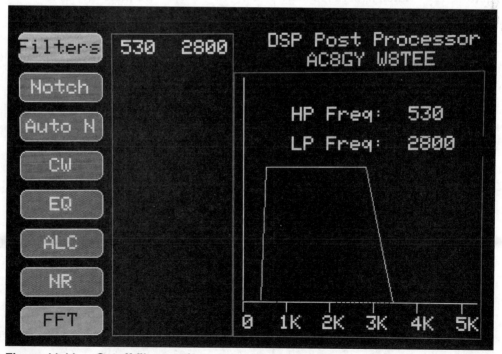

Figure 11.11 — Cutoff filter setting.

## 1) Cutoff Filters (LP and HP)

Low-pass (LP) and high-pass (HP) filters limit the audio passband to improve speech recognition. High-frequency noise, above 2000 Hz, may mask speech, reducing clarity. Low frequencies, under 400 to 500 Hz, generally add little to speech recognition and may reduce intelligibility. The cutoff filters allow the operator to tailor the spectrum to reduce noise.

Both the HP and LP filters are characterized by their –3 dB cutoff frequencies. The encoders allow real-time adjustment of the cutoff frequencies while listening to the result. The HP filter is set with the primary (left) encoder and the LP cutoff is set by the second encoder. Figure 11.11 shows the filter setting screen (note the 530 and 2800 numbers in the setup area of the display, representing 530 Hz cutoff for the HP filter cutoff and 2800 Hz for the LP filter). Simply rotate the encoders while listening to the signal through the speaker or headphones until the desired result is achieved. The –3 dB cutoff frequency is shown both on the graphical display and in the status display.

The HP filter is adjustable from 150 Hz to 1000 Hz, and the LP filter goes from 1000 Hz to 5 kHz. The HP filter has a final slope of 24 dB/octave and the LP final slope is 48 dB/octave. Given that each operator has their own audio "signature," the DPP can make it easier to listen to someone comfortably.

**Figure 11.12** shows a frequency response plot of the filters at several cutoff frequency settings. Note the extremely sharp falloff above and below the cutoff points. Several LP frequencies are shown for one HP filter setting. Also note that the signal "tops" are relatively flat.

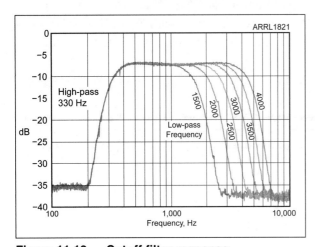

Figure 11.12 — Cutoff filter response.

## 2) Notch Filter

The purpose of the Notch filter is to remove a single pure tone, while preserving the ability to recognize the desired signal. The Notch setting screen is shown in **Figure 11.13**, with a typical notch filter setting at 1 kHz.

The notch filter frequency is continuously variable using the left encoder. To use it, set the Notch filter frequency (for example, 1000 Hz) with the Frequency encoder to eliminate the tonal noise while listening to the signal.

The Notch filter has two parameters: The reject frequency and the Q value, which determines the "sharpness" of the filter. The frequency is displayed next to the graphical display of the Notch. Q values are set in the program source code file, which the user may change to suit. A *#define* value in the .h file defines the Q constant, which may be altered, and the code uploaded again. Q values less than 2 result in a broad filter shape, whereas values above 5 result in quite sharp filters.

If you wish to alter the Q setting, pick a value that eliminates the noise but doesn't reduce the usable audio. The default is set to Q=20. In practice this value eliminates the pure tone but does not interfere with speech recognition. **Figure 11.14** shows plots of the actual Notch response with no other filters and in conjunction with the cutoff filters active.

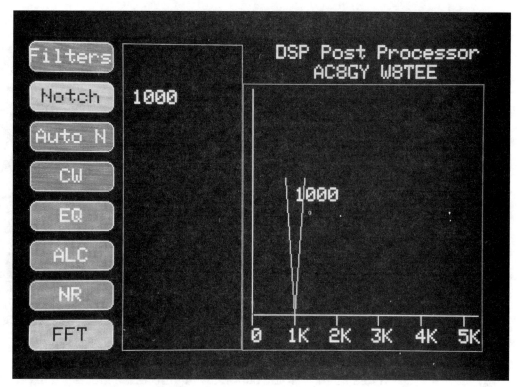

Figure 11.13 — Notch filter setting screen.

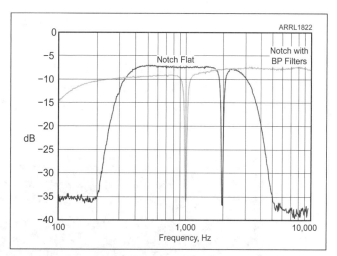

Figure 11.14 — Notch filter response.

### 3) Automatic Notch Filter

When the Auto Notch function is started, the routine searches for any stationary tones with significant amplitude. The Notch filter is then "tuned" to the tone frequency, removing the noise tone from the output. If the tone frequency changes the notch filter tracks the tone.

The Auto Notch function requires a certain level of tone amplitude to be effective. If the Auto Notch "lock" is lost periodically during a QSO, the "hold" function may be used. (The "Hold" button is only available when the Auto Notch option is selected.) Once the tone frequency is located, the "hold" function can be used to lock the frequency. This reduces the chance that the lock will be lost.

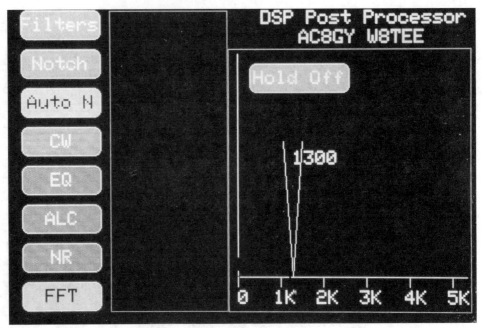

**Figure 11.15 — Auto notch setting screen.**

Pressing the "Hold" button again turns off the lock. **Figure 11.15** shows the Auto Notch setting screen.

## 4) CW Filters (CW)

There are three preset CW filters, with –3 dB bandwidths of approximately 60 Hz (CW1), 100 Hz (CW2) and 230 Hz (CW3).

The filter center frequency is selected with the *Frequency* encoder and the filter width is selected with the *Secondary* encoder. **Figure 11.16** shows the CW filter setting screen and **Figure 11.17** shows the actual responses of the three filters.

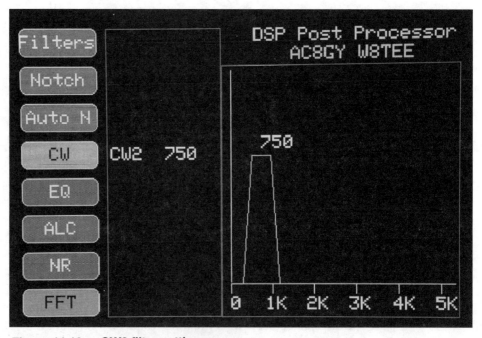

**Figure 11.16 — CW2 filter setting screen.**

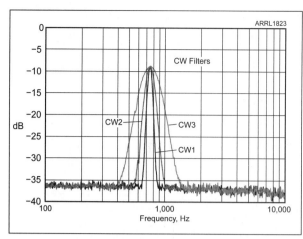

**Figure 11.17 — CW filter response plot.**

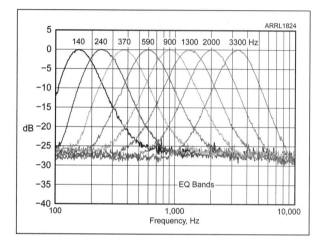

**Figure 11.18 — EQ band responses.**

## 5) Equalizer (EQ)

The Equalizer (EQ) has eight bands: 150, 240, 370, 590, 900, 1300, 2000, and 3300 Hz. **Figure 11.18** shows the individual band responses.

In the EQ Setting screen, the bands are graphically depicted on the screen, with two parameter selections, as shown in **Figure 11.19**. The *Frequency* encoder selects the band and the *Secondary* encoder sets the level for that band. Each band has a range of +12 dB to −12 dB. Note that there is interaction between adjacent bands, since the individual bands have finite skirt slopes and the filters overlap. In practice this does not affect the performance, because the amount of adjustment to achieve "natural" sound is typically not large.

The best way to set the EQ profile is to listen to a typical SSB signal while adjusting the individual bands. Alter the band levels to achieve the most "natural" sound. Don't rely on just one QSO, but try several receptions to get the best result. The EQ may be used in conjunction with any of the other Filter functions.

A response plot of the EQ is shown in **Figure 11.20** with the Notch filter, along with a second response without the Notch. If the SSB signal still sounds like Donald Duck after band adjustments, you're doing something wrong! For instance, perhaps you did not tune the SSB signal exactly.

**Figure 11.19 — EQ setting screen.**

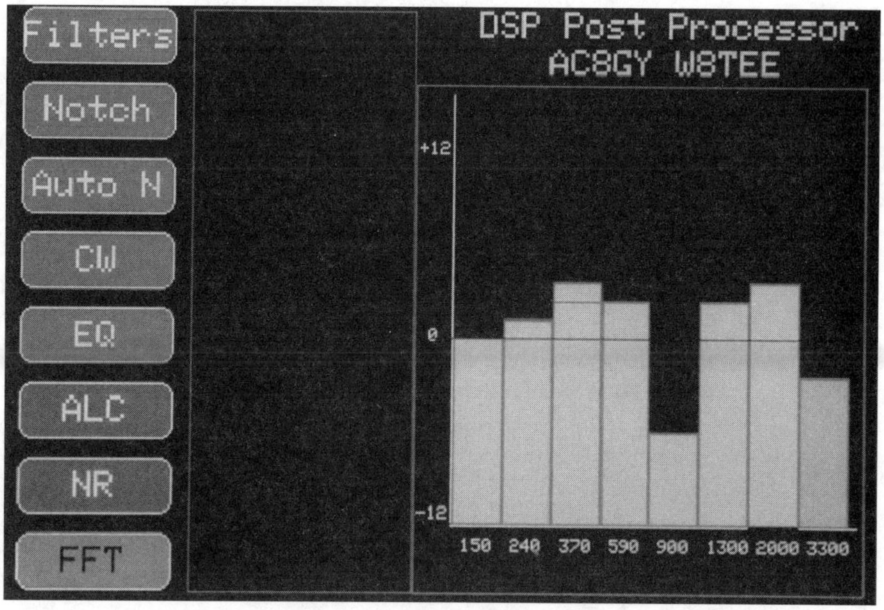

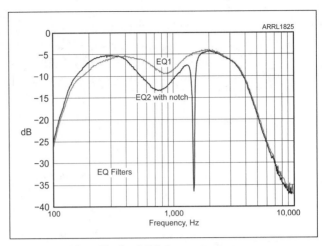

Figure 11.20 — Typical EQ frequency response.

### 6) Automatic Level Control (ALC)

Audio level varies not only from QSO to QSO, but also because of fading and band variations. Most transceivers have automatic level controls that may be effective or leave something to be desired. The ALC function here augments the built-in automatic level controls by boosting low volume levels and limiting the maximum volume levels. The Threshold level, at which the Auto Level control starts to act, is adjustable using the first encoder. The Threshold is variable from 0 dB, relative to full scale, to about −50 dB. A typical level for speech is about −18 dB.

### 7) Noise Reduction (NR)

A simple DSP Noise Reduction algorithm is activated by pressing the NR on/off button. There are no parameters to be set for this function.

### 8) FFT

A Fast Fourier Transform (FFT) display is available during operation, showing the frequency spectrum from 100 Hz to 5 kHz. Its 128 frequency bands clearly show the frequency content of the audio. You can use this information to make informed changes in the equalizer bands. The display is refreshed approximately every 500 ms.

To activate the FFT, press the FFT button. FFT is only available during operation, not while making settings. All other functions may be used with FFT enabled, so the effect of Filters and other noise reduction functions can be observed. **Figure 11.21** shows a typical FFT spectrum of white noise.

## Software Code Description

The DSP Post Processor performs a number of functions to modify and improve the audio signals from your receiver. Some of these functions are similar to circuit functions in the analog world, some are not. In any case, all of our processing is done digitally by performing mathematical operations on the audio signals to alter frequency response, change levels and reduce noise. The speed of

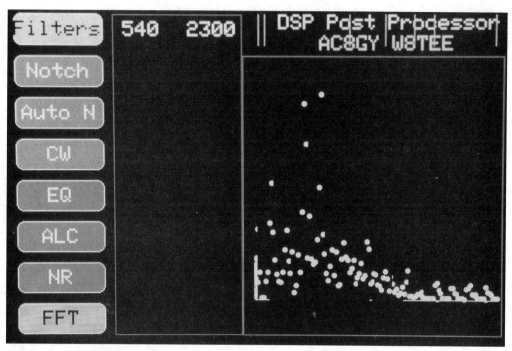

**Figure 11.21 — FFT display.**

the T4 is sufficient to do all of this seamlessly.

Much of the DPP code is like that of previous projects. What is new is the use of the Teensy Audio Library to implement DSP functions. The library functions consist of a number of modules that are connected to form a signal path. Each module also has a set of individual functions with parameters. The software design consists of using the online graphical design tool to lay out the data flow for the audio DSP and then copying the automatically generated code to the Arduino IDE. The graphical design tools is at **www.pjrc.com/teensy/gui/**. (The design software is very slick, but takes some time to master. It's worth the effort if you want to extend our DPP.) Function parameters are set with the code to tailor the DSP responses to suit. The flow diagram for Channel A is shown in **Figure 11.22**.

In the case of the various frequency domain filters we use, there are analog circuit equivalents. For instance, the low-pass filters are equivalent to a cascade of RC (resistor-capacitor) filters with gain. Each single analog RC stage yields a frequency roll-off of –3 dB/octave. Our eight cascaded digital filters have a total roll-off of 48 dB/octave or 160 dB/decade, which is the equivalent of 16 stages of RC filters! Imagine wiring up such an array, with appropriate gain stages. We do it all in software.

The compressor, or Automatic Level Control, is the equivalent of a variable gain stage, controlled by the signal level — again a complex circuit. We do all of that with a few lines of code and a very nice audio library that works with the Teensy Audio Adapter.

The filters we use are called biquad or biquadradic cascaded filters, which are a variety of recursive linear filters based on a mathematical formula that has feedback incorporated. The interested the reader can find out more at **peabody.sapp.org/class/350.838/lab/biquad**. In essence, the filters we use have a grouping of four filters in a module that can be cascaded with additional modules to obtain

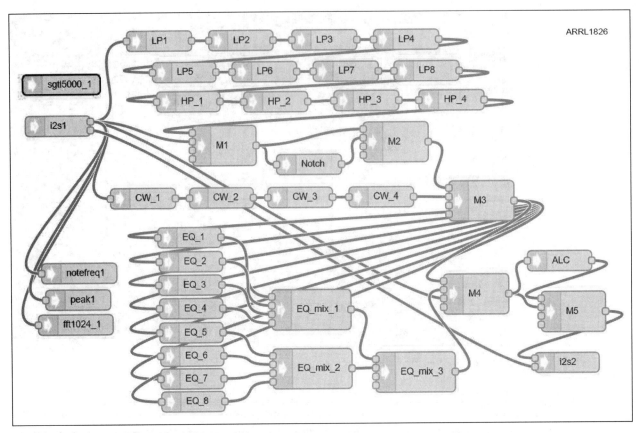

**Figure 11.22 — Single channel DSP signal flow diagram.**

better performance. The BiQuad filters can be configured as low-pass (LP), high-pass (HP), notch, or band-pass (BP) filters. We use all of those configurations in our DPP.

In addition to digital filters, we also use mathematical operations to calculate Fast Fourier Transforms (FFT) to give us real-time spectrum plots of our audio signals. Once again, the T4 is fast enough to calculate all of the filter function operations and the FFT at the same time. A variant of the FFT is used to find single tones in the audio signs for an Auto-Notch function.

## Signal Flow

At the input in Figure 11.22, an audio capture module labeled "I2S1" is the starting place for the audio signal. This connects the analog audio input to the DSP functions. Each module is interconnected using a (graphical) digital "connector" wire.

Starting at the top, we see the LP1 – LP8 blocks which are DSP low-pass filters cascaded to achieve steep roll-off after the cutoff frequency. The eight LP filters attach to the four HP low-frequency cutoff filters for the high-pass function. Each of the major filter sections is interconnected using a "Mixer" function (labeled M1, M2, EQ_mix1, and so on) that allows setting gains and turning individual functions on or off, as well as bypassing that section.

After the HP/LP filters there is a Notch filter in between two mixers that allows for bypassing the notch function. Auto Notch uses the same filter, but uses a routine with "notefreq1" function that finds the frequency of the tone to be suppressed and sets the notch frequency automatically.

In a parallel circuit, the CW band-pass filters are used instead of the LP/HP combination to give very narrow filtration for CW work. Three CW bandwidths are available, and the center frequency of the CW filter can be altered using the Setting function.

Next in line we have the eight-band equalizer. Signals are sent to eight different biquad filters each with a different frequency. Human hearing is approximately logarithmic with frequency, so the spacing is approximately set to be 1.5× the previous band frequency, which gives an equal spacing on a log scale. The eight bands cover the range of 150 Hz to 3300 Hz in eight steps. The output of each filter is sent to a series of three mixers, in which the individual band gains are set. The eight outputs are then combined into one composite signal. The spectrum is shaped by setting the individual band gains from +12 dB to –12 dB. Because the filter skirts drop off at a finite value, there is interaction among the filters, limiting the sharpness of peaks and dips. The purpose of the EQ section is to gently shape the spectrum to eliminate humps or dips in the ultimate frequency spectrum, compensating for microphone variations on the other end. Note that the filters create phase changes, so the gains alternate + and – indicating the appropriate phase reversals.

Automatic Level Control (ALC) is computed in a special section using an augmented Audio library. As with the other functions, ALC is turned on or off using mixer gains.

In the software, there are other code routines that perform various functions. For instance, there is an ongoing routine that reads the on-screen touch-activated buttons to determine when one was pressed. Then each Setting function has three major routines: 1) Draw the graphical representation on the screen, 2) Read the encoders and set the appropriate biquad filter parameters and/or the mixer gains, and then, 3) Turn the functions on or off using the front panel pushbuttons.

In the code there are also the usual menu functions, encoder interrupt functions, screen management, and so on, all of which has been adequately described elsewhere in the previous chapters. The code is broken up into 10 different files for convenience of both coding and debugging. Adequate comments throughout should help anyone wishing to make extensions or improvements.

Finally, a simple Noise Reduction routine has been included to reduce random high-frequency noise using a FIR filtering technique. There are many other more comprehensive mathematical routines such as convolution filtering that could be implemented but were beyond the scope of the current project function set. This is left to the reader to add later.

Possible extensions for the adventurous include:

- Independent filter functions for Channel A and Channel B. This is primarily a UI coding exercise, adding setting and selection routines for each channel.

- A more rigorous implementation of the NR routines.

- Additional filtering and/or averaging of the Auto Notch signals to improve signal-to-noise ratio.

- Adding real-time spectrum during setting in order to observe the spectral response variations while making changes.

## Conclusion

In summary, the DPP presented here can bring an older transceiver up to modern signal processing standards and could even be a valuable adjunct to lower-end modern rigs that are perfectly adequate in all other ways. There are a lot of good, inexpensive, QRP rigs that could benefit from our DPP, and the improved audio can make your listening experience that much more enjoyable.

# CHAPTER 12

# DSP Audio Mic-Processor

While most modern transceivers have extensive digital signal processing (DSP) functions for microphone audio signal conditioning, older rigs do not. Also, many inexpensive transceivers have no DSP capability. Regardless of their shortcomings, many amateurs still like to use their older (vintage) or less expensive units on a regular basis.

Just like the Chapter 11 DSP Post Processor (DPP), we have created a DSP-based audio microphone signal conditioner based on the Teensy line of microcontrollers. As with the Post Processor, we elected to use the newer Teensy 4.0 unit, primarily for its Audio Library DSP capabilities. The Teensy 4 (to be known as the "T4" from now on) is a real powerhouse, clocked at 600 MHz, with 1024 KB of RAM and 2048 KB of flash memory. (As we go to press, PJRC has announced the Teensy 4.1, which is the same as the T4, but with a higher pin count and larger footprint, and about $5 more expensive.) The T4 initially is being offered in a reduced-size development board package but it has sufficient GPIO pins for our application. Note that the Chapter 11 Post Processor required additional GPIO pins from the underside of the T4 board, while the Mic-Processor does not.

So, what can an audio mic-processor do for you? Some of the possibilities are:

- Automatic speech compressor
- Multi-band equalizer
- FFT display
- Interface both dynamic and capacitor/electret mics to an older transceiver.

There are several commercial units with some of the functions mentioned above, but typically at relatively high prices, some starting at several hundred dollars. The unit we are describing does all of the specs listed for commercial units and costs less than $70, if you build it yourself. In addition, the software, which is the heart of the unit, is open-source, so you can extend and tailor the processor to your needs. We also present some ideas of ways to augment the features of our DSP Mic-Processor (DMP), if you choose to do so.

## User Interface (UI) Overview

The stuff under the hood of our DMP is quite interesting, but everyday interaction with the user interface is what makes the high-powered DSP technology accessible. We have spent a significant amount of time and countless prototypes streamlining the way our DSP unit interfaces with you and the functionality it brings to the table.

A UI has several aspects, most important of which are: 1) the control/data input mechanism, 2) feedback to the user, and 3) the ease with which the user can make changes. We have elected to present data about the state of the processor using a high-resolution TFT color LCD screen. Data input/control uses the touch capability of the TFT screen as well as rotary encoders and an analog-to-digital volume pot. Our experience has shown that this combination of physical controls and TFT touch-screen input is quite efficient and minimizes the number of steps necessary to make changes and set the state of the unit.

**Figure 12.1** shows one of the front panel menu options we elected to use,

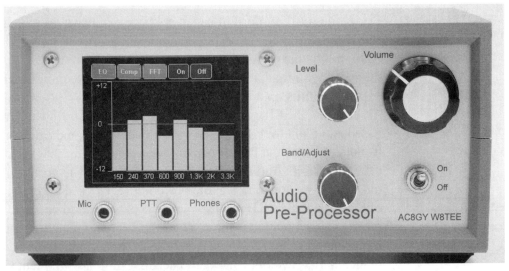

Figure 12.1 — DSP Audio Mic-Processor.

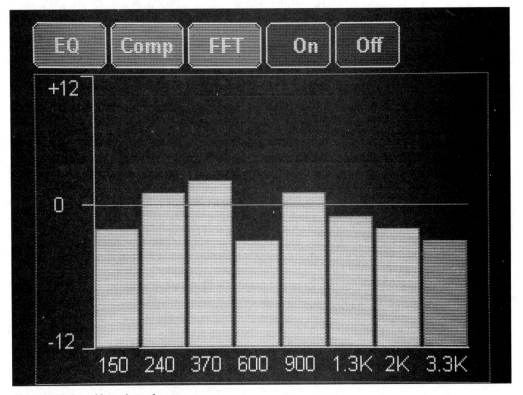

Figure 12.2 — User interface.

combining a TFT display, volume control, two rotary encoders, and three input/output jacks — there are more jacks on the back panel. With the TFT display, the user gets graphical feedback as to the state of the unit's functions and can select functions and set parameters. A real-time FFT display shows the spectral content of the various signals, after DSP. Because of the less stringent graphics requirements, this project uses a slightly lower resolution touch screen (320×240 pixels) than was used in the DPP.

**Figure 12.2** shows a close-up of the TFT screen with the graphical display graphic equalizer function shown. Here the user can select bands and band levels, while monitoring the effect using the headphones. Real-time interaction makes fine tuning the unit a breeze. The UI is explained in detail later in this chapter.

## Circuit Description

This project is about 25% electronic hardware and 75% software, so let's get the circuit out of the way first and then dive into the software.

**Figure 12.3** shows a block diagram of the circuit. The heart of the unit is the Teensy T4 microcontroller. The second most important component is the Teensy Audio Adapter (TAA), which contains all of the audio I/O, such as ADCs and DACs necessary to capture the audio signals. It also sends the digital data to the T4 and converts and outputs the processed signals back to analog audio again. The TAA has 16-bit conversion and operates at a 96 kHz conversion rate, enough for full 16-bit 48 kHz CD-quality audio. The TAA can directly piggyback onto the T4, too, significantly simplifying the circuit wiring. Finally,

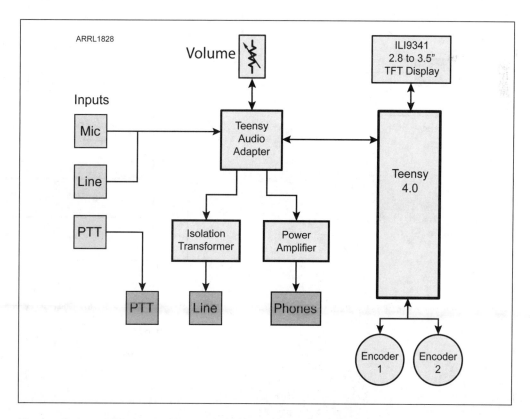

Figure 12.3 — DSP Audio Mic-Processor (DMP) block diagram.

the T4 has enough horsepower to run a real-time FFT processing at the same time as it is doing the DSP filtering on the audio channels. Note that there is a specific version of the TAA for the T4. The older TAA module does not work correctly with the T4.

Starting at the input, we see that provision can be made for two inputs which are to be routed to the computational channels. At this time, only the left channel is utilized. However, it would be very easy to add the DSP functions to the right channel if you wish. We did not need two independent DSP channels, so adding a second channel remains as an exercise for the reader. The microphone is connected to the TAA Mic input, which has extra gain to accommodate the low-level signals from the microphone. The Line input operates at line levels and is suitable for high-level line signals. This allows you to input

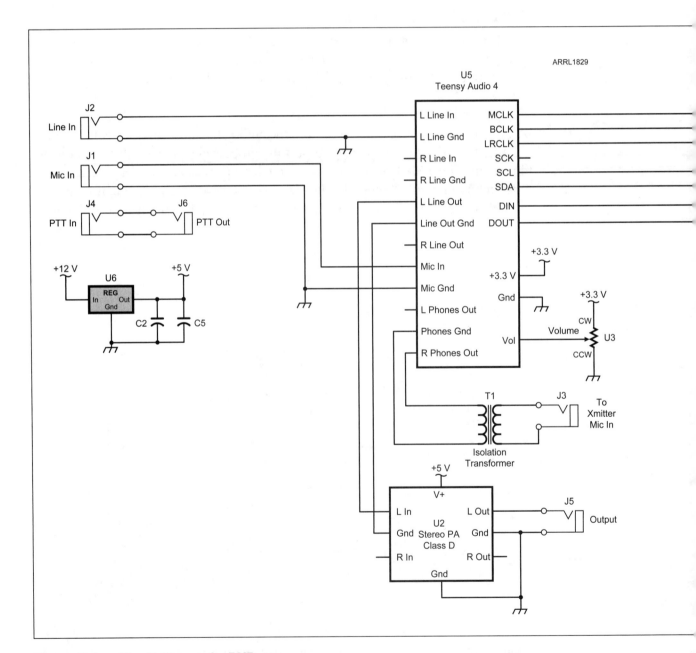

Figure 12.4 — Circuit diagram for DMP.

signals from an audio mixer source that outputs line-level signals. Selection of the source must be done in the setup portion of the Comp Screen.

Signals from the TAA "headphone" output go to an isolation transformer. The isolation transformer eliminates the possibility of ground loop noise and isolates the grounds of the transceiver and Audio Mic-Processor.

The low-level mic inputs accommodate either dynamic, capacitor, or electret mics. Capacitor and electret mics require a "bias" dc voltage, which is provided by the TAA. A Mic Gain setting allows the specific level characteristics of your mic to be accommodated.

The outputs include a line-level output as well as an amplified headphone output for monitoring. This amplified output is capable driving a small speaker as well, depending on the application.

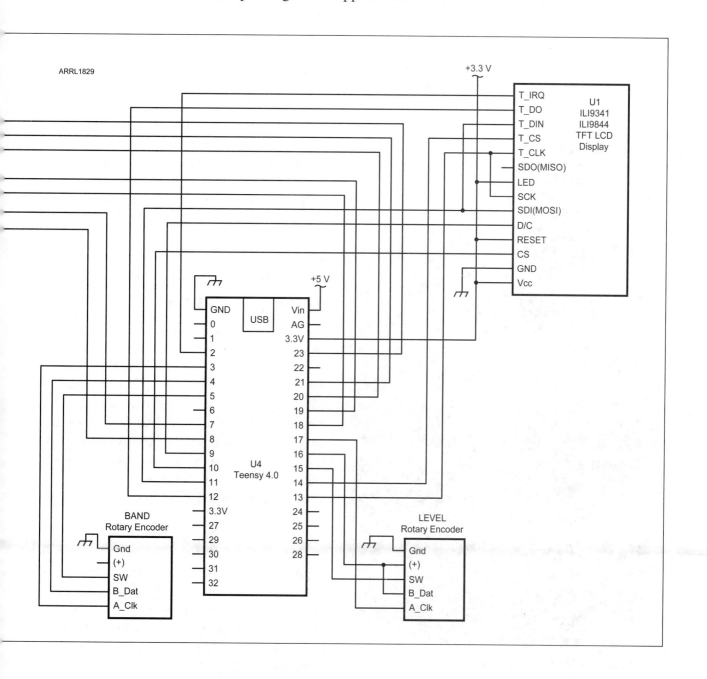

**Table 12.1**
**Teensy 4 Pin Assignments**

| Teensy 4 Pin | Attachment | Teensy 4 Pin | Attachment |
|---|---|---|---|
| 2 | Touch IRQ | 17 | Encoder2 A |
| 3 | Encoder1 A | 18 | Audio Adapter |
| 4 | Encoder1 B | 19 | Audio Adapter |
| 5 | Encoder1 SW | 20 | Audio |
| 6 | N/C | 21 | Audio |
| 7 | Audio | 22 | |
| 8 | Audio | 23 | Audio Adapter |
| 9 | LCD D/C | 24 | |
| 10 | LCD CS | 25 | |
| 11 | LCD, Touch MOSI | 26 | |
| 12 | Touch MISO | 27 | |
| 13 | LCD, Touch SCK | 28 | |
| 14 | Encoder2 SW | 29 | |
| 15 | Volume Pot | 30 | |
| 16 | Encoder2 B | 31 | |

The detailed circuit diagram, including pin assignments is contained in **Figure 12.4** and **Table 12.1**. Note that some Teensy pins, such as USB, are not available because of system use. One voltage regulator provides power for the Teensy, display LED, and power amplifier. The 5 V fixed regulator drops the 12 V to the T4 Vin pin. 3.3 V is provided from the T4 on-board regulator.

## Building the DMP

One advantage of building circuits for the audio frequency range is the relaxed requirements on layout. The DMP can be constructed using any of the popular methods. We built our prototype on a modified breakout board. The Audio Adapter plugs directly onto the Teensy, so control wiring is minimal. We always use sockets for the microprocessors for ease of insertion and removal, if necessary. The other components are connected using IDC connectors and ribbon cable for convenience.

No special wiring precautions are required. We like to use #30 AWG solid hookup wire because of the close 0.1-inch hole spacing in the proto-board (same for most perf boards.) The main power connections use #24 AWG or larger wire to minimize resistive losses.

**Figure 12.5 — DMP circuit board (top).**

Figure 12.5 shows the top of the main circuit board, with the Teensy and Audio Adapter installed. Note the use of multi-pin IDC connectors for display, front panel connections, and encoders. Figure 12.6 shows a view of the bottom of the circuit board. Figure 12.7 shows the front and back panel wiring, with

**Figure 12.6 — Bottom view of circuit board.**

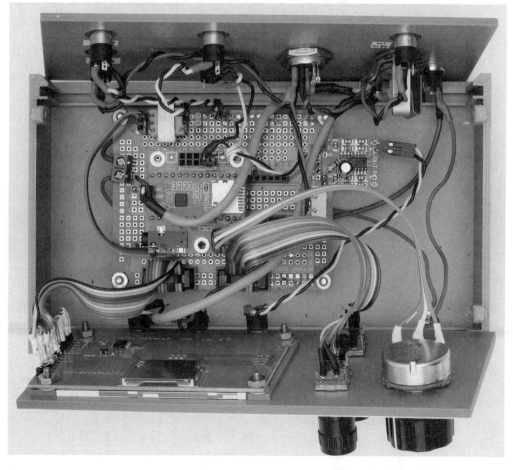

**Figure 12.7 — Front and back panel wiring.**

**Figure 12.8 —
Top view of
complete unit.**

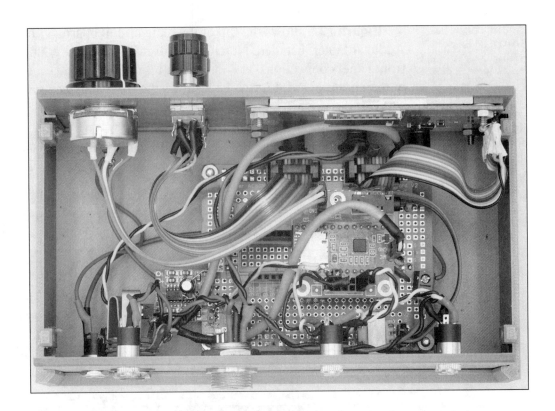

**Figure 12.9 —
The DMP at
home in its
case.**

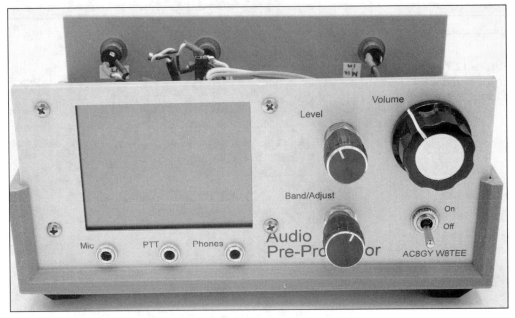

the ribbon cables and female IDC connectors.

Finally, a top view of the assembled unit is shown in **Figure 12.8**. Shielded cables are used to carry the low-level audio signals from the front panel to the Audio Adapter and from the inputs.

For most of the projects in this book we have assumed that the user has ample 12 V or 13.8 V power available from rig power supplies. This project doesn't require much power and could be battery operated in a pinch. If an adequate supply is not available, a small switch-mode 12 V supply capable of a couple of amps works just fine, or you can build the Utility Power Supply

Figure 12.10 — The DMP back panel.

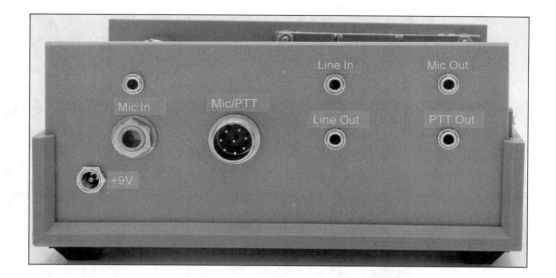

described in Chapter 5. Be sure to add adequate filtering of the incoming 12 V supply to minimize noise from getting into the audio path.

We made a 3D printed case with a custom faceplate. **Figure 12.9** shows the unit with the top lid off and **Figure 12.10** shows the back panel.

## Operating the Mic-Processor

Most of the difficult parts of using the processor are in the setup and, as we indicated previously, we spent a lot of time trying to make the user interaction as easy and intuitive as possible. The operational steps are a simple 1-2-3 process:

1) After hooking up the DMP to your rig and mic, you must set the parameters for the functions you want to use. To enter the "Setup" mode for any function, touch the on-screen function button to bring up the setup routine for that function. Using the two encoders, set the parameters for that function. (A detailed description of each setup is given below.) If displayed, touch the "Set" button to store the parameters in (emulated) EEPROM.

2) Turn each function on or off using the on-screen buttons at the top of the display. The status of the function is displayed by the color of the button for that function.

3) Adjust the output level using the volume control. You are now ready to go!

## Function Detail

There are only three functions available for the DMP

### 1) Equalizer (EQ)

The EQ utilizes eight bands: 150, 240, 370, 590, 900, 1300, 2000, and 3300 Hz. **Figure 12.11** shows the individual band responses.

In the EQ Setting screen, the bands are graphically depicted on the screen, with two parameter selections, as shown in **Figure 12.12**. The *Band/Adjust* encoder selects the band and the *Level* encoder sets the level for that band. Each band

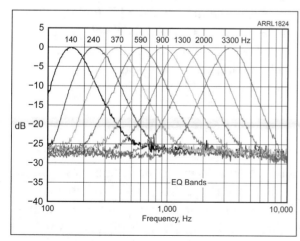

Figure 12.11 — EQ band responses.

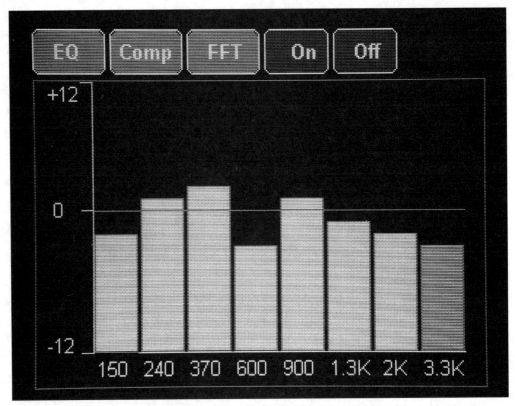

**Figure 12.12 — EQ setting screen.**

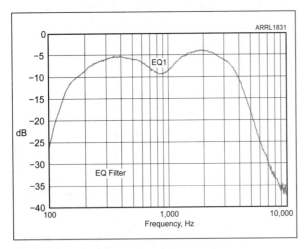

**Figure 12.13 — Typical EQ frequency response.**

has a range of +12 dB to –12 dB. Note that there is interaction between adjacent bands, since the individual bands have finite skirt slopes and the filters overlap. In practice this does not affect the performance, because the amount of adjustment to achieve "natural" sound is typically not large.

The best way to set the EQ profile is to monitor your speech while adjusting the bands and levels by altering the band levels to achieve the most "natural" sound. The EQ may be used in conjunction with any of the other functions. We suggest recording your speech and listening to it later. This eliminates the aural feedback present when you speak. Just hook an audio recorder to the headphone output as you speak.

A typical response plot of the EQ is shown in **Figure 12.13**.

### 2) Automatic Level Control (ALC) — Compressor

Speech audio levels vary considerably during the typical QSO. Perception of the strength of your SSB signal depends on having the modulating level more nearly uniform than is usual with typical speech. The ALC function here augments the perceived levels by boosting low volume levels and limiting the maximum volume levels. Both the upper Threshold level, at which the auto level control limits the top volume and the lower Threshold level, at which the gain is increased, are adjustable using the first encoder. Simply touch the button for the Threshold parameter to set and turn the encoder. When the level is where you want it, press the "Set" button to save the values in EEPROM.

Mic Gain may also be set on this screen, to account for differences in microphone sensitivity. **Figure 12.14** shows the Comp setting screen.

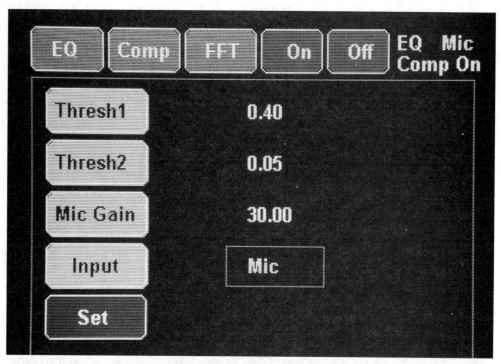

Figure 12.14 — Comp Parameter setting screen.

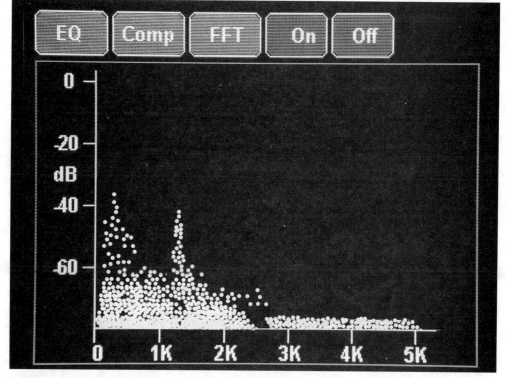

Figure 12.15 — FFT display.

**DSP Audio Mic-Processor**  12-11

## 3) FFT

A Fast Fourier Transform (FFT) display is available during operation, showing the frequency spectrum from 100 Hz to 5 kHz. 128 frequency bands clearly show the frequency content of the audio. The display is rapidly refreshed for a real-time depiction of the spectrum.

To activate the FFT, press the FFT button. FFT is available during operation, but not while changing settings. All other functions may be used with FFT, so the effect of EQ and Compression can be observed. **Figure 12.15** shows a typical FFT spectrum. The vertical axis is a log scale in dB.

## Software Code Description

Much of the DMP code is like that of the Post Processor, and a general introduction to digital signal processing (DSP) was presented in Chapter 11. Again, the Teensy Audio library is used to implement DSP functions. The library functions consist of a number of modules that are connected to form a signal path. Each module also has a set of individual functions with parameters. The software design starts by using an online graphical design tool to lay out the data flow for the audio DSP and then copying the automatically generated code to the Arduino IDE. The graphical design tool is available from **www.pjrc.com/teensy/gui/**. (The design software is very slick, but takes some time to master. It's worth the effort if you want to extend our DMP.)

Function parameters are set with the code to tailor the DSP responses to suit. The flow diagram is shown in **Figure 12.16**. The DSP Mic-Processor performs a number of functions to modify and improve the audio signals going to your transmitter. Some of these functions are similar to circuit functions in the analog world, some are not. In any case, all of our processing is done digitally

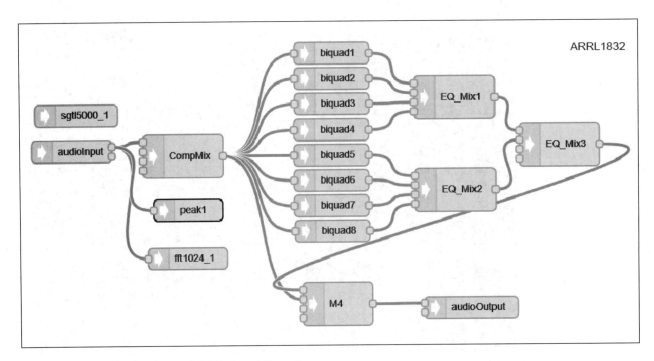

Figure 12.16 — Single channel DSP signal flow diagram.

by performing mathematical operations on the audio signals to alter frequency response, change levels and reduce noise. The speed of the T4 is sufficient to do all of this seamlessly (which is why we selected the more expensive T4).

There are analog circuit equivalents to some of the frequency domain filters we use. For instance, the low-pass filters we used in the Post-Processor are equivalent to a cascade of RC (resistor-capacitor) filters with gain. Each single analog RC stage yields a frequency roll-off of –3 dB/octave. Our eight cascaded digital filters have a total roll-off of 48 dB/octave or 160 dB/decade, which is the equivalent of 16 stages of RC filters! Imagine wiring up such an array, with appropriate gain stages. The band-pass filters used in the DMP are ever more complex to implement in hardware. We do it all in software.

The Compressor or Automatic Level Control is the equivalent of a variable gain stage, controlled by the peak signal level — again a complex circuit. We do all of that with a few lines of code and the very nice audio library that works with the Teensy's Audio Adapter.

The filters for the EQ we use are called biquad or biquadradic cascaded filters, which are a variety of recursive linear filters based on a mathematical formula that has feedback incorporated. The interested the reader can find out more at **peabody.sapp.org/class/350.838/lab/biquad/**. In essence, the filters we use have a grouping of four filters in a module that can be cascaded with additional modules to obtain better performance. The biquad filters can be configured as low-pass (LP), high-pass (HP), notch, or band-pass (BP) filters. We use the band-pass configuration in our DMP.

In addition to digital filters, we also use mathematical operations to calculate Fast Fourier Transforms (FFT) to give us real-time spectrum plots of our audio signals. Once again, the T4 is fast enough to calculate all of the filter function operations and do the FFT at the same time.

## Signal Flow

At the input in Figure 12.16, an audio capture module is the starting place for the audio signal. This connects the audio input to the DSP functions. Each module is interconnected using a (graphical) digital "connector" wire.

Starting at the left in Figure 12.16, we see the CompMix module, which provides the variable gain necessary for the Compressor. The gain is set in the *Compressor( )* routine, which samples the input signal peak values and adjusts the gain according to the sampled level. The response is fast enough to react to changing levels but averages the control signal sufficiently to ensure natural sounding speech. The speech levels are sampled using the "Peak" function.

Next in line we have the eight-band equalizer. Signals are sent to eight different biquad band-pass filters (biquad1-biquad8) each with a different center frequency. Human hearing is approximately logarithmic with frequency, so the center frequency spacing is set to be approximately 1.5× the previous band frequency, which gives an equal spacing on a log scale. The eight bands cover the range of 150 Hz to 3300 Hz in eight steps. The output of each filter is sent to the inputs of two mixers, in which the individual band gains are set. The eight outputs are then combined into one composite signal by another mixer. The spectrum is shaped by setting the individual band gains from +12 dB to –12 dB.

Because the filter skirts drop off at a finite value, there is interaction among the filters, limiting the sharpness of peaks and dips. The purpose of the EQ section is to gently shape the spectrum to eliminate humps or dips in the ultimate frequency spectrum, compensating for microphone variations. Note that these filters create phase changes, so the mixer gains alternate + and − to account for the appropriate phase reversals.

In the software, there are other code routines that perform various functions. For instance, there is an ongoing routine that reads the on-screen touch-activated buttons to determine when one was pressed. Each Setting function has three major routines:

1) Draw the graphical representation on the screen;

2) Read the encoders and set the appropriate biquad filter parameters and/or the mixer gains; and then

3) Turn the functions on or off using the front panel push buttons.

In the code there are also the usual menu functions, encoder interrupt functions, screen management, and so on, all of which has been adequately described elsewhere in the previous chapters. The code is broken up into nine different files for convenience of both coding and debugging. Adequate comments throughout should help anyone wishing to make extensions or improvements.

Possible extensions for the adventurous include:

- Independent functions for Channel A and Channel B. This is primarily a UI coding exercise, adding setting and selection routines for each channel. This would allow for more than one mic to be attached and compensated separately, or a high-level mixer input could be added to the second channel.

- A digital recorder could be incorporated to allow the operator to review speech quality off-line. The Teensy Audio Library has the necessary components to implement such a recorder.

- Adding real-time spectrum monitoring during setup in order to observe the spectral response variations while making changes.

## Conclusion

In summary, the DSP Mic-Processor presented here can bring an older transceiver up to modern signal processing standards and could even be a valuable adjunct to lower-end current rigs that are perfectly adequate in all other ways. If you do enhance the DMP, we hope you'll post your work on the SoftwareControlledHamRadio website!

# CHAPTER 13

# Signal Generator

There are several indispensable pieces of test equipment every serious ham experimenter needs: 1) a multimeter, 2) an oscilloscope, and 3) a signal generator. While there are lots of commercial signal generators (SG from now on) available, there are few amateur radio specific SGs with features you might need that are not common on reasonably-priced units costing less than several hundred dollars. These extra features include: 1) AM modulation with waveforms other than sine waves, 2) selectable modulation frequencies for frequency response testing of the complete receiver RF/IF chain, 3) double-sideband, suppressed-carrier modulation to work with SSB receivers, 4) external modulation inputs, 5) ability to create two-tone signals for receiver 3rd order intermodulation testing, and 6) easily selectable step output attenuation as well as a wide frequency range and the other usual requirements.

**Figure 13.1** shows our finished SG in its case. This is probably the most complex project in this book, so we are making a PCB available to make construction easier for the builder.

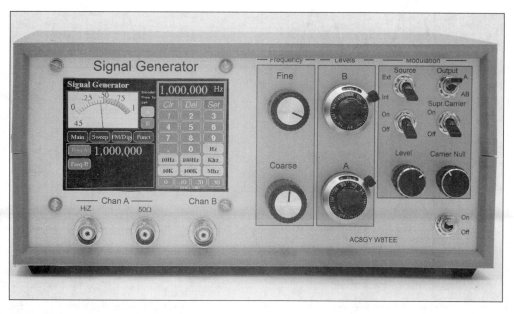

**Figure 13.1 — Signal generator.**

**Signal Generator    13-1**

## What Does a Signal Generator Do?

Simply stated, an SG provides the simplest forms of signals, such as the sine waves, in a controllable, repeatable, and stable manner. Other waveforms, such as square, triangle, or arbitrary waveforms, also may be available, but to start, our focus is on sine waves.

What is so special about sine waves? A unique property of a sine wave is that it is comprised of just one frequency, whereas other waveforms are made up of multiple sine waves, harmonically related or not, all superimposed upon each other. **Figure 13.2** shows several waveforms commonly encountered in ham radio. With sine waves, we can readily control and measure the effect of signals on our radios and other instruments.

What are the common requirements for a signal generator to be useful in amateur radio applications?

1) The output frequency should be known, stable, accurate, and controllable.

2) The frequency should be drift-free and accurate to less than 1 part per million.

3) The SG should have low distortion (less than 1% harmonic distortion). That is, it should be just one sine wave to better than one part in 100.

4) The amplitude of the sine wave should be controllable in both stepped and variable increments.

5) The output level should be accurately displayed in volts or dBV.

Now, let's see how this chapter's SG project stacks up to those requirements.

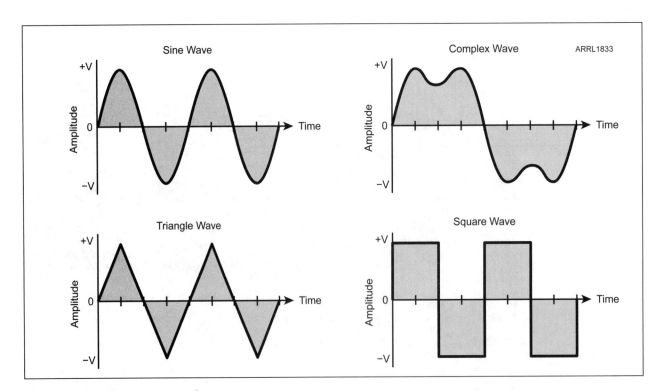

Figure 13.2 — Common waveforms.

## Table 13.1
**Signal Generator Feature Set**

Feature Details

| | |
|---|---|
| Frequency range | 10 Hz to 40 MHz |
| Two channels | Channel A – 10 Hz to 40 MHz, sine waves only<br>Channel B – 10 Hz to 10 MHz<ul><li>Sine</li><li>Square</li><li>Triangle functions</li></ul> |
| Modulated or unmodulated output | AM modulation – Channel A modulated by Channel B<ul><li>Modulation level adjustable from front panel 0 to 100%</li><li>Suppressed carrier, double sideband modulation – carrier null adjustable</li><li>Internal or external modulation signals</li><li>Modulating frequency range – 10 Hz to 10 MHz</li></ul>FM modulation<ul><li>Adjustable modulating frequency in preset steps</li><li>Adjustable frequency span in preset steps</li></ul> |
| Outputs | Sine<ul><li>Channel A — Max 2.5 V RMS impedance output</li><li>Max 1 V RMS into 50 W</li></ul>On-screen attenuator selection: 0, –10, –20, –30 dB<br>Function (Channel B) output 1 V RMS |
| Sweep function – Channel A | Start frequency<br>End frequency<br>Sweep time<br>Number of points in sweep<br>External trigger |
| Attenuators | Fixed: 0, –10, –20, –30 dB<br>Variable: 0 to –40 dB |
| Frequency | Resolution: 1 Hz<br>Stability: Better than 0.3 parts per million<br>Channel A sine wave distortion: Less than 1% up to 15 MHz<br>Channel B sine wave distortion: Less than 0.5% |
| User Interface | Touch Screen<br>3.5-inch TFT color LCD – 480×320 pixels<br>Direct keypad frequency input with selectable multiplier<br>On-screen step attenuator<br>Fine and coarse frequency adjustments<br>On-screen digital readouts<br>Frequency – 1 Hz resolution<br>Channel A output level displayed on on-screen analog meter |
| Power Supply | 12 V dc |
| Frequency Calibration | Totally software dependent – user calibration in code |
| Expandable | Possible future additions:<ul><li>DC Sweep ramp output for scope</li><li>Trigger out</li><li>High speed sweep for alignment</li></ul> |

## SG Specifications

What follows is a list of the specifications for our SG. As we said before, this is not the easiest project to build. However, compare the specs you see here with any reasonably-priced commercial SG and you will discover that the SG built in this chapter is a bargain. Even better, you'll have the satisfaction of actually building it!

The specs are as presented in **Table 13.1**. This SG should satisfy even the most rigorous demands of the experimenter.

## Circuit Description

### Block Diagram

A block diagram of the SG circuit is shown in **Figure 13.3**. Starting on the upper left you can see the heart of the circuit, which is an AD9851 DDS (direct digital synthesis) module, creating sine waves from 10 Hz to more than 40 MHz for Channel A. AD9851 control is provided by an ESP32 microprocessor through a Serial data connection provided by a special AD9851 Library. Next in line is a wide-band RF amplifier that increases the signal level to about 2.5 V RMS into a hi-impedance load, which can supply 1 V RMS into 50 Ω for RF work.

The Attenuator block has both fixed attenuation in 10 dB steps and a variable attenuator for fine level control. The output level is measured using an AD8397 log amp, which reads the output level and outputs a dc signal proportional to 20 log(voltage level), in dBV. This dB signal is displayed on an on-screen analog type meter.

A second Channel is constructed around the AD9833 DDS, which has sev-

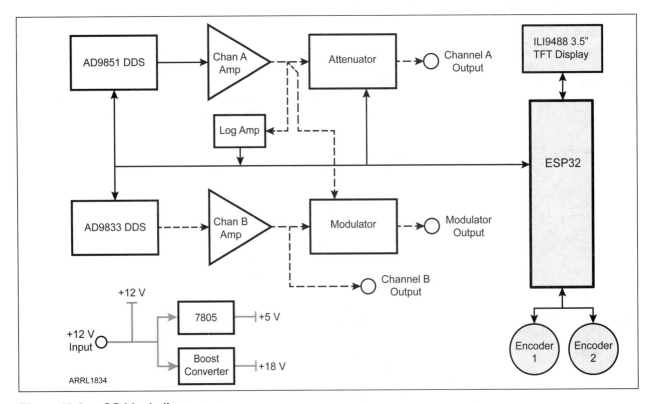

Figure 13.3 — SG block diagram.

eral important features. While it does not have the wide range of the AD9851, the AD9833 can output triangle waves as well as square waves. In addition, the sine wave output is very low distortion, typically less than 0.5%. The limitation is that its range is only up to about 10 MHz. For the intended function, this is not a real problem, however, since the primary use of Channel B is to provide variable modulating signal to be combined with Channel A to yield high quality AM and suppressed-carrier outputs.

Output from the AD9833 is amplified and sent to both a separate Channel B output and to the Modulator circuit. We had some ambitious expectations for our modulator, specifically: 1) that is should provide low distortion AM signals capable of being modulated throughout the audio range, and 2) that a suppressed carrier mode be provided to use as input to an SSB receiver using various signal sources.

True SSB modulation circuits are complex and beyond the scope of this project, but a close approximation can be created using a double-sideband suppressed-carrier mode. The signal is just like conventional SSB, except both sidebands are present, but no carrier. This would not do for actual SSB transmissions because of the extra power required to transmit the signal and because it would intrude on adjacent SSB channels. For testing, however, this is usually not a limitation. The receiver readily rejects the opposite sideband, passing a signal equivalent to regular SSB reception. Thus, complex tests on an SSB receiver are possible to easily characterize receiver bandwidth, distortion, and other characteristics.

The modulator we selected is an "oldie but goodie" chip, the MC1496. It works extremely well but requires more support circuitry than more recent integrated chips. The problem is there aren't other alternatives for what we want to do in a reasonable price range, so we had to add the extra resistors. The good news is that the flexibility of the MC1496 makes it ideal for our purposes. The modulation mode is selected using a simple front-panel switch and the suppressed carrier mode is fine-tuned with a single pot, also on the front panel.

Modulating signals are provided by either the Channel B output or an external input. Output levels are adjustable, as is the degree of modulation. Channel A provides the carrier frequency. Other Modulation modes also are available, primarily FM, which is done in software.

Continuing in the block diagram, we see the microcontroller, an ESP32 Node MCU. We have elsewhere commented on the variations of ESP32 available and which one to select. The ESP32 is a very capable 32-bit unit, running at 240 MHz with lots of memory. It is interfaced to a high-resolution 3.5-inch TFT LCD screen, which uses the ILI9488 controller chip. The ILI9488 chip is significant because the correct library must be selected for the combination of the microcontroller/display, as described in Chapter 4.

Finally, we use two rotary encoders and the touch screen to interface with the ESP32. At the bottom of the block diagram there are several voltages shown that need to be provided. Some of the amplifier circuits require +12 V, while +18 V is needed by the Channel A amplifiers to be able to output up to 2.5 V RMS. The ESP32 and other circuits need 5 V, which is stepped down from +12 V using a 7805 linear regulator. The +18 V is provided by a switching boost converter that creates the higher voltage from the +12 V input.

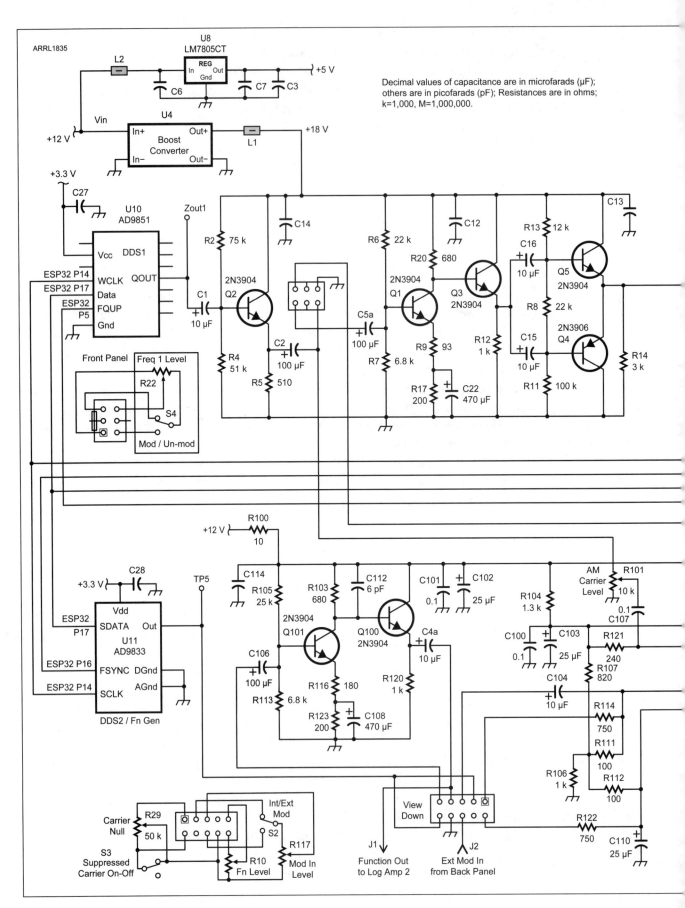

Figure 13.4 — SG detailed schematic diagram.

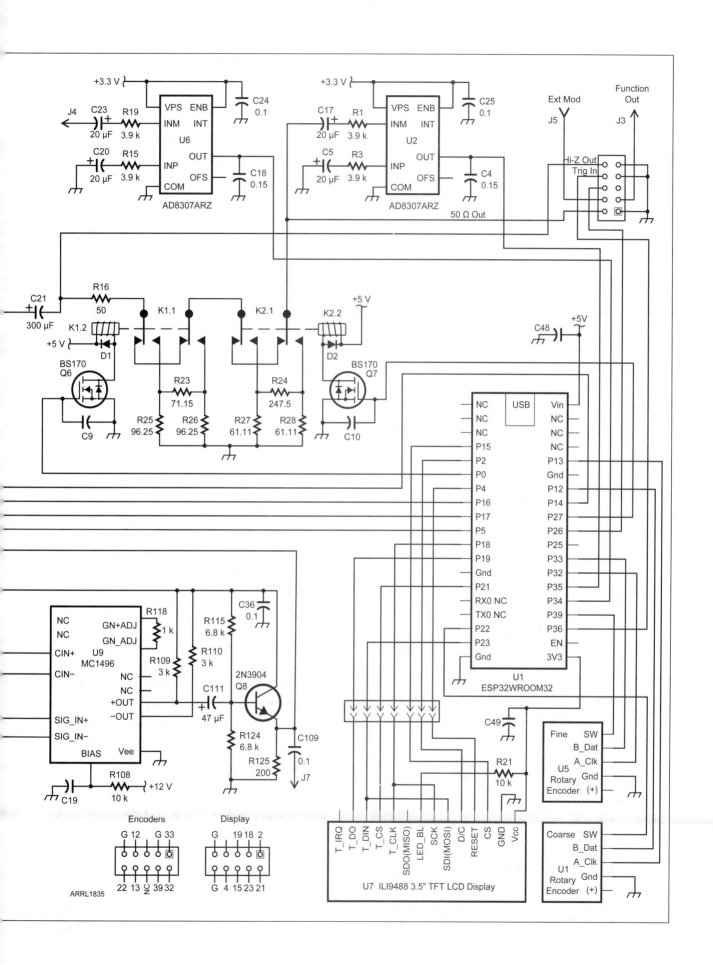

Signal Generator

## Detailed Schematic Diagram

The details of the circuit are shown in the schematic diagram in **Figure 13.4**. Most of the circuit elements are readily evident, but we will cover a few more unusual items in detail.

First, after the AD9851 output, which is about 400 mV RMS, we see the wide-band RF amplifier. This amp has two basic functions: 1) to increase the output voltage to about 2.5 V into a high-impedance load, and 2) to drive a 50 Ω load up to 1 V out to 40 MHz. The 50 Ω output is important when dealing with frequencies above about 10 MHz because of the effects of cable capacitance and length on signal transmission. Standing waves and resonances in the connection circuit will cause uneven frequency response of the signal if there are significant impedance mismatches. For this reason, most RF frequency tests are done with 50 Ω sources and cables, hooked up to 50 Ω loads, effectively minimizing impedance mismatch issues.

Our amplifier consists of several dc coupled stages, followed by a power amplifier capable of driving 50 Ω loads. The output of the AD9851 DDS is directed to a potentiometer to give fine control of the levels. The pot output is connected to an emitter follower stage that provides the level control pot and DDS with a high impedance. The emitter follower is direct coupled to the common emitter voltage gain stage, also dc coupled to another emitter follower, providing some power gain into the last stage. The last stage is NPN/PNP complimentary-symmetry push-pull output stage capable of driving the 50 Ω loads as required, without the need for transformers, up to 40 MHz. This stage is fed from a 18 V dc source to provide required the larger voltage swings. The tricky part of this amplifier is getting the gain/bandwidth necessary to output up to 2 V or more at 40 MHz, which requires careful biasing and the 18 V supply. The amplifier is essentially flat from about 20 Hz to above 20 MHz, as shown in **Figure 13.5,** which is a frequency response plot, or Bode plot, of the RF amplifier alone.

Following the Channel A amplifier is a step attenuator controlled by the ESP32. It consists of two DPDT relays and the associated relay drivers used to select 50 Ω pi attenuators. Attenuation steps of 10, 20, and 30 dB are offered. Additional attenuation steps would require another relay, which is left to the reader as an enhancement. The relay drivers are MOSFETs attached to ESP32 digital pins.

The modulation circuit was touched on earlier and can be operated in either a conventional AM mode or as a suppressed-carrier double-sideband modulator. Two rotary encoders serve to provide Coarse and Fine Frequency control through the ESP32.

Channel B uses an AD9833 DDS, which was selected for low cost and ability to output different functions under microprocessor control using an I2C interface. The AD9833 DDS has exceptionally low distortion and additional

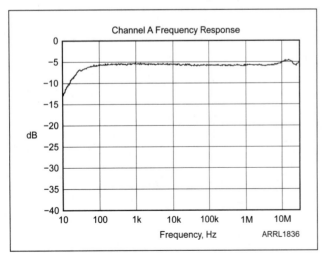

Figure 13.5 — Channel A RF amplifier frequency response.

available waveforms. The DDS output is amplified by a simplified version of the Channel A RF amp, consisting of a two-stage amplifier. Stage 1 provides gain, while a Stage 2 emitter follower presents a low impedance to the Modulator. Gain is approximately 10 dB and the output is both directly available for separate Channel B Function Generator out and as input to the Modulator. Attenuation is by means of pot at the input to the gain stage.

The modulator is a MC1496 balanced modulator, a general purpose wide bandwidth balanced modulator IC that can be configured in several ways. We use two of the available modes: an AM modulator with Channel A providing high-frequency carrier and Channel B providing the modulating frequency. This modulator is characterized by low distortion, giving a faithful representation of modulating input, and allows full control over output levels and percent modulation from 0% to 100%. The second configuration is suppressed carrier double sideband.

SSB modulation is difficult to achieve in a variable frequency circuit configuration, requiring either narrow-band filtration to remove the second sideband or a special quadrature circuit with constant 90 degree phase shift. Both of those were beyond the requirements of this instrument, so we took a simpler approach.

Creating SSB signals with a single modulating frequency is easy — just off-tune the receiver. But, if you want to test the receiver with more complex signals, something additional is needed. For that we simply provide a double sideband, suppressed carrier signal. The SSB receiver rejects the second sideband, allowing the SG to augment the test procedure with complex waveforms to test IF and audio stages for intermodulation distortion and other characteristics.

As in the case of the AM modulation, the input to this circuit can be from internal or external sources. The suppressed carrier null is adjusted from the front panel for minimum carrier level. The suppressed carrier feature may be turned on or off from the front panel.

Near the top of the circuit diagram we see the two log amps, only one of which is initially implemented. One is used to measure the Channel A output for the on-screen display, and the other can be connected as the user would prefer. With some additional coding, the SG could be configured as a simple Bode plotter (without phase). This is left to the reader as an exercise.

## Building the SG

Our SG is built on a custom PC board using surface mount devices (SMDs), so it is rather compact, as shown in **Figure 13.6**. Yeah, we know... "Oh, No! SMDs!" Far too many people are afraid to try working with SMD parts, and that's a shame for many reasons. First, after about 30 minutes of working with SMD parts, you'll be throwing rocks at PCBs that use through-hole components. Second, SMD parts are relatively inexpensive. You can buy an assortment of 200 large (size 1206) resistors for $1.80, including shipping! Third, it reduces the number of pico acres needed for a given PCB, which makes the project even less expensive to build. Fourth, unsoldering an SMD mistake is much easier than correcting a through-hole soldering mistake. And finally, you can look down your nose at people who say they can't work with SMD parts. (Actually, it would be better if you tried to convince them that SMDs are their friend.)

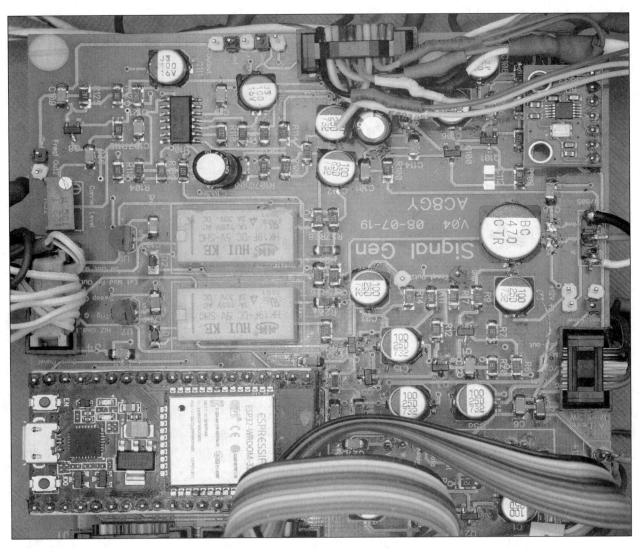

Figure 13.6 — Populated SG PC board.

If you haven't already done so, we hope you'll give SMD components a try.

The power supply linear regulator is mounted separate from the main board, as is the boost converter for convenience and because the PCB was getting rather crowded. Good RF practices are required for this build, since we are handling frequencies up to 40 MHz. The topic of component and wire placement in RF circuits is discussed elsewhere in this book, including Chapter 16, so we won't repeat all of that here.

We used a custom case and followed the procedure that is also described in Chapter 16 on construction. The interior of our case is shown in **Figure 13.7.**

If the builder wants to "roll" his own case and boards, we suggest dividing the circuit up into three parts: digital circuitry, Channel A and Channel B with modulator, and power supplies. This division makes it easier to lay out everything without crowding and to keep inputs and outputs separated from each other and from the digital signals.

Be sure to use adequate power supply bypassing and isolation with lots of caps and ferrite beads. Shielded cable throughout keeps the cross-talk and interference to a minimum.

Figure 13.7 — Interior layout.

## Performance

In this section we show how well the SG performs, by presenting screen shots of various outputs in different modes. These were primarily taken at 1 MHz for Channel A and at 1 kHz modulating frequencies with both internal (Channel B) and external modulation signals.

### Channel A Outputs

Channel A can output sine waves from 20 Hz to 40 MHz. The AD9851 DDS has distortion levels of less than 1%. It does create some non-harmonic "spurs" at higher frequencies, which are effectively filtered out in the DDS module. **Figure 13.8** shows the output at 1 MHz at full 2.4 V RMS level. A frequency spectrum of that output level is shown in **Figure 13.9**. Note that harmonic and non-harmonic content above 1 MHz are more than 45 dB below 2.4 V. Harmonic distortion is less than 1%.

This SG can be used for audio as well as RF applications. The 40 MHz upper range covers the major amateur HF bands.

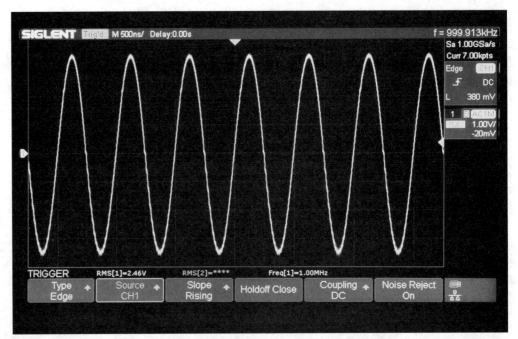

Figure 13.8 — Channel A output, 1 MHz at 2.5 V RMS sine wave.

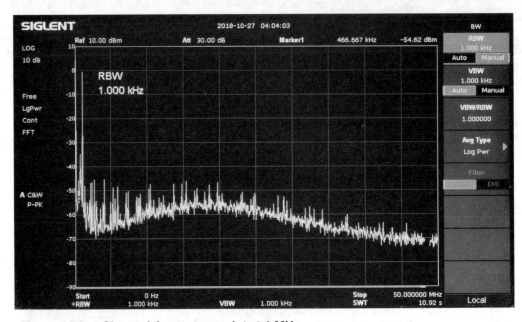

Figure 13.9 — Channel A spectrum plot at 1 MHz.

## Channel B Outputs

Channel B serves two purposes. First, the second channel can output sine, square, and triangle waveforms as a second independent source. The AD9833 DDS is capable of low distortion output up to about 10 MHz. This DDS was selected for its multi-waveform capability and low cost.

The second use is as a variable modulating signal for the Channel A RF signals. An example of a Channel B sine wave is shown in **Figure 13.10**.

In addition to sine waves, the AD9833 is capable of outputting square and triangle waves. These waveforms are shown in **Figure 13.11** and **Figure 13.12**

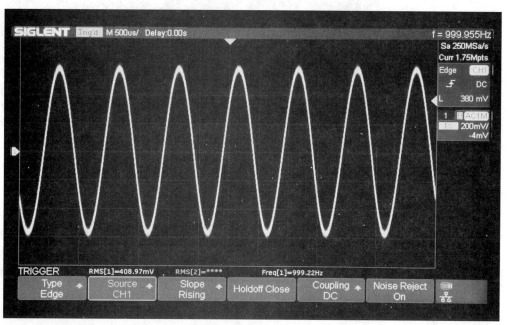

Figure 13.10 — Channel B sine wave 1 MHz,

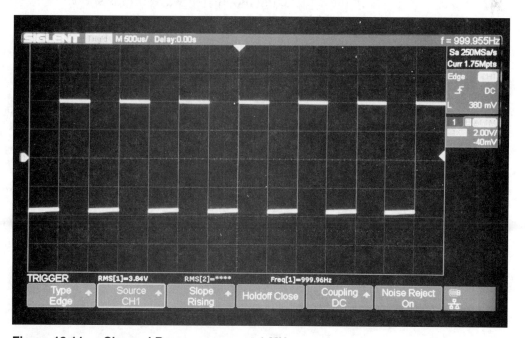

Figure 13.11 — Channel B square wave at 1 MHz.

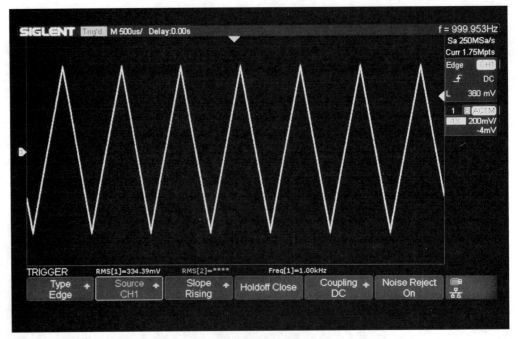

Figure 13.12 — Channel B triangle wave, 330 mV at 1 MHz.

## Modulation Outputs

One of the key features of the SG design is its ability to provide high quality modulated signals — AM, suppressed carrier double sideband, and FM. FM is created in software, whereas the other modulation modes are produced using the MC1496 balanced modulator circuit. The AM modulation results are shown in **Figures 13.13** through **13.18**.

In these plots, the modulated waveform is shown at the top and the modulating waveform at the bottom, both synced to the modulating signal frequency. The modulation percentage is adjusted using front panel controls.

**Figure 13.13** shows sine wave modulation of about 15%. **Figure 13.14** presents modulation of around 50%, and finally **Figure 13.15** has 100% modulation. A frequency spectrum of the 50% modulation signal is shown in **Figure 13.16**, in which the carrier and AM sidebands are clearly visible.

Modulating waveforms are not limited to sine waves. Internal triangle and square waves may be used, as well as external inputs. **Figures 13.17** through **13.20** show additional modulation examples, including triangle waves, square waves, along with external step waves and pulses. External sources could also include voice signals, or other waveforms.

Finally, the suppressed-carrier double-sideband spectrum is shown in **Figure 13.21**. Note the reduced level of the 1 MHz central carrier. The carrier suppression null is adjustable from the front panel.

**Figure 13.13 — AM modulation, low % modulation.**

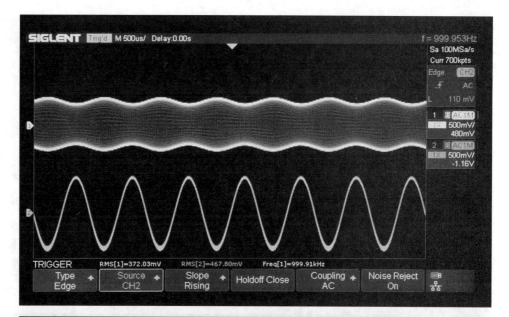

**Figure 13.14 — AM, approximately 50% modulation.**

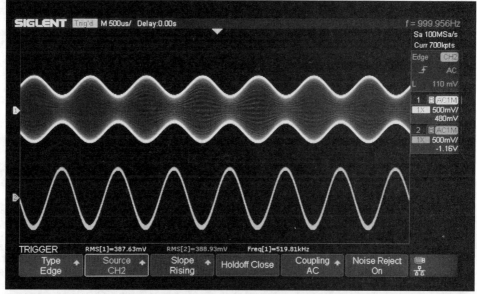

**Figure 13.15 — 100% AM modulation.**

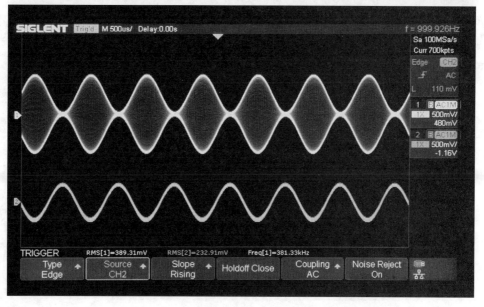

**Signal Generator 13-15**

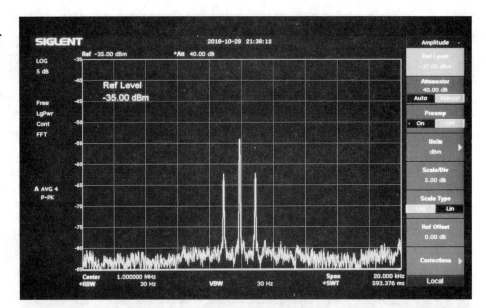

Figure 13.16 — Modulated carrier spectrum.

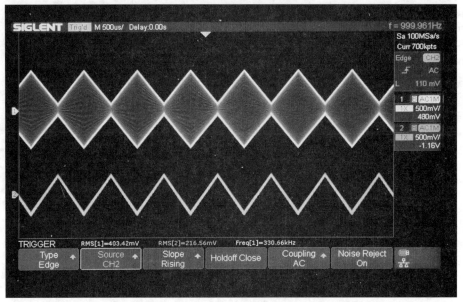

Figure 13.17 — AM modulation — triangle wave modulation.

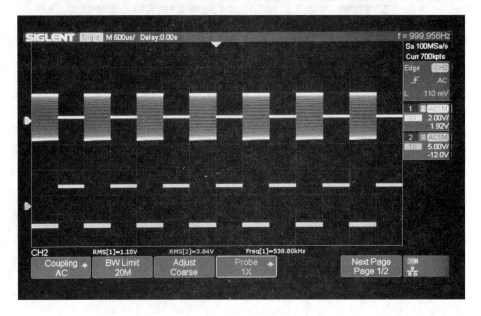

Figure 13.18 — AM modulation — square wave modulation.

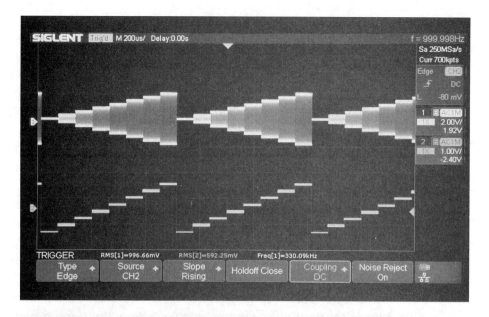

Figure 13.19 — Modulated carrier step wave from external input.

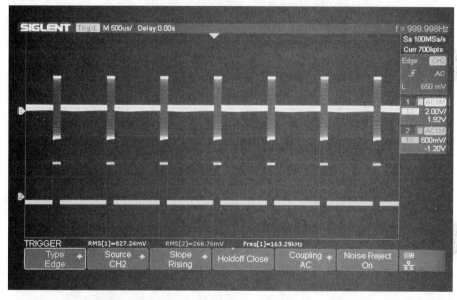

Figure 13.20 — Modulated carrier pulse wave from external input.

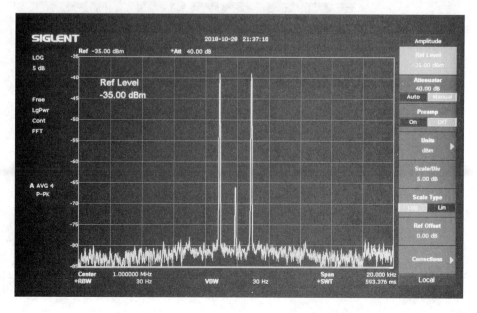

Figure 13.21 — 1 MHz suppressed carrier spectrum.

Signal Generator   13-17

## User Interface

Frequency selection with the SG is accomplished either by using the on-screen keypad or the encoders. To input a frequency, first select a multiplier from the lower section (the buttons marked with the Hz labels). Then touch the digit buttons to input the frequency. Note that your input is "scaled" by your multiplier selection. For example, touching the 1 button after selecting the MHz button results in 1,000,000 being displayed in the frequency window at the top-right corner of **Figure 13.22**. Once the frequency displayed above the keypad is correct, press the Freq A or Freq B button at the lower-left side of the screen. The Clr and Del buttons may be used to edit the inputs. Once the Freq A or Freq B button is pressed, the appropriate DDS begins outputting the signal.

The encoders may now be used to fine tune the frequency. The previously selected multiplier applies, and the coarse and fine encoders may be used to modify the frequency. The Fine tuning encoder is one-tenth the multiplier rate. **Figure 13.23** shows Freq A set to 1 MHz and Channel B set to 1 kHz. To change the multiplier, simply touch a different multiplier button. The encoders may be used to set either Channel A or Channel B frequencies, by touching the A or B buttons in the center next to the meter.

The Channel A output level is shown on the analog style meter and may be altered using the front panel encoder controls or the attenuator. The attenuator is only active on the Channel A on the 50 Ω output.

Other functions are available by selecting the appropriate mode button. The Channel A frequency may be swept using the Sweep function. **Figure 13.24** shows the sweep setting screen. Note that the lower left panel now shows the Sweep parameters. The Start, End, Time, and number of Points are set using the on-screen keypad.

Finally, frequency modulation is available using the FM button. Both modulating frequency and span may be selected. In **Figure 13.25**, the 12K span option has been selected, as indicated by the light background on the 12K Span button at the bottom of the screen.

Figure 13.22 — Channel A, 1 MHz output selection.

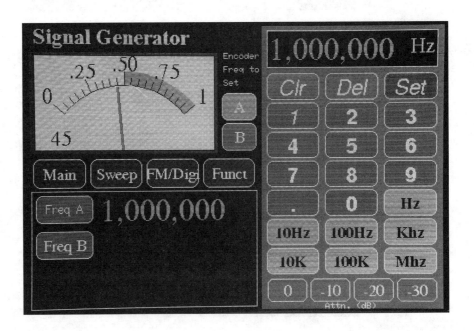

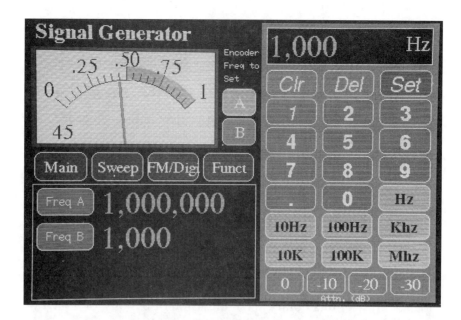

**Figure 13.23 —** Both Channel A and Channel B selected.

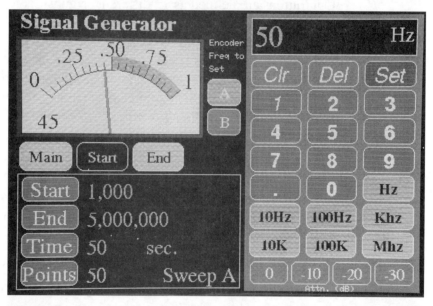

**Figure 13.24 —** Sweep function.

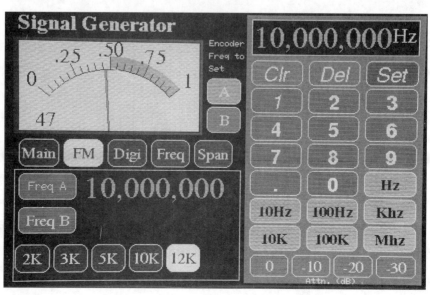

**Figure 13.25 —** FM modulation setting screen.

Signal Generator    13-19

A placeholder "Digi" button has been provided for user-supplied digital mode modulation code activation. We thought we would give the builder some motivation to enhance the SG capabilities.

## Signal Generator Software

At this stage in your project building you have encountered most of the relevant software facets we have to describe, but there are a couple of interesting routines in the Signal Generator that merit description.

The SG code is divided into 12 different files, based on function. Of these, the Keypad.cpp, Modulation.cpp, and Meter.cpp code are unique to this project. We also briefly discuss DDS and screen calibration.

First a bit about the graphical UI. We are using a 3.5-inch TFT LCD display with ILI9488 controller, with a 4-wire SPI interface. The graphics library specific to the combination of the ESP32 and ILI9488, in this case called TFT_esp.h, found at **github.com/Bodmer/TFT_eSPI**. One of the nice features of this library is the inclusion of buttons, of which we have made extensive use. Our buttons are defined as 12 different array variables in the .ino file. Using arrays to create and interrogate the buttons saves lots of tedious code writing, as we will demonstrate with the keypad. The button parameters such as labels are defined as character arrays, again for convenience. Key colors are also similarly defined as arrays. Using arrays in this fashion gives the user lots of flexibility and makes changes to the parameters very easy, simply by modifying the parameter arrays, rather than tracking down each instance of a specific key color, for example.

## Keypad

The onscreen keypad is drawn by the *drawKeypad()* function, called in *setup()*. There are 21 buttons in the Keypad, which would take 42 statements to initialize and draw if done individually. Instead, by using the array variables, we can draw the entire keypad with just eight statements. Note the use of the constants KEYPADXOFFSET, KEY_X. KEY_Y, KEY_H, KEY_W, and KEY_SPACING_X, as well as the arrays for color and key labels.

```
for (uint8_t row = 0; row < 7; row++) {
  for (uint8_t col = 0; col < 3; col++) {
    uint8_t b = col + row * 3;
    if (b < 3 and b < 14) tft.setFreeFont(LABEL1_FONT);
    else if (b > 3 and b < 14)tft.setFreeFont(LABEL2_FONT);
    else if (b >= 14)tft.setFreeFont(FSB9);
    key[b].initButton(&tft, KEYPADXOFFSET - 3 + KEY_X + col * (KEY_W + KEY_SPACING_X + 3),
        KEYPADYOFFSET + KEY_Y + row * (KEY_H + KEY_SPACING_Y - 3) - 12,
        KEY_W + 6, KEY_H, TFT_WHITE, keyColor1[b], keyColor2[b],
        keyLabel[b], KEY_TEXTSIZE);// x, y, w, h, outline, fill, text
    key[b].drawButton();
  }
}
```

Two nested *for* loops are used, one for the horizontal buttons, the other for the vertical. One benefit of this approach is the ability to resize and relocate the entire keypad by changing a couple of constants. It also makes reuse of the code with other displays very simple. Clearly, a considerable timesaving approach.

In a similar fashion, the keys are read when pushed by using a simple polling routine that determines when a key is pushed and output the key value. Again, use is made of the keypad array variables to output the value in just a few statements:

```
for (uint8_t b = 0; b < 21; b++) {
   if (pressed && key[b].contains(t_x, t_y)) {
     key[b].press(true);  // tell the button it is pressed
   } else {
     key[b].press(false);  // tell the button it is NOT pressed
   }
 }
 // Check if any key has changed state
 for (uint8_t b = 0; b < 21; b++) {
   if (b < 3 and b < 14) tft.setFreeFont(LABEL1_FONT);
   else if (b > 3 and b < 14)tft.setFreeFont(LABEL2_FONT);
   else if (b >= 14)tft.setFreeFont(FSB9);
   if (key[b].justReleased()) key[b].drawButton();      // draw normal
   if (key[b].justPressed()) {
     key[b].drawButton(true);  // draw invert
     switch (b) {
       case 0:
         numberIndex = 0; // Reset index to 0
         numberBuffer[numberIndex] = 0; // Place null in buffer
         freqOut2 = 0;
         displayKbdOut();
         break;
       case 1:
         numberBuffer[numberIndex] = 0;
         if (numberIndex > 0) {
           numberIndex--;
           numberBuffer[numberIndex] = 0;//' ';
           displayKbdOut();
```

The function *displaykbdout()* puts the button press values in the desired place for screen display. Using the Button library functions and array variables greatly simplifies the code and makes it much more portable.

## Modulation

AM and suppressed carrier double sideband modulation functions are accomplished primarily in hardware. FM modulation and any digital modes the user might add, on the other hand, use a software approach.

The AD9851 DDS output frequency is set using a function *sendfrequency(freqFM)*, that writes any desired frequency to the DDS. Both the

ESP32 and the DDS are sufficiently fast to be able to change the DDS frequency rapidly enough for FM. The process is to calculate a sine wave, which alters the DDS frequency according to the sine value. A delay function is used to set the period of this modulating sine wave, and its amplitude sets the span or deviation amount of the frequency modulation around the carrier frequency. While sine waves are used here, other routines could be added to create different FM waveforms.

The code snippet below shows the calculation of the modulating sine wave. Note that each cycle of the sine wave contains 32 points, ensuring a reasonably low distortion sine result. The variable *fmDelay* sets the duration of each point, effectively setting the modulating frequency.

```
for (int j = 0; j < 32; j++) {
        float sinArgument = (6.28 / 32) * j;
        sineWave2[j] = sin(sinArgument);
    }
    //Serial.print("span=  "); Serial.println(span);
    //Serial.print("fmDelay=  "); Serial.println(fmDelay);
    while (FM_FlagOff != 1) {
        Serial.print("FM_FlagOff= ");Serial.println(FM_FlagOff);
      for (int i = 0; i <= 31; i++) {
        freqFM = freqOutA + span * (sineWave2[i]);
        sendFrequency(freqFM);
        delayMicroseconds(fmDelay);
      }
    }
```

## Meter

One of the unique graphical features of the SG is the on-screen analog meter. Digital outputs are very convenient, but don't we all fondly remember the old analog meters that graced our vintage equipment? Well, we have provided a digital version of those old meters. Actually, creating an on-screen analog meter isn't too difficult, but takes a fair amount of coding. The process has two parts: first drawing the main meter elements such as the face and numbers, and second, drawing the moving needle in real time. We won't go into great detail on this code but suggest that you study the documented code. The tricky part of the code concerns the needle, which must be refreshed at about 10 ms intervals, by first erasing the old position and then drawing the new. Because the needle moves in an arc, transcendental *sin()* and *cos()* functions are required to map the analog values to be displayed to screen positions. Nothing profound, just a lot of detail.

## Sweep

The frequency Sweep mode is accomplished simply by determining the start, end, and frequency increments and then using the *sendfrequency(freqFM)* function to output the frequencies one at a time for the desired duration. Notice how we reuse functions to carry out quite different tasks such as frequency modulation and frequency sweep. Creating your own set of functions allows you to rapidly and reliably build new software based on tried-and-true routines you and others have developed.

## Frequency Calibration

The main DDS frequency calibration is set by a constant defined in the .h file toward the top of the file:

```
#define DDS_CLOCK 180000000 // DDS
```

If the frequency output of your unit is found to be slightly off, the DDS_CLOCK constant may be varied to bring the output into specification.

## Touch Calibration

Touch screens must be calibrated before use. For this purpose, we have included a function, *touch_calibrate()*, that is run the first time a new processor is deployed. Just follow the on-screen directions. When complete, the routine saves the screen calibration parameters to EEPROM for use any time the system is restarted. If recalibration is required, the system file SPIFFS must be deleted, and the system restarted to initiate calibration.

## Conclusion

The SG is one of the more complex projects in this book, but one that is well worth the effort expended to build it. It has a number of features that simply are not available on moderately priced commercial SGs. That said, this is one project where we do urge you to get the PCB for the project. It simplifies building the SG significantly.

# CHAPTER 14

# Double-Double Magnetic Loop: A Luggable Portable HF Antenna

## Inspiration

It appears that portable HF antennas, especially for 40 meters and lower frequency bands, are always a compromise. Either they are too big, only work at low power levels, or require trees or other tall supports to work well. Because of their small size, magnetic loops (ML from now on) should be good candidates, but commercially available units are either for QRP power levels only or are too big. Usually 40 meters requires a 2-meter diameter or larger loop — way too big to fit in my car. Many hams who face homeowner association (HOA) restrictions are also interested in unobtrusive, yet effective, HF antennas. The question: Is there a way to get reasonable power capability (up to 100 W), but in an unobtrusive size that fits in a car trunk, that would be easy to set up, and that operates with reasonable efficiency? Those are the design goals we have set. A search of the literature indicated a definite maybe.

Our specific design requirements were for a 1-meter diameter (or smaller) ML capable of at least 100 W input, with reasonable efficiency at 40 meters, and the ability to work on the 40, 30 and 20 meter bands using one variable capacitor. The loop would also have to be portable — or at least luggable — and not too heavy, and it would need to be constructed using commonly-available materials. Lots of constraints! To jump ahead a bit, the final loop we came up with, the Double-Double Magnetic Loop (DD-ML) is shown in **Figure 14.1**. The journey that took us to that final design is next.

**Figure 14.1** — Double-Double Magnetic Loop (DD-ML).

## Mag Loops (ML)

Just to refresh everyone's memory, magnetic loop (ML) antennas consist of three (and optionally four) basic parts.

### 1. The Loop

Loops are commonly made of a conductive material such as coaxial cable, copper or aluminum tubing, or in some cases, microwave waveguides. The shape of the loop can be round, octagonal, square, or some other approximately circular configuration, although theory indicates that the circular shape is most efficient.

### 2. Tuning Capacitor

Because the ML is to be a resonant circuit, a capacitor is needed to tune the system to the desired frequency. High voltages are involved in such a tuned circuit at resonance when driven to moderately high powers. A 100 W input to the ML, for example, can produce in excess of 5 kV of circulating RF with high amperage at resonance. This high voltage is created by the very high Q factor of a resonant system that has low resistive losses. A ½-inch (1.25 cm) copper tube that is 3 feet (1 meter) in diameter has a resistance in the milliohm range at RF frequencies. If the capacitor also has low losses, Q factors of greater than 500 are not uncommon. The peak input RF voltage of a 100 W signal can reach 100 V, which, when amplified by the high Q resonant circuit yields kilovolt potentials across the capacitor. (By the way, this does *not* mean the power is amplified by the Q factor as well. The circulating currents are not in phase with the voltages everywhere, so power-in still equals power-out.) However, not all of this RF power going into the antenna is radiated, since some of the energy is

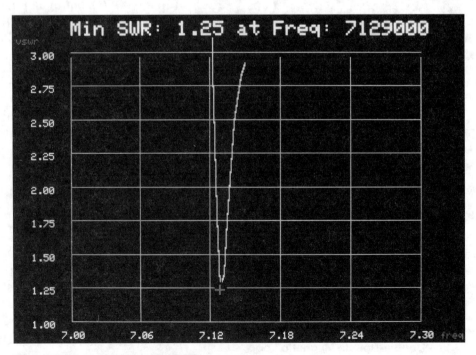

Figure 14.2 — 40-meter ML SWR.

dissipated as heat in the ML itself or in its surroundings, hence concerns with efficiency.

**Figure 14.2** shows a typical SWR plot of a mag loop at 40 meters. Note the very narrow curve near resonance. The high voltages across the capacitor mean that special attention must be paid to the capacitor plate spacing to prevent arcing between the plates. The *ARRL Handbook* gives some spacing guidelines for applied RF voltages. **Figure 14.3** shows voltage tolerance vs. plate spacing. For 100 W on 40 or 20 meters, a spacing of about 3.5 mm is adequate. *High voltages also mean care should be taken when placing the ML so no person can come near it during operation.*

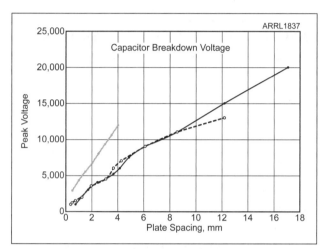

Figure 14.3 — Recommended air capacitor plate spacing (based on information from the 2012 *ARRL Handbook*).

The best capacitors to use for mag loops are old butterfly capacitors, but they are extremely rare on the used market in the capacitance and spacing ranges needed (we've looked). Second choice is a vacuum variable capacitor (VVC), but they are expensive, even used. See **Figure 14.4**.

Next are the split-stator air variable capacitors (AVC) usually salvaged from old tube-type transmitters. These are readily available at flea markets at reasonable prices. These AVCs are hooked up by connecting to the stators only, which are electrically coupled through the rotating plates, fixed to a common shaft. The capacitance is cut in half, but the voltage rating is approximately doubled. A bonus of this arrange-

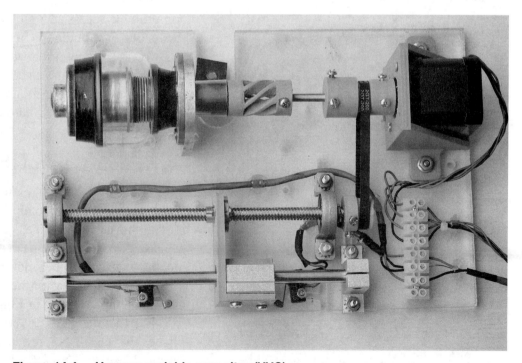

Figure 14.4 — Vacuum variable capacitor (VVC).

**Double-Double Magnetic Loop: A Luggable Portable HF Antenna    14-3**

**Figure 14.5 — Split stator air variable capacitor (AVC).**

**Figure 14.6 — Faraday coupling loop.**

ment is that there are no rotating connections to add losses. The wiper at each end of the rotor is a sliding connection, which can add resistance and thus losses. By not having to connect through any sliding connections, we decrease losses and increase efficiency. If you can find such a unit in good condition, the capacitance of each section should be approximately 20 pF to 200 pF. Hooked in series, the values are then about 10 pF to 100 pF. (Look out for broken ceramic parts.) **Figure 14.5** shows a typical AVC split stator capacitor. The least desirable AVCs are the single rotor/stator type, primarily because of the resistive losses from sliding contacts between the frame and rotor.

## 3. RF Coupling Loop

The third major component is the RF coupling element, which transforms the inherent high impedance of the loop to 50 Ω for connection to the transmitter. Lots of information is available in the literature on various coupling configurations. We experimented with several of them, and selected the tried and true Faraday loop for our design for three reasons: 1) it is easy to construct from copper tubing, 2) it works on multiple bands, and 3) It can easily be adjusted to vary the coupling to give the best compromise across several bands. **Figure 14.6** shows our Faraday loop (FL) in place on the DD-ML. A Faraday loop/ML combination can be considered an impedance matching transformer. Notice that our mounting system for the FL is such that we can rotate it around the vertical support axis.

## 4. Remote Control Unit (Optional)

The fourth (optional) element of the system is a mechanism for remotely controlling and moving the capacitor vanes or vacuum variable cap's shaft position. We used a NEMA 17, a simple, inexpensive stepper motor and gear assembly to move the capacitor for the tests described here. Our test setup is shown in **Figure 14.7**.

Note that the gear reduction drive accomplishes two tasks: 1) it reduces the stepper motor increment by about 2.5:1, and 2) it electri-

Figure 14.7 — Stepper motor with belt reduction drive.

cally isolates the stepper motor from the capacitor's high voltage. If a reduction drive is not used, a high-voltage, non-conducting, coupler must be used.

All of the parts for mounting the NEMA 17 motor and the gear assembly were printed on a 3D printer, as were the two gears. The drive belt is a standard small rubber gear belt. The gear teeth and belt minimize slippage when the NEMA motor is active. The stepper motor has 200 steps per revolution, giving 1.8° of arc for each step. The stepper motor controller can theoretically provide up to $\frac{1}{32}$ of a step, although we are only using half that granularity.

## Design

Various ML design programs indicate that a 1-meter diameter loop at 40 meter would have an efficiency of about 9%, which is 10 dB down from an ideal antenna. That loss is a lot of wasted power, all going to heat up the surroundings. (For reference, one S-unit is approximately 6 dB.)

Efficiency is defined as a ratio involving radiation resistance (Rrad) and loss (non-radiating) resistance (Rloss), or:

Eff % = (Rrad / (Rrad + Rloss) × 100.

We can use the Mag Loop Design Calculator program from Simone Mannini, IW5EDI (**www.iw5edi.com/software/magnetic-loop-calculator**) to characterize the efficiency of a single loop. For a copper loop running 100 W, **Table 14.1** shows the results for loops that are 10 feet (≈3 meters) and 20 feet (≈6 meters) in circumference.

Clearly the 20-foot circumference loop has better efficiency (almost 90%), but the diameter, at 6.1 feet, is much larger than is convenient for portability in

### Table 14.1
**Comparative Statistics for Loops with Diameters of 10 and 20 Feet at 40 and 20 Meters**

|  | 10-Foot Diameter | | 20-Foot Diameter | |
| --- | --- | --- | --- | --- |
| Operating frequency (MHz) | 7 | 14 | 7 | 14 |
| Loop circumference (feet) | 10 | 10 | 20 | 20 |
| Loop circumference (meters) | 3.05 | 3.05 | 6.1 | 6.1 |
| Conductor diameter (inches) | 0.5 | 0.5 | 0.5 | 0.5 |
| Bandwidth (kHz) | 6.1 | 16.7 | 10.8 | 84.6 |
| Capacitor value (pF) | 164 | 34.1 | 79.7 | 9.3 |
| Capacitor voltage (kV) | 3.9 | 4.7 | 4.0 | 2.8 |
| Inductance (μH) | 2.995 | 2.995 | 5.512 | 5.512 |
| Loop Q value | 1139 | 840 | 646 | 165.5 |
| Radiation resistance (Ω) | 0.005 | 0.082 | 0.082 | 1.316 |
| Resistive loss (Ω) | 0.053 | 0.075 | 0.105 | 0.149 |
| Efficiency (%) | 8.9 | 52.5 | 43.8 | 89.9 |

an ordinary car. The best the 10-foot loop can manage is well under 50% for 20 meters and a dismal 9% on 40 meters. Not good.

The problem with the smaller loop is the very low radiation resistance at 7 MHz, which, coupled with losses in the antenna and ground effects, reduces the amount of power radiated as useful signal. Over 90% of the signal disappears as heat.

Glenn Gardner, AA8C, in an article titled "A More Efficient Magnetic Loop Antenna," suggested that creating multiple parallel loops increases the radiation resistance proportional to the number of loops. The radiation resistance is given by

$$Rrad = [31171 \times ((N \times A^2) / \lambda^4)]$$

where N is the number of loops, ignoring the capacitive coupling between the loops, A is the loop area, and $\lambda$ is the wavelength. Coupling is minimized by spacing loops far enough apart. This means that increasing the number of turns to two doubles the efficiency by increasing the radiation resistance.

Now, with two 3-foot loops, we have efficiencies of 16% (–8 dB) at 7 MHz and more than 90% (–0.5 dB) at 14 MHz. Better, but not yet good enough. Note that increasing the number of turns above two has diminishing returns, probably because of coupling effects between the loops. We also limited our experiments to two parallel turns for ease of construction and size.

It appears that the efficiency at 7 MHz is thus improved from 8.9% for a single 3-foot diameter loop to 16% for a 3-foot diameter two-turn loop. Progress, but are there any other means by which the efficiency can be improved? How about reducing the non-radiating losses or further increasing the radiation resistance?

Other references indicate that folding the larger 6-foot loop into a multi-turn configuration improves the efficiency over the single, smaller loop. This loop folding creates two smaller 3-foot diameter loops in series if we use two turns. Using the Multi-turn Magnetic Loop calculator from Jakub Jalowiczor (**comtech.vsb.cz/mlacalc/**), we get the results shown in **Table 14.2**.

### Table 14.2
**Multi-turn ML Statistics**

| Frequency (MHz) | 7 | 14 |
| --- | --- | --- |
| Diameter (meters) | 1 | 1 |
| Number of turns | 2 | 2 |
| Pipe diameter (inches) | 0.5 | 0.5 |
| Spacing (inches) | 7 | 7 |
| Radiation resistance (Ω) | 0.157 | 0.3 |
| Loss resistance (Ω) | 0.983 | 0.1 |
| Efficiency (%) | 13.8 | 64.2 |
|  | (–8.6 dB) | (–1.9 dB) |

**Table 14.3**

**Approximate Double-Double Performance Expectations**

|  | 7 MHz | | 14 MHz | |
| --- | --- | --- | --- | --- |
|  | Efficiency* | ML Power Loss (dB)* | Efficiency* | ML Power Loss (dB)* |
| Single Loop | 8.9% | −10.5 | 52.5% | −2.8 |
| 2 Series Loops | 13.8% | −8.6 | 75% | −1.5 |
| 2 Parallel Loops | 16.0% | −8.0 | 82% | −0.86 |
| Double-Double | 38.6% | −4.1 | 90% | −0.41 |

*Power Loss compared to 100% efficient antenna

Suppose we combine the two configurations into a loop consisting of two series-coupled loops of two parallel loops each — four loops total (our Double-Double). If we assume the parallel-series combination has increased radiation resistance because of the parallel combination proportional to the number of loops, using the **comtech.vsb.cz** calculator, we get the efficiencies and power reductions (compared to a 100% efficient antenna) shown in **Table 14.3**. (By the way, these were untested assumptions at the time we were designing our DD-ML. Our empirical tests are reported later.)

As yet another way to approach the two-turn loop efficiency calculation, Steve Adler, VK5SFA (**members.iinet.net.au/~sadler@netspace.net.au/tmla.html**) suggests in his article that by going to a two-loop series-connected configuration, the following is obtained:

- Radiation resistance increases proportional to N squared, or for N = 2 Rrad × 2 = 4 × Rrad
- For a 3-foot diameter single loop, Rrad = 0.005,
- Two turn loop, Rrad = 0.005 × 4 = 0.02
- Efficiency = 0.02 / (0.02 + 0.053) × 100 = 27.4%
- If we use double parallel loops, Efficiency = 0.02 × 2 / (0.02 × 2 + 0.053) × 100 = 43.0%, which is −3.7 dB compared to ideal.

Note that these approximations assume the Rloss is unchanged. In practice, the extra loops should reduce the skin effect losses as well. If we assume the parallel combination is equivalent to using 0.75-inch pipe, the Rloss is reduced by about 0.67.

- Efficiency = (0.04 / (0.04 + 0.053 × 0.67)) × 100 = 53%

a 2.75 dB reduction from ideal. Note these calculations are approximate because effects such as proximity coupling between the loops have been ignored or minimized. We also make big assumptions about combining the series and parallel loops. The objective in doing the calculations is to give guidance in designing a practical small loop for 40 and 20 meters. (Operating tests verified that this design approach was reasonable.)

The data presented in Table 14.3 suggest the results are in the ballpark for our design objectives. Adding more parallel loops might improve the efficiency somewhat, but at the expense of weight, complexity, and cost. AA8C also suggests that adding loops in parallel over two has diminishing returns.

You might ask: Well, if adding one series turn improves the efficiency, how

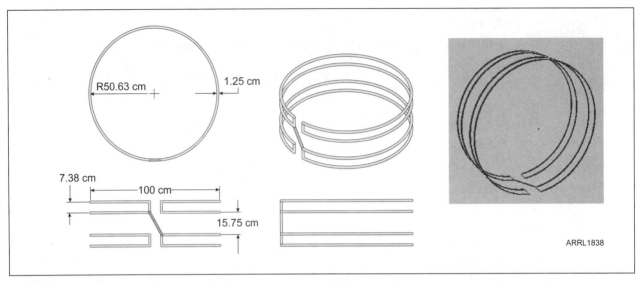

Figure 14.8 — Proposed Double-Double Mag Loop (DD-ML) dimensions (in centimeters).

Figure 14.9 — Tubing Bender.

about more than one? The answer is yes it will improve, but unfortunately the loop inductance also increases, meaning you need an even lower capacitance to tune to the top of the 20-meter band. Since we are already at the practical lower capacitance limit with AVCs at around 15 pF, this is not a practical way to improve efficiency. Other ways to improve things even more would be to increase the pipe diameter to 0.75 inch, but the efficiency for the two-loop series combination only goes from 13.8% to about 18% more at double the cost and about double the weight. We elected to live with a slightly lower efficiency for more convenience and lower cost. The proposed loop dimensions in centimeters are shown in **Figure 14.8**.

## Building the Loop

Once we had some design parameters, we moved on to build a prototype. We elected to construct the first Double-Double Mag Loop from readily available plumbing materials — ½-inch type M copper pipe for the loop and PVC pipe for the supports. Copper pipe is available in both soft and rigid types in ½ inch, ¾ inch, and larger sizes. We

Figure 14.10 — Mag loop PVC pipe structure.

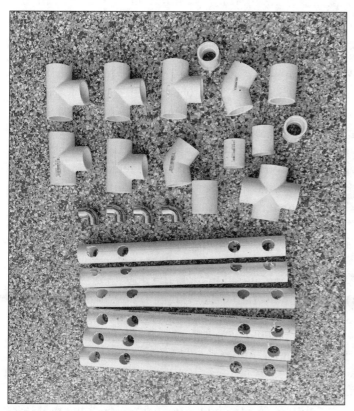

Figure 14.11 — PVC fittings and pipe supports.

first tried soft copper, but getting a nice round circle proved to be a real challenge, so we elected to try the rigid pipe. To bend the pipe into a circle we got access to an inexpensive tubing bender from Harbor Freight (see **Figure 14.9**). Others who have done this suggested annealing the pipe first to soften it, but this just got us back to a soft bendable tube, so we ventured to just try the rigid pipe. After some trial and error, we got nice round circles, as shown in Figure 14.9. Our Double-Double design requires four loops, so we used four pipes, each 10 feet long.

We also needed to consider a structure to hold it all together. Our prototype structure was made from ¾-inch and 1-inch PVC schedule-40 pipe, because it is readily available, is easy to cut and join, and is relatively inexpensive. It also has dielectric properties that won't increase the residual capacitance too much. **Figure 14.10** shows the structure and **Figure 14.11** shows the fittings and pipe supports.

The copper pipe fittings were all carefully soldered using good plumbing practice for copper pipes. We used regular 60/40 solder, but slightly better results might come from silver solder. We also made a simple tripod from the PVC pipe to hold everything upright for testing, as shown in **Figure 14.12**.

The Faraday loop is attached to the vertical support in such a way as to allow rotation of the coupling loop to tune the system for best SWR. We elected to put the small loop at the top and the capacitor and stepper motor at the bottom for weight balance. The offsets in the top and bottom PVC structure are to allow for pipe cross-connection at the bottom and full rotation of the Faraday loop at the top. The whole assembly is shown in **Figure 14.13**.

After some experimentation, the Faraday loop diameter was set to about 11 inches. This was a compromise that gave very low SWR readings at all frequencies of interest.

Once the capacitor was attached, reso-

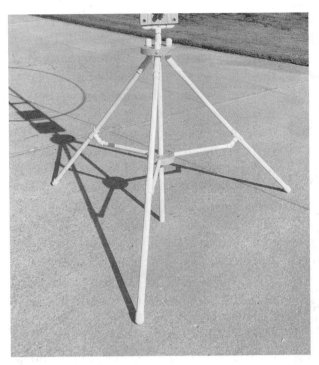

Figure 14.12 — Tripod.

Figure 14.13 — ML assembly.

Figure 14.14 — 3D printed structure parts assembled.

nance tests showed that the inductance of the loop assembly was a bit too high to be able to cover the entire 20-meter band with our capacitor. The fix for this limitation was to shorten the loop circumference by 4 inches to reduce the inductance. This allowed the SWR to be minimized across both the 20 and 40 meter bands, as well as 30 meters in between. The Faraday loop rotational position was used to obtain a compromise for which the minimum SWR was less than 1.2 across both the 20 and 40 meter bands.

We measured the DD-ML Q-factor using an SWR bridge with our Siglent SSA3021x spectrum analyzer. We found that with a ground plane of eight radials, each 10 feet long, the Q was about 590, which indicates our resistive losses in the piping and capacitor are very low.

After the unit was all assembled, we decided the configuration looked pretty good, so 3D printed versions of the main support elements were designed and fabricated, as shown in **Figure 14.14**. The whole 3D printing process took about 50 hours of print time, but made the assembly very easy, much sturdier, and nicer looking.

After the capacitor and stepper motor assembly were attached, it was time for some real tests.

## Initial On-Air Tests

The first objective of the testing was to see if the unit could really accept 100 W RF for a sustained period. Indeed, we ran the power up to 200 W without arcing, but conservatively rate the ML at 100 W. The limiting factor is the capacitor's voltage rating.

One major concern was whether the increased losses at 7 MHz would cause resistive heating in any of the elements. The copper pipe was all carefully soldered together and the two mechanical connections to the capacitor were nice and tight, but this aspect needed to be checked. On an unused 40-meter frequency (we checked carefully) a 100 W CW signal was broadcast for 10 minutes continuously. Using an infrared temperature gun, no excess heating was observed. One of the mechanical connections warmed a couple of degrees, which certainly was not a problem. That connection was tightened again. Subsequent tests showed that the problem had been fixed.

Figure 14.15 — Field Day setup.

## ARRL Field Day

One of our initial goals was to try the new "portable" configuration during the June 2019 ARRL Field Day. We set up a Yaesu FT-950 transceiver with the Double-Double ML and operated phone on 40 and 20 meters for a portion of the official Field Day time period during the daylight hours. (Our goal was not a high QSO count, so we used a hunt-and-pounce approach to see if

Figure 14.16 — 2019 Field Day contacts with mag loop.

we could make contacts while also measuring and recording our experience.) The antenna setup that was used is shown in **Figure 14.15**. Unfortunately, we oriented the antenna to favor north-south directions, so far western US was not in the strongest lobe. We made about 50 contacts, mainly in the eastern half of the US, from our Cincinnati, Ohio, location. Band conditions appeared to be rather poor during our operating time, but we got good signal reports overall... even a "great signal" report from one contact! A map of our contacts is shown in **Figure 14.16**. The farthest contact was to Colorado.

## Digital Mode Tests

We felt a good way to test our new antenna was to compare it to an existing antenna with real quantitative signal reports from reverse beacons like WSPR or by using FT8 signal reports. The comparison antenna is an 80 - 10 meter, multi-band end-fed from MyAntennas, which is resonant on both 40 and 20 meters with relatively low native SWR. The end-fed antenna is situated in an East-West orientation about 40 to 50 feet above the ground. It is generally in open space, except for the attachment to the peak of the roof. A good ground system consisting of several 8-foot ground rods completes the setup.

### FT8 Reports

During our club's special event commemorating the first man on the moon, we ran FT8 for about four hours total, spread over a Saturday and Sunday. We made well over 150 contacts, with an average signal report of –5.7 dB. Subjectively, it appeared that the ML was performing very nicely, just like our FD experience.

**Figure 14.17** shows the distribution of our FT8 contacts from PSK Reporter. **Figure 14.18** presents the FT8 signal reports by distance.

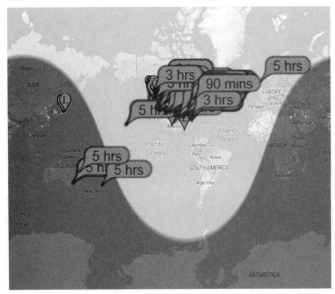

Figure 14.17 — FT8 signal reports using DD-ML.

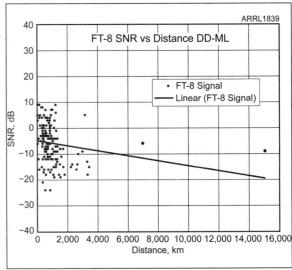

Figure 14.18 — FT8 signal reports by distance.

## WSPR Tests

We also did an extensive series of moderate power WSPR tests over several days, alternating between the DD-ML, end-fed half-wave, and a 6-foot single-loop ML. Each antenna was run for a full 24 hour series of tests on succeeding days. It would have been better to run the comparison tests simultaneously, but our equipment setup precluded this.

We ended up with lots of data from many reporting stations all over the globe (more than 10,000 signal reports over four days.) These are only from stations responding to our CQ. The farthest contact was about 18,000 km, while most were within North America. A sample WSPR map is shown in **Figure 14.19**. Most of the activity is within the continental US, but some nice DX was reported as well.

The data were analyzed by plotting reported SNR versus distance for each antenna. For each time period, two plots are shown, one for North America and the other including the full distance contacts. A trend line for each antenna is also shown. **Figure 14.20** is a comparison of all 40-meter reports out to 18,000 km. The DD-ML is compared to the end-fed and a full size, 6-foot diameter single loop ML as well. **Figure 14.21** shows the North America results. Finally,

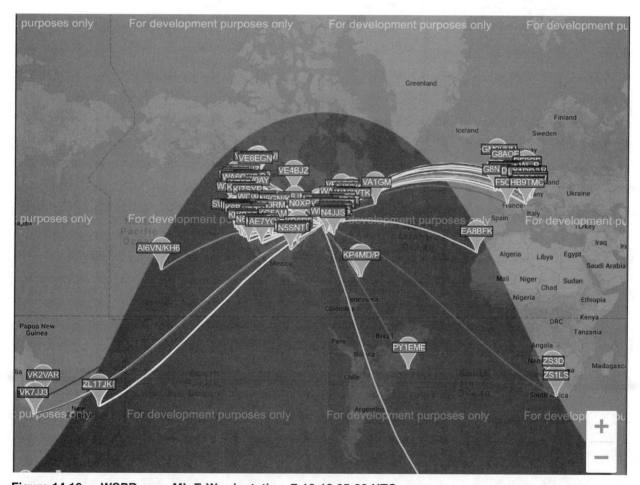

Figure 14.19 — WSPR map, ML E-W orientation, 7-13-19 05:22 UTC.

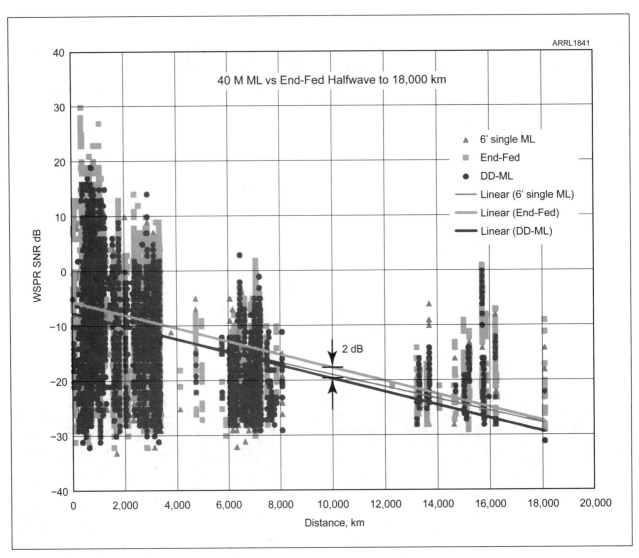

Figure 14.20 — WSPR SNR, 40-meter DD-ML vs end-fed, all distances.

### Table 14.4
### 40 Meter Results Summary

|  | All Distances (dB) | NA only (dB) | DX only (dB) |
|---|---|---|---|
| End-Fed | −9.70811 | −7.11517 | −18.217 |
| 6-ft Single Loop ML | −11.5137 | −9.57447 | −18.8155 |
| DD-ML E-W | −11.7015 | −9.01532 | −18.8645 |
| | | | |
| Differences: | | | |
| DD ML vs. 6-ft ML | −0.18772 | 0.55914 | −0.0489 |
| DD ML vs. End Fed | −1.99336 | −1.90016 | −0.64647 |

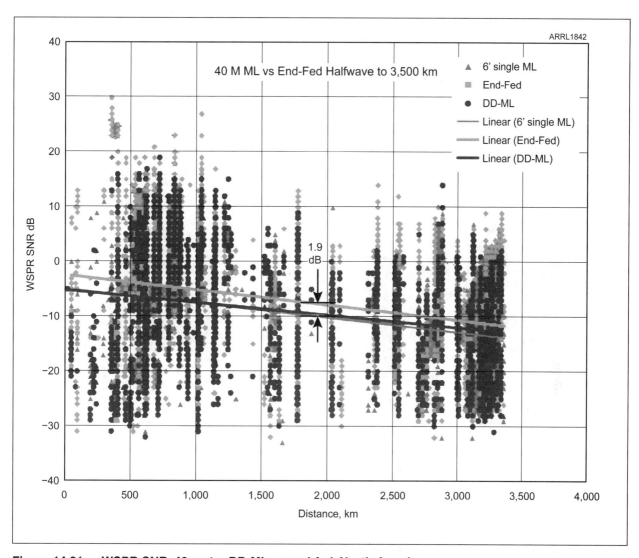

Figure 14.21 — WSPR SNR, 40-meter DD-ML vs end-fed, North America.

### Table 14.5
### 20 Meter Results Summary

|  | All Distances (dB) | NA only (dB) | DX only (dB) |
|---|---|---|---|
| End-Fed | −11.6053 | −9.68693 | −18.8666 |
| DD-ML E-W | −12.8327 | −12.0476 | −19.1659 |
| Differences: | | | |
| DD-ML vs. End Fed | −1.2274 | −2.36066 | −0.29931 |

**Double-Double Magnetic Loop: A Luggable Portable HF Antenna**

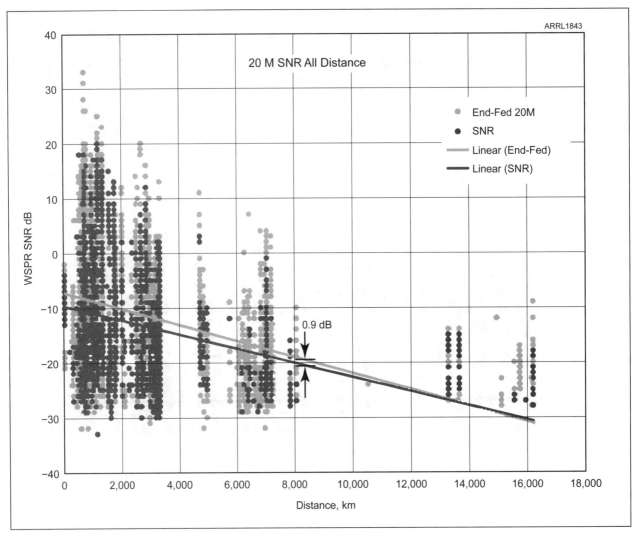

Figure 14.22 — WSPR SNR, 20-meter DD-ML vs end-fed all, distances.

## Table 14.6
### Differences Calculated vs. Measurement

|  | *40 Meters* | *20 Meters* |
|---|---|---|
| Calculated Difference (Mag Loop vs. Ideal) dB using IW5EDI | −4.1 dB | −0.41 dB |
| Calculated Difference (Mag Loop vs. Ideal) dB using VK5SFA | −2.75 dB | N/A |
| Measured Average Difference | | |
| North America (Mag Loop- End-Fed) | −1.9 dB | −2.3 dB |
| All Measured Difference (Mag Loop vs. End-Fed) | −2 dB | −1.2 dB |

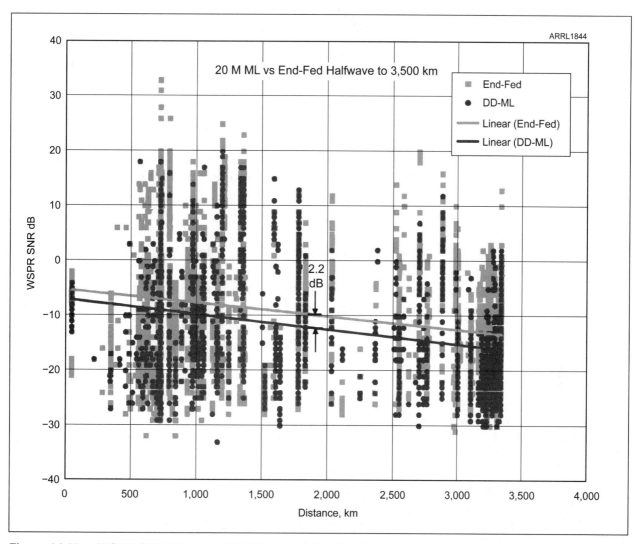

Figure 14.23 — WSPR SNR, 20-meter DD-ML vs end-fed, North America.

**Figure 14.22** shows the 20-meter SNR results for all distances and **Figure 14.23** contains the 20-meter SNR results for just North America.

To determine the relative performance between the antennas, the differences in the SNR from the trend lines were calculated. As a final check, averages of all SNR reports for the antennas were compared. These are just the raw SNR reports plotted against distance. As you can see, there is lots of variance, but with about 4000 data points in each data set, I think we can infer that on balance we have characterized each antenna's behavior fairly.

The comparison results are summarized in **Table 14.4** (40 meters) and **Table 14.5** (20 meters). It is worthwhile noting that the full-size 6-foot ML and the 3-foot DD-ML yielded nearly identical results on 40 meters. (Because of time constraints, 20 meters on the 6-foot ML was not tested.)

Finally, we compared the observed results with our calculations as shown in **Table 14.6**. The DD-ML performed better than the calculations, but the comparison is to a real antenna, not an ideal. Differences could be attributed to some combination of reduced efficiency of the end-fed reference antenna, radiation pattern differences, and the reduced resistive losses of the parallel ML configuration.

Looking at the worldwide results, for both 20 and 40 meters, it is evident that the SNR difference among the three antennas is nearly constant with respect to distance. At only about 2 dB down from a 130-foot end-fed antenna, for those confronted with antenna restrictions (lack of space or HOA rules), the DD-ML seems to be a viable choice.

One note on the data analysis: Multiple reports from identical stations were received in each measurement period. There was a typically a large variance in the reports from the same station in the same time period. Some attempt was made to correlate the reports from the same station, but it appeared that the large variation made this less reliable than we at first thought. It was then decided to take the simpler approach of simply comparing averages over a large number of samples instead of pairing results from individual stations. The authors are fully aware that propagation is a very complex subject and the limited tests might not fully characterize the DD-ML performance. However, our experience while listening and making contacts with the antennas tends to reinforce the findings presented. We think an imperfect real-world test is better than no test at all.

## Ease of Use

Anyone who has used a ML has faced the challenge of re-tuning the loop each time you make a frequency change. Because of capacitive effects, many operators use a non-conductive stick from a short distance to adjust the tuning capacitor when changing frequencies. This is cumbersome because it either requires the ML to be in close proximity to the rig, or walking to the ML's location, making a "stick" adjustment, and then going back to see if the change was within the necessary range for operating at the new frequency. For those reasons, the Double-Double ML uses remote tuning via a stepper motor and controller board.

**Figure 14.24** shows how the NEMA 17 stepper motor is used to control an air variable capacitor. Limit switches (microswitches) control the end travel of the capacitor. **Figure 14.25** shows one of the microswitches mounted on a plastic bracket that is attached to the plate that holds the capacitor assembly. The limit switch is mounted just above the stepper motor, and the other microswitch is

Figure 14.24 — Air-variable capacitor with stepper motor and limit switches.

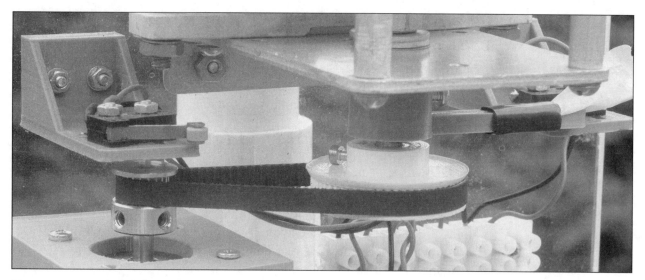

Figure 14.25 — Close-up of the limit switch mounting arrangement.

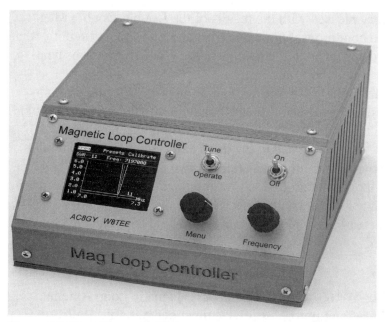

Figure 14.26 — ML controller.

mounted on the opposite side of the assembly. There's a plastic arm attached to the capacitor rotation shaft that swings through a 180-degree arc. When one limit switch is pressed by that arm, the capacitor plates are fully open, and at the other extreme they are fully closed. The control wires at the lower right of Figure 14.24 run to the Main Controller unit that uses the STM32F103 ("Blue Pill") microcontroller. The STM32 was chosen because it's cheap (under $5), has enough memory and I/O resources to do the job, and can be programmed from within the familiar Arduino IDE.

The board also uses an AD9850 chip as part of an SWR bridge, two rotary encoders, and a small (2.2-inch) TFT display, which together provide an easy way for users to control the ML. **Figure 14.26** shows the display in a project case. The remote control feature was used during the Field Day tests and worked well.

Of course, the DD-ML can be operated without stepper motor control and you can lower the construction costs by eliminating the remote-control feature. However, do you really want to run to the ML each time you change your operating frequency? Given that a lot of HOA-bound operators would benefit from our ML and will likely place it in the attic, that seems like a lot of steps for a frequency change. If you value your time at more than two pennies an hour, you're probably ahead to build the remote control unit.

The construction of the ML controller and its software are the subjects of the next chapter.

## Observations

Based on the WSPR data averages, we make the following observations:

- At all distances tested, the Double-Double Mag Loop does quite well on both 20 and 40 meter bands. Only a few dB separate the averages when compared to a conventional end-fed antenna.
- The DD-ML performance is nearly identical to a full-sized 6-foot single ML.
- The DD-ML appears to perform about as the calculations would suggest. No account was taken of radiation patterns, reference antenna performance, or ground effects, which might explain any observed differences.
- None of these tests should be taken as definitive scientific results. They are just meant to be indicative of what can be achieved in a real-life situation.

## Conclusion

Overall, we are very pleased with the results of the Double-Double Mag Loop under real-world conditions, both using SSB and FT8 modes on both 40 and 20 meters. Also, it appears that the much smaller DD-ML performs nearly identically to a 6-foot single loop ML.

The size of the final Double-Double configuration is clearly suited to portability. While not as small as some of the QRP antennas that use coaxial cable for the loop elements, the ability to handle 100 W or more is a real bonus.

A tripod assembly may not be necessary in all cases. On a recent camping trip, we threw a rope over a tree limb and attached two ropes to the ML and then tent-staked the ropes to the ground to prevent the antenna from spinning. It worked exactly the same as you would expect.

The DD-ML will clearly accompany us on future field trips and we hope you'll give it a try, too.

## References

Steve Adler, VK5SFA, "A 160 Metre Transmitting Magnetic Loop Antenna Design": **members.iinet.net.au/~sadler@netspace.net.au/tmla.html**

Frank Dorenberg, N4SPP — a great summary of everything ML: **www.nonstopsystems.com/radio/frank_radio_antenna_magloop.htm**

Glen E. Gardner, Jr., AA8C, "A More Efficient Magnetic Loop Antenna": **gridtoys.com/glen/loop/loop3.html**

Paul Hamilton KE7UAE, "Tuned Magnetic Loop Antennas": **www.seapac.org/seminars/2018/SEA-PAC2018-tuned-magnetic-loop-antennas-ke7uae.pdf**

Small Transmitting Loop Antenna Calculator: **www.66pacific.com/calculators/small-transmitting-loop-antenna-calculator.aspx**

Mag Loop Design Calculator program from Simone Mannini, IW5EDI: **www.iw5edi.com/software/magnetic-loop-calculator**

Multi-turn Magnetic Loop calculator from Jakub Jalowiczor: **comtech.vsb.cz/mlacalc/**

Leigh Turner, VK5KLT, "Small Loop Antennas": **www.qsl.net/vk5bar/**

## CHAPTER 15

# Controller for Double-Double Mag Loop

The magnetic loop (ML) antenna you constructed in Chapter 14 represents a viable HF antenna for many hams, especially those who face homeowner association (HOA) restrictions or would like an effective "luggable" antenna for portable operations. True, you're not going to throw this ML into your suitcase for portable use on a business trip, but our Double-Double Magnetic Loop (DD-ML) may be a good alternative for a vacation where you are driving to that location.

Any small loop-based antenna is a compromise when compared to full-sized antennas. It is also true that there are a number of small MLs that are easier to tote around, but most suffer from the following limitations:

- Operation at QRP power levels only
- Manual tuning
- Audio noise-level fine tuning

To us, the most severe limitation is the manual tuning. Because of the high-Q nature of ML antennas, even small frequency changes require retuning of the ML. If you're confronted with HOA restrictions, our guess is that you'll get tired running from your basement shack to the attic each time you want to make a 10 kHz change in frequency. While many commercial ads show the small ML sitting on a picnic bench within easy reach for retuning, we would only suggest this proximity to the ML with QRP power levels. Even then, it's possible to develop high voltages near the ML. (For more information, see ARRL's RF Exposure resources online at **www.arrl.org/rf-exposure**.)

Finally, many MLs require you to listen to the background noise on your receiver to determine when you are properly tuned to the new frequency. Some hams have constructed small LED indicators to give a visual clue about the current tuning state of the ML. While these work, they are a little fuzzy in terms of their ability to tune the ML precisely. If you cannot tune the ML with fairly high precision, most of that precious power is wasted heating the environment.

For all of these reasons, we designed a companion remote tuning unit to be used with the Double-Double Magnetic Loop antenna. The design is a semi-automatic electronic package using a microprocessor and micro-stepper con-

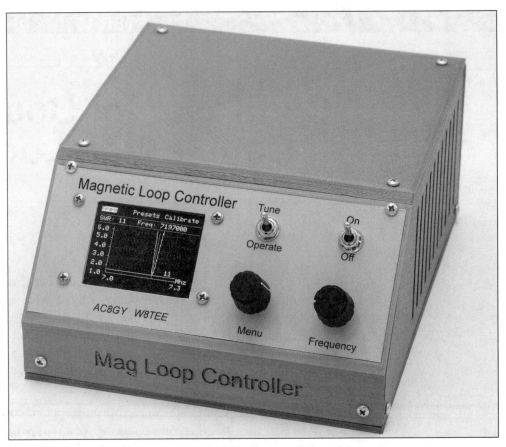

Figure 15.1 — The prototype DD-ML tuning unit.

troller that controls a NEMA 17 stepper motor. **Figure 15.1** shows our finished automatic tuning unit. The primary design goal was to allow remote tuning of the DD-ML, provide a convenient readout of the SWR while tuning, and to automate the process of retuning during operation as much as possible — all at reasonable cost. The remote tuning unit provides:

- SWR readout on a small TFT color display while tuning,
- Fine tuning during operation,
- Stepper motor presets, and
- Remote band switching.

In all cases, we have been able to get the SWR for each band to less than 1.2:1.

## Circuit Description

A block diagram of the DD-ML remote control unit is shown in **Figure 15.2**. The schematic for the unit is shown in **Figure 15.3**.

The easiest way to visualize the ML plus controller is as a system consisting of two parts: 1) the physical (and remote) ML antenna unit; and 2) the main controller unit of the system that resides in the shack. Each of these parts is discussed in the following sections.

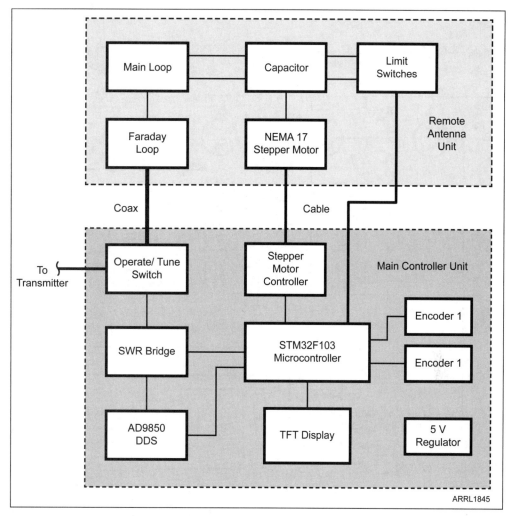

Figure 15.2 — A block diagram of the DD-ML remote control unit.

# 1. The Physical (Remote) ML Antenna Unit

Chapter 14 covers the elements of the ML is greater detail, while this section provides a refresher. For purposes of discussion, we subdivide the remote ML antenna itself into five parts.

### a. Main Loop

The first part of the remote antenna unit is the main loop of the antenna, which you know from Chapter 14 is actually four loops connected together to (electrically) form one single loop. This main loop acts as an inductor that works in concert with the tuning capacitor to form a tunable antenna system.

The main loop and the tuning capacitor form a resonant circuit, which can be tuned to the desired transmit frequency. The main loop nominal inductance at 7 MHz is about 5 µH and the capacitor value for the ML varies from less than 15 pF to 95 pF. There is also a residual capacitance in parallel with the tuning capacitor of about 9 pF, yielding a range of about 24 pF to 105 pF. The calculated resonance frequency range is then about 7.0 MHz to 14.5 MHz, just what is required for the ML's designed coverage.

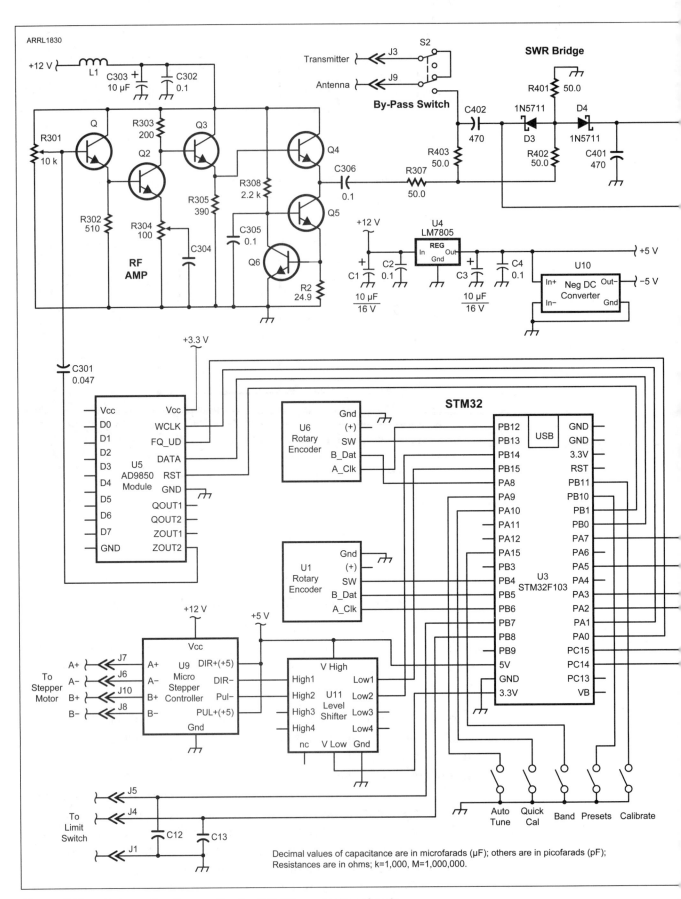

Figure 15.3 — Schematic diagram for the DD-ML controller circuit.

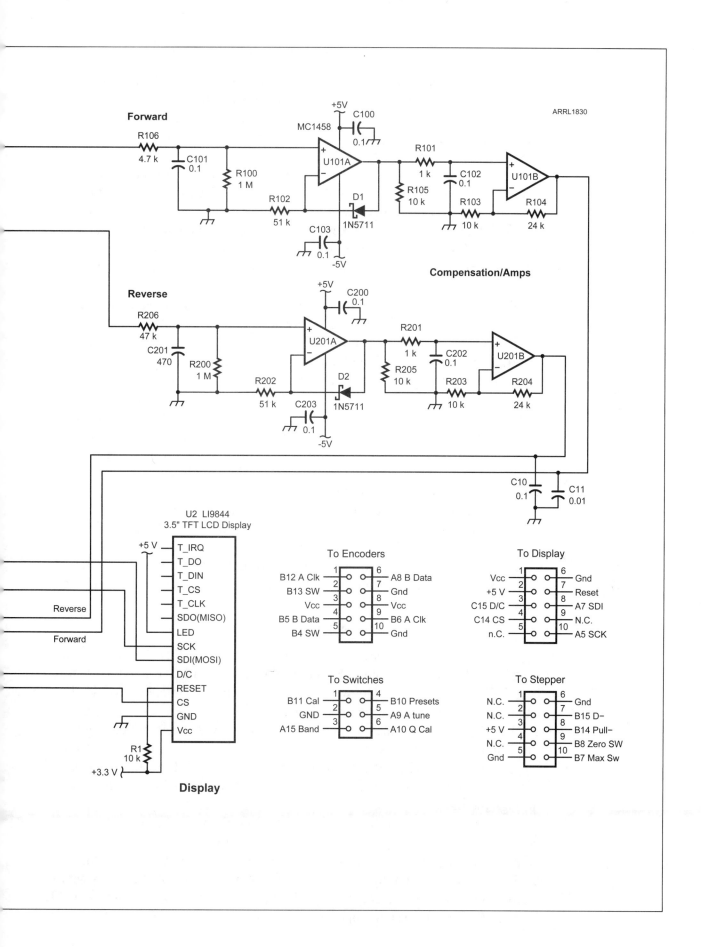

Controller for Double-Double Mag Loop 15-5

### b. Tuning Capacitor

The second part of the ML system is the *tuning capacitor*. The actual capacitor you need depends upon a number of considerations, including expected operating power level and cost. (Chapter 14 discussed these considerations in greater detail.) As already mentioned, the purpose of the tuning capacitor is to complete the resonant circuit and to tune the ML to the desired frequency. Capacitors have inherent losses, which add to the system losses and, if too high, can reduce the ML efficiency significantly. Most old tube-type transmitting capacitors in good condition have low losses. If you are buying a used capacitor, it's a good idea to check for broken insulators, corroded connections, rust anywhere, and dirt. Clean it up as best you can if any part is suspect. As previously discussed, be certain the capacitor plate spacing is adequate for your power levels.

The wiring for your capacitor of choice may vary slightly depending upon the type of capacitor you are using. There are a number of options:

**1. Single stator, single rotor.** This is the most common type. The stator is wired to one side of the ML and the rotor to the other.

**2. Dual stators with common rotor**. These are often found used in good condition and were probably rescued from an old tube-type transmitter. Commonly, the two-stator sections are the same. To connect this type of capacitor to the circuit, connect one end of the ML to one stator and the other end to the other stator. *No connections are made to the rotor assembly.* (If the capacitance values or range are not correct, the capacitor may be modified to change the values. This is discussed later in this chapter.)

**3. Vacuum variable capacitors (VVC).** These are easy to connect. Just connect in a manner like the single stator units. However, VVCs are not cheap and are probably overkill for QRP operation.

In all cases, make sure you make good physical and electrical connections between the tuning capacitor and the ML. Poor connections severely affect efficiency, and not in a good way!

### c. Faraday Loop

The third part is the *Faraday loop*. The Faraday loop (FL) electrically connects the ML to the transmitter. Exact placement of the FL is critical to obtain best SWR results. This is why we allow the FL to be movable, both vertically and relative to the plane of the main loop.

The FL element can be thought of as part of a simple transformer, whose function is to convert impedances. The FL is attached to the transmitter through the Operate/Tune switch of the remote control unit. The ML is the other half of the transformer. The impedance of the ML is matched to 50 Ω at resonance through the FL.

Note that not every location or position of the FL gives the desired impedance match. That is, the plane formed by the FL with respect to the ML does have an impact on the impedance match. Again, this is why we have designed the FL mounting system to allow both up/down positioning of the FL, as well as rotation around the main axis plane of the ML. (More on how to position the FL in the Setup and Test section later in this chapter.)

### d. Stepper Motor

The fourth part is the *stepper motor*. The function of the stepper motor is to rotate the tuning capacitor rotor. By changing the capacitance of the resonant circuit in conjunction with the DD-ML, we can tune the circuit for minimum SWR at the transmit frequency.

We elected to use a NEMA 17 stepper motor to control the variable capacitor. These stepper motors are reasonably priced yet have enough power to rotate almost any tuning capacitor. The NEMA 17 stepper motors have four wires marked A+, A–, B+, and B–. If your unit not does not identify which lead is which, there are a number of online videos that explain how to identify the leads for your stepper motor. Often the vendor also has a spec sheet for the stepper. Finally, keep in mind that *the stepper motor must be electrically isolated* from the capacitor, often through an insulated shaft coupling, because of the high voltages on the capacitor.

### e. Limit Switches

The fifth and final parts of the remote antenna unit are the *limit switches*. The purpose of the limit switches is to allow the controller to sense when the capacitor's end of travel has been reached. Limit switches are absolutely necessary for vacuum variable capacitors. Trying to turn a VVC past its end stops can physically damage the capacitor. Some air variable caps can only be turned to their physical stops, while butterfly caps could be rotated until the cows come home. However, while the absence of limit switches on air caps may not be fatal to the cap, they do help prevent overheating of the stepper motor and unnecessary roll-through. The lower limit switch is also used to set the zero position for calibration.

Limit switches must be wired in the normally-open (NO) position. How to attach and position the switches was covered in Chapter 14.

## 2. The Main Control Unit

The second component of the DD-ML system is the main control unit, pictured in Figure 15.1. The control unit is placed in the ham shack, presumably near the operator so it is within easy reach. As shown in Figure 15.2, the STM32F103 (Blue Pill) microcontroller is used to control the functions of the remote control unit. The Blue Pill is powerful enough to get the job done, yet costs under $3 and can be programmed in the familiar Arduino IDE using C. The STM32 controls: 1) the small TFT color display via an SPI interface, 2) the stepper motor controller, and 3) the SWR bridge. Some additional attention has to be paid to the power sources, since various parts of the control unit need +12 V, +5 V, –5 V and +3.3 V.

Two rotary encoders are used in the remote control unit. One of the encoders is used to control menu selection and other elements of the user interface. A second encoder is used to control data input (such as frequency or motor position). Both encoders are interrupt-driven to make them as responsive as possible. If you modify our code, avoid any code routines in the interrupt service routines that use their own interrupts to avoid blocking, such as the common *delay()* function. Also note that some encoders such as the KY-040 are mounted on their own circuit board and often include SMD 10 kΩ pull-up resistors. You should remove those resistors. Those resistors are clearly marked on the back of the board. Simply heat them up with a soldering iron and slide them off the board. (We are shying away from the KY-040 encoders, as their quality seems to be eroding. We find the Bourns encoders to be very good yet reasonably priced.)

The encoders we use have built-in switches, which we do use in the software. While you could use external switches, the cost of buying encoders that include the switch is about the same as those that don't include the switch. Also, built-in switches use less front-panel real estate. On the down side, the encoders require enough force that you will likely have to hold the unit with one hand while you press the encoder with the other. (You might consider replacing the encoder switches with separate normally-open pushbutton switches. If so, there's no sense buying encoders with the switch built in.) One more thing: Be careful when ordering the encoders. A lot of vendors sell encoders that do *not* have a threaded shaft housing, which makes them a lot less convenient to mount on a panel.

We chose the AD9850 to generate the RF signals for the SWR bridge. The reason for this choice over some obvious alternatives (such as the Si5351) is that the AD9850 produces a sine wave as output. The alternatives generate squares waves which are, as you know, loaded with harmonics. It would be virtually impossible to adequately filter the square wave signal without a rather elaborate circuit, beyond the scope of this project.

The output from the AD9850 is amplified by the RF Amp to provide about 1 $V_{RMS}$ to the antenna. The SWR is determined using a resistive SWR bridge with diode sampling. The dc out from the bridge is amplified by the two compensating amplifiers, one each for the Forward and Reverse standing waves. The bridge amplifier uses use Schottky diodes in the feedback circuit to reduce the issues of diode nonlinearity. We also use Schottky signal diodes in the bridge to get the best low-level performance.

The remote control unit does present a number of graphical output screens, including one for the SWR bridge. The SWR tuning depiction, however, is not the standard "V-shaped" plot you're accustomed to seeing. The reason is because we are not interested in varying frequency to find the minimum SWR. Once the minimum SWR has been achieved, a plot of the SWR vs frequency is shown, allowing visualization of the antenna bandwidth.

Instead, tuning involves the user selecting a desired frequency and the software then attempts to minimize the SWR for that *given* frequency by adjusting the capacitor. Once you specify the desired operating frequency, the remote unit automatically minimizes the SWR of the ML for that frequency. The remote unit typically finds the minimum SWR in a few seconds, although the exact time depends upon a host of exogenous factors over which we have no control (stepper motor, controller, gearing, type of capacitor, and so on). When the minimum SWR is found, the SWR and Freq fields in **Figure 15.4** are updated with the values from your frequency selection. (While Figure 15.4 shows a 1.00 SWR, it is likely closer to 1.004 and the result is rounded to fit the display field dedicated to the SWR value. Details on using the Freq menu option are discussed later in the chapter.)

While the code does provide for further fine tuning, our experience is that the auto-tune does a good job of finding a low SWR for the frequency selected.

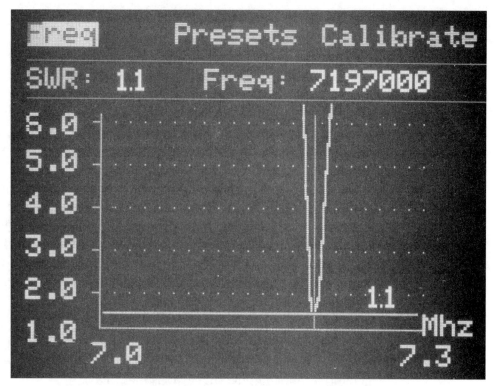

Figure 15.4 — Display after running the Freq menu option.

(We've done some additional testing and have found that fine tuning makes virtually no difference in the SWR...0.05 at most. Given that the auto-tune can find the "sweet spot" in less than a couple of seconds, we may remove the fine tuning since it does little other than slow down the frequency change process.)

There are five front-panel pushbutton switches which allow quick access to the following functions: Quick-Cal, AutoTune, Band Selection, Presets, and full Calibrate. Each of these switches, when pressed, activates a specific functional section of the code. Briefly, these functions serve the following purposes:

- Quick-Cal — does a fast calibration of the selected band only, refining the band edge definitions.
- Auto-Tune — initiates auto-tuning at the selected frequency.
- Band — Quick access to band selection.
- Presets — Alternative way to access the Presets Menu.
- Calibrate — Shortcut to the Calibrate Menu function.

The front panel also provides an Operate/Tune switch. The operator should *always place the Operate/Tune switch in the Operate mode when the transmitter is to be used*. Otherwise the transmitter is not connected to a 50 Ω load...probably not a good idea.

## Constructing the Remote Control Unit

As is true with other projects in this book, you have choices in building the remote control unit:

1) Prototype circuit board style using: perf-board, Manhattan-style, dead-bug style, or another method that you prefer.

2) The PCB designed specifically for this project.

Our prototype used perf-board construction since this is probably the most commonly-used construction method. We initially elected to use two smaller boards, one for the STM32 and the other for the SWR bridge circuits. We used a proto-board that Al designed for the STM32, but a simple perf-board works just the same. Dividing the circuits into two pieces allows additional flexibility in mounting — for instance, the two boards could be mounted vertically to save space. On a perf-board where ground plane shielding is impractical, the two separate boards isolate the RF and digital signals.

Since the original prototype, we have designed and built a PC board that includes all functions, with isolation of signals built in, but either approach yields excellent results. Note that the perf-board example uses a different RF amp and diodes amps. The accompanying schematic diagram and the final PC board use the upgraded circuit, which was partially done to eliminate some difficult-to-get components. The perf-board construction is shown for reference only.

## Wire up the STM32 board

Our prototype board for the STM32 board is shown in **Figure 15.5** (top view) and **Figure 15.6** (bottom view). We suggest you follow a similar layout and build it following the suggestions that follow.

There are several things to notice in our construction of the boards. First, we use sockets for the STM32. While a lot of people say that sockets add a point-of-failure to the circuit, we do not find this to be a problem. You can "build" a

**Figure 15.5 — Top view of STM32 prototype board**

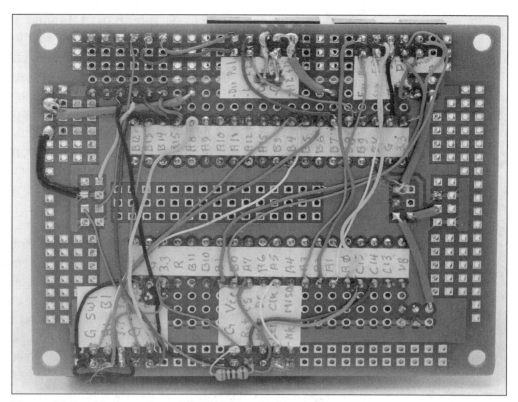

Figure 15.6 — Bottom view of STM32 board with wiring

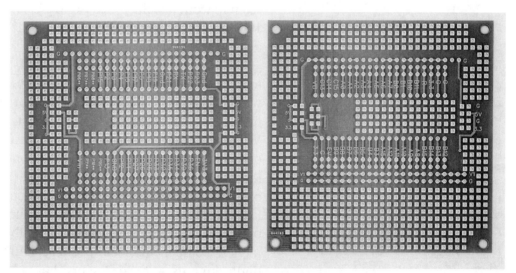

Figure 15.7 — Front and back view of Al's prototype board.

20-pin socket from smaller sockets, but you should file the end of any socket that needs to "touch" the end of another socket. For example, if you only have 10-pin sockets, you will need to file one end on each socket and place those filed ends together for a good fit. If you don't file the ends, pins 10 and 11 will be too far apart for a normal fit. You can also buy 20 pin header sockets online for $3 for 20 sockets. It's worth it. If you ever toast the STM32 and need to replace it, you will *really* appreciate the fact you used sockets. Removing a 40-pin IC from a soldered circuit is like trying to nail a blob of Jello to the wall. Use a socket.

**Figure 15.7** shows the prototype board we used. When I was working on

Figure 15.8 — Ribbon cable female IDC connector.

**Table 15.1**
**Pins for Component Wiring**

| Display Pin | STM32 Pin |
|---|---|
| TFT_CS | PC14 |
| TFT_DC | PC15 |
| MOSI | PA7 |
| MISO | n/c |
| SCK | PA5 |
| LED | +5V |
| Reset | 3.3V thru 10K |

| DDS Pin | STM32 Pin |
|---|---|
| Clk | PA1 |
| FQ_UP | PA0 |
| Rst | PB1 |
| Data | PB0 |

| Limit Switches Pin | STM32 Pin |
|---|---|
| Zero | PB8 |
| Max | PB7 |

| Encoder Pin | STM32 Pin |
|---|---|
| EnSW1 | PB4 |
| PinA1 | PB6 |
| PinB1 | PB5 |
| EnSW2 | PB13 |
| PinA2 | PB12 |
| PinB2 | PA8 |

| Micro-stepper | STM32 Pin* |
|---|---|
| Pul+ | PB15 |
| Dir+ | PB14 |
| Pul- Dir- | +5V |

| SWR Br. Pin | STM32 Pin |
|---|---|
| Forward | PA3 |
| Reverse | PA2 |

| Pushbutton SW | STM32 Pin |
|---|---|
| Quick-Cal | PA10 |
| Auto-Tune | PA9 |
| Band | PA15 |
| Preset | PB10 |
| Calibrate | PB11 |

* Level Shift Required

the Morse Code Tutor project, I found myself flipping the TFT front-to-back just short of a bazillion times as I was building the project. Al was watching and said: "Why don't you take a piece of scrap sticky-label paper and write the pin IDs on it?" Duh! Another flat forehead moment. As you can see, we did the same thing on the proto board by adding sticky labels to identify pin numbers. This will save you a couple of hours of construction time, reduce the likelihood of errors, and probably save you some hair follicles. That discovery led Al to design prototyping boards for both the STM32 and the ESP32 boards. Both are similar to that shown in Figure 15.7 (note the silk screen).

The PC boards use SMD components for most elements. These are 1206 size, which are pretty easy to work with after a bit of practice. The SMD approach saves lots of time and makes the final circuit much more compact.

We use IDC connectors and ribbon cable for the encoders, the TFT display, micro-stepper, and all external connections to the remote control boards. The connectors are not exactly cheap, but we buy them in 50 lots, which lowers the cost considerably. If you share your order with others, the cost isn't too bad. We do buy these overseas and, so far, they seem to work just fine. (See **Figure 15.8**.) We also found a 100-foot roll of 10-conductor, color-coded, ribbon cable at a flea market for $12 which is perfect for these IDC connectors.

The boards can include the LM7805 voltage regulator, too. These regulators get a bit warm and should be used with a heatsink. You can find a fistful of heatsinks at hamfests for about a dollar...seems like cost-worthy insurance. With the perf-board, once you have all of your components on hand, we suggest you do a trial fit of each component on the board. It is very frustrating to solder everything in place only to find that you need one more pico-acre to have that last part fit on the board. There's very little reason to skimp on the board size. We suggest a board that is at least 2 × 3 inches. Components should be placed to separate the RF and digital circuits — hence the two-board approach.

Wire placement on the proto-boards is not critical. Just use general point-to-point wiring techniques. We use #30 AWG solid wire with different colored insulation for the digital connections. You might think that stranded wire would be better, but with these small gauges, solid copper is a better choice if for no other reason than you can tuck it somewhere and it tends to stay there.

Once all the components appear to fit with ample space on the board to route the wires, start attaching the major components. Keep in mind that other wires may be attached to the same pins.

First, wire up the STM32 to the encoder connector, followed by the display connector. Then wire the connectors to the SWR board and the micro-stepper connector through the level shifter. **Table 15.1** shows the pin connections for each of the major components. This table should also be used to check connection continuity after the wiring is complete.

## Wire Up the SWR Bridge Board

**Figure 15.9** shows the layout we used to wire the SWR board. If you use the perf-board approach, place the discrete components on the board as shown in Figure 15.9, one at a time. Notice that we have used some prefab modules in building the SWR board. We use a AD9850 DDS board, a –5 V dc-to-dc switching converter to get –5 V for the op amps, and a level-shifter to convert the STM32 3.3 V signals to 5 V for the micro-stepper controller. **Figure 15.10** shows the PC board with all components installed. We use 1206 SMD components on the PCB, which are large enough to be soldered in place easily. The use of the PCB really makes life easier and we recommend getting one if possible. Our source is listed in the Appendix.

Carefully wire any discrete components together using #30 AWG wire as needed. Be especially careful not to overheat the diodes when soldering them in place. (Solder one end, wait for a short while, then solder the other end...it's a rather-safe-than-sorry thing.) Now solder the AD9850 sockets and the other daughter boards. Finally, wire a ribbon cable to the SWR board for connection to the STM32 board. We suggest using header pins to connect the ribbon cable, soldering all connections. We like to use a bit of hot glue on the ribbon cable ends for strain relief.

The PC board has space for IDC connectors for the display, encoders, stepper, and front panel switches. Use ribbon cable to attach these external components.

Wire the connections to the remote control unit connector using #24 AWG solid wire because of the heavier currents encountered on this connector. The +12 V and +5 V connections also should use the heavier gauge wire.

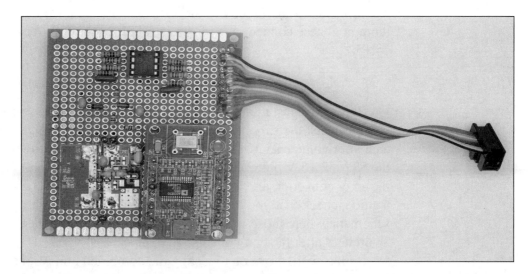

Figure 15.9 — Placement of SWR components on prototype perf-board.

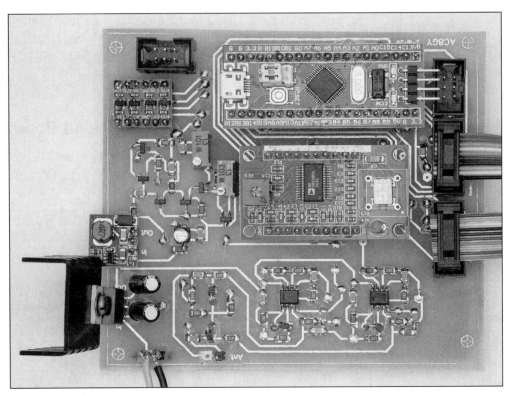

**Figure 15.10 — PC Board with components installed.**

Once everything wired, check for bad solder joints and check out the wiring connections. The best way to check the wiring is to use a printed copy of the schematic diagram (Figure 15.3) with the connection table (Table 15.1) and trace the connections one-by-one, marking off each connection on the diagram with a yellow marker. If everything was done correctly, the schematic should be a sea of yellow and all component and connections should be highlighted.

Now, using ribbon cable, wire up the TFT display and the encoders. We recommend wiring directly to the TFT display and encoders. Use heat shrink tubing on these connections for strain relief, as shown later in Figures 15.14 and 15.15.

## Power Supply Check

Before you install any plug-in components such as the STM32 or AD9850 boards, check the supply voltages that are coming into the STM32 board. Attach your voltage source for the +12 V inputs and measure the voltages at: 1) the STM32 +5 V pin, 2) the HV connection to the level shifter, 3) the input to the −5 V converter, and 4) the Display LED. Using the schematic diagram as a guide, check that you are seeing either +12 V or +5 V at those points on the schematic where those voltages are expected. You will check these voltages again later after the components are in place.

If all is well, move on to testing the combined units and assembling everything in its case.

## Remote Cables

Two data cables are required to connect the remote control unit with the DD-ML antenna. The first cable is between the main controller and the stepper motor. This cable consists of four wires between the respective A and B pairs of wires on the stepper motor and the micro-stepper controller.

The second cable connects the limit switches for the variable capacitor to the main controller. Ethernet cable is a good choice for these cables. Cat 5 or Cat 6 cables have four twisted pairs of #24 AWG wire. In order to minimize the voltage drops that can occur over long cable runs, we suggest using one twisted pair hooked together for each of the four stepper motor wires. The limit switch current requirements are very low, and Cat 5 cable is cheap and readily available at flea markets, so use it for these connections as well. We have tested this cable arrangement up to 75 feet. For much longer runs, we would suggest heavier gauge wire such as #18 AWG zip cord or similar.

The coax connection is nothing special, but low-loss coax is a good investment in this case. *Do not* use RG-174 or any other small diameter coax. The small coax simply has too much loss, and we are fighting for every dB we can keep. If you're running QRP power levels, you might have more invested in your antenna and feed lines than you do in the transceiver. It would be kinda silly to be pound-foolish on the coax.

We suggest that you use polarized connectors on both the antenna side of the cables and the remote side, as well. Using polarized connectors means they cannot be hooked up incorrectly, thus avoiding one of those campfire stories your "friends" love to tell. There are seven total data wires, so one multi-pin connector can take care of everything. We happened to have some DB15 15-pin connectors in the junk box, so we used those on each end. This way there is no possibility of plugging in the cables wrong. If female connectors are used on the main controller and the remote unit, the interconnect cable with male DB15 connectors can be used either way. We have also tried 8-pin locking connectors with success.

After you have everything all wired up, once again double check the connections with the circuit diagram and wiring chart. Now finally, it is time to test everything together.

## Testing the Completed Units

Prior to putting everything in a nice case, it is a good idea to test the components. We realize there's an awful temptation to simply hook everything up and try it out, but resist the urge. Don't get too eager to plug it all in and fire it up. You need to test things in stages, so as not to fry everything and ruin your investment.

First, if you have not already done so, load the software onto the STM32. As with the other projects, the software can be downloaded from our website mentioned in the introduction. We suggest you create a new directory for the software. We like to put major projects off the root directory: C:\MagLoopSoftware\.

Download the software and unzip it into your new directory. Look in the header file (MagLoop.h) to make sure there are no new libraries that need to be installed. As always, we put a comment after any *#include* in the source code header file that gives the URL where you can download the library. Once downloaded, install the library as discussed in Chapter 2. Note you must shut down the IDE after any new library is installed, and then reload the IDE. The reload forces the new library to be registered with the IDE. Now load the ML software, compile, and upload as usual.

Once the software has been loaded successfully, proceed with testing the rest of the main board.

## Testing the Main Board

First, adjust the dc offset trimmer resistor in the RF amp to mid-point (about 5 kΩ). This will be fine tuned later.

Connect the STM32 board and the SWR board with the IDC connectors and ribbon cable. Do not plug the STM32 or AD9850 into their sockets. Also, do not connect the TFT display, encoders, or remote unit just yet. First, you should double check the polarity of the +12 V power source and then attach it to the +12 V input to the board. Measure the voltages at the following points:

- Power input: +12 V
- Output pin of 5 V regulator: +5 V
- 5 V pin of STM32: +5 V
- Power input of micro-stepper controller: +12 V
- Output of the dc-to-dc converter: –5 V

If all these voltages are correct, you can now begin plugging in the components.

First, turn off the power and plug in the STM32 into its socket. (You did use a socket, right?) Now measure the following voltages

- 3.3 V pin of STM32: +3.3 V
- Vcc pin of AD9850: +3.3 V
- Vcc pin of TFT connector: +3.3 V

If all these voltage readings are correct, turn off the power and attach the TFT display and the encoders to the board. If everything is correct and the software was loaded properly, the splash screen should appear, followed in a few seconds by the main menu.

Note that the code attempts to reset the stepper motor to position zero. If the stepper motor is not connected, the zero limit switch cannot be read properly and the code might "hang" waiting for the zero switch to close. If this happens, you can bypass the stepper reset by adding comment characters in front of the reset function call. Somewhere around line #250 in the project's .ino file, you comment out the function call by adding the double-slash comment pair:

```
// ResetStepperToZero();
```

This at least allows you to see the initial display screen. Don't forget to uncomment the line when the stepper motor is powered up.

Turn the encoders and watch for appropriate changes on the screen. Obviously the menu encoder should move the menu selection highlight bar according to the direction you turned the encoder. If for some reason a clockwise turn produces a counterclockwise movement, you have two choices: 1) rewire the encoder (messy); or 2) reverse the encoder pin numbers in the Mag_Loop.h header file. For example, suppose the menu encoder is moving the wrong way. Around line #65 (this number will likely change over time) in the header file, you will see:

```
#define MENUFREQENCODERSWITCH      PB4      //MENUFREQENCODER switch
#define MENUFREQENCODERPINA        PB6      //MENUFREQENCODER
#define MENUFREQENCODERPINB        PB5      //MENUFREQENCODER
```

Simply reverse the two pin numbers so it becomes:

```
#define MENUFREQENCODERSWITCH      PB4      //MENUFREQENCODER switch
#define MENUFREQENCODERPINA        PB5      //MENUFREQENCODER
#define MENUFREQENCODERPINB        PB6      //MENUFREQENCODER
```

Changing the code is a lot easier than reversing the encoder wiring.

If there is any other problem at this stage, it is usually a bad connection or poor solder joint. Go back and re-check everything.

## Testing the RF Output

This step is best done using a scope to look at the RF signals, but can be done with just a good voltmeter. First set the output of the controller to about 7 MHz. Look at the output of the AD8950 on the Zout2 pin of the AD9850. You should see a sine wave of about 400 mV at 7 MHz. Then view the output of the RF amp at C306. The voltage should be about 1 to 1.5 V. If it is more or less, adjust the gain trim pot to get between 1.2 and 1.5 $V_{RMS}$. Now adjust the Offset trimmer to get the cleanest sine wave, with little or no distortion. You may have to reduce the output level a bit to get there. When you have achieved a 1 to 1.5 V sine wave, this part of the setup is finished.

## Testing the Stepper Motor and Controller

Now connect the micro-stepper controller to the main board. Connect the cables between the completed main controller unit and the remote unit. Turn on the power to the main controller. If the stepper motor is not at its left-most stop, the stepper motor should immediately begin to move to the "zero" position. Once the stepper motor reaches the zero position, it should hit the limit switch and the stepper should stop. If not, there's a hiccup somewhere. Check your switches and their wiring.

Depending upon the exact stepper motor you are using and perhaps the gearing you are using, the current being drawn from the power source should be about 1 A at 12 V. If the current is much over 1.5 A — *stop* and check for a short or incorrect connection...white smoke is a dead giveaway!

If the motor does not move or just "hums," the stepper motor wires may be hooked up wrong. If the stepper motor goes backward from the expected movement, the A and B pairs are reversed.

If the motor does not run at all, try depressing the limit switches. If the motor now starts, the limit switches are set to NC instead of NO. If the stepper motor does not want to stop at the end of travel, turn off the unit and check the limit switches. Are they wired correctly? Test for +3.3 V at the switches.

Assuming everything works so far, turn the Frequency encoder clockwise. The stepper motor should begin to move clockwise. Reversing the position encoder should move the capacitor back counterclockwise.

It you got to this point without error, it looks like all is well so far.

## Initial Tests of the SWR Bridge

Complete testing requires the DD-ML loop and capacitor be connected to the system, but for now we can check the SWR bridge with a couple of resistors. If you built the Mini Dummy Load project presented in Chapter 6, this would be a good place to use it. Place a 50 Ω resistor (or the dummy load) across the input coax connection. The SWR should read 1.0.

Now place a 100 Ω resistor across the input of the coax connection instead of the 50 Ω load. Depending upon the exact value of resistor you selected, the SWR should read about 2.0. If your SWR readings are very high, your connections to the STM32 analog inputs may be reversed.

## Tests of the SWR Bridge with ML Remote Assembly

Fully testing the SWR bridge requires connection to the completed ML, with the capacitor assembly also connected. First, you need to check the ML/capacitor range. To do this, set up the ML in a vertical position and away from metal objects as far as possible.

Don't attach the coax cable between the main and remote units yet. If you don't have access to an antenna analyzer or SWR meter, we feel this test is sufficiently important that you should see if you can borrow one from a club member. You could also consider building your own antenna analyzer. (QRP Guys sells the PCB and has the assembly instructions for Jack's (W8TEE) Antenna Analyzer.) Attach the data cables between the units, but don't turn on the main unit yet.

You need to adjust the position of the Faraday loop (FL) at the top of the main ML support. You can see the small copper FL near the top of the support structure in **Figure 15.11**. As a starting point, place the FL in the same plane as the ML. Now, manually position the capacitor in the fully-meshed position. (You should move away from the ML when testing/tuning.)

Figure 15.11 — The complete DD-ML antenna.

Using the antenna analyzer that you managed to beg, borrow, or steal, read the SWR on the meter with the capacitor in the fully-meshed position. A typical ML SWR plot is shown in **Figure 15.12**. Note the extremely sharp null.

The SWR should be below 10:1 or so at a frequency around 7 MHz. If the lowest frequency for a null is above 7.0 MHz, your capacitor may not have enough capacitance to tune low enough. (See the section on modifying your capacitor later in this chapter.)

If you get a definite minimum in the SWR, but it is well above 1.5, slowly rotate the Faraday loop in small increments out of the ML plane and observe the SWR reading. Note that your body capacitance affects the reading, so change the FL position and back away to make each reading. Continue moving the FL until the SWR reading is below 1.2 or as low as possible.

If the SWR meter does not indicate a null around 7 MHz, check for ML loop connection issues, or FL loop problems, including bad solder joints or faulty cables. When all is well, move on to the next test.

Now position the capacitor to the fully-open position. The SWR should be below 3.0 at a frequency of greater than 14.35 MHz. Note the present position of the FL and rotate the FL until the SWR at around 14 MHz is as low as possible.

Going back and forth between fully meshed and fully open, and locate a compromise FL position that gives SWR reading below 1.5 at both extremes. If the frequency extremes are not less than 7 MHz and greater than 14.35 MHz, refer to the section on "Modifying Your Capacitor" later in this chapter to be able to get the full frequency range.

Now we test the SWR bridge in the main unit. Connect the coax between the antenna and the remote unit. Turn on the power and allow the capacitor zero position to be initialized. This resets the capacitor to its stop-limit position with

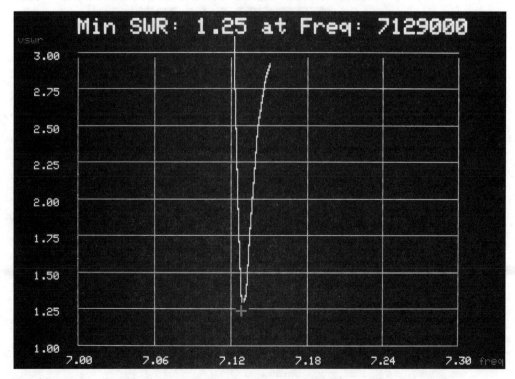

Figure 15.12 — An SWR plot of the antenna on the W8TEE antenna analyzer.

the capacitor plates fully meshed. The SWR reading will likely be very high — this is normal.

Using the Frequency encoder, set the frequency to 7.0 MHz. Then rotate the Position encoder clockwise until the SWR begins to decrease. Continue until a minimum is reached, which should be around the value observed in setting the compromise FL position. Looks like the SWR Bridge is working. Congratulations!

At this point follow the instructions for the Calibrate function, presented later in the Software and Menu Selection section of this chapter, to complete the testing.

## Assemble the Main Controller

Once all the tests have been completed, it is time to assemble everything into a case. The exact details of the final construction depend on the case selected. We have given many examples of how to do this in previous chapters, so at this point we will just make a few suggestions to make the project a success.

We designed a case that uses a sloped front which we 3D printed. The case is approximately 150 × 170 × 90 mm. **Figure 15.13** shows the stepper motor controller, the AD9850 module, and the STM32 board module mounted on the base plate of the case. Don't plan on going to the library and using their 3D printer, as the case is fairly complex and required about a quarter of a spool of PLA line and 30 hours to print! If you don't have a 3D printer, ask your club members if you could use theirs or simply use an off-the-shelf case.

In retrospect, we should have mounted the STM32 board just a smidgen further to the right than pictured in Figure 15.13. Its present placement makes it a little difficult to plug in the USB cable for updating the software. Because of

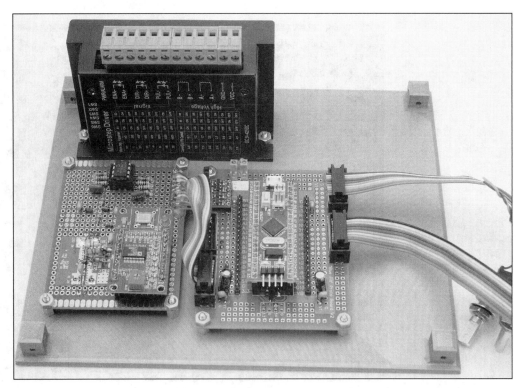

Figure 15.13 — Major components on bottom panel — perf-board prototype.

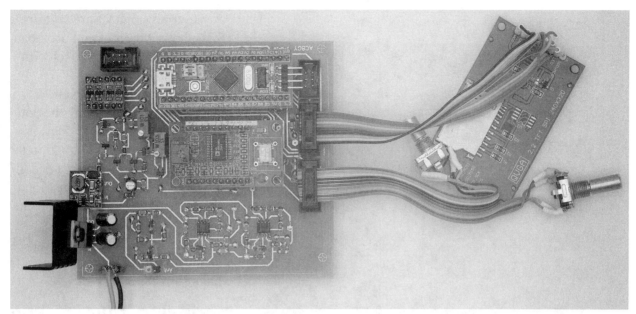

**Figure 15.14** — PCB version of the wired controller with encoders and display.

**Figure 15.15 — Front and rear panels added — PCB version.**

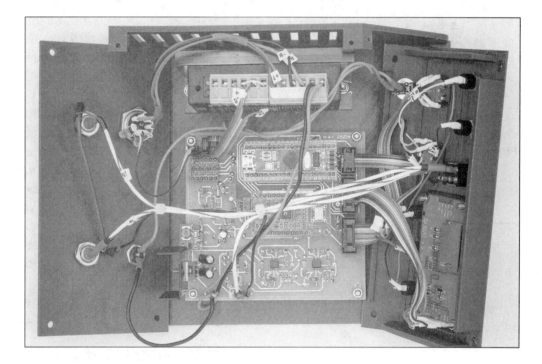

that difficulty, we relocated the STM32 board closer to the front of the case to allow easier access, as shown in Figure 15.15. An even better option is to spend about $6 and add a USB extension cable from the STM32 to a position on the back of the case. See **Figure 15.14**. While we all know Jack writes perfect code (ahem), there may come a time when you'll want to update the software. The USB extension cable makes it possible to upgrade the software without taking the case apart.

**Figure 15.15** shows the base panel and the front panel of the PC board version connected together with the ribbon cables and the back panel with its power supply connections. **Figure 15.16** shows the front and back panels attached to the base panel with small self-tapping screws.

Follow normal construction practices when assembling the components, keeping leads as short as possible. Pay particular attention to the length of the display and encoder cables. They should be just long enough to allow you to unscrew the front panel to have easy access to their connectors. Add the top and side panels and you end up with what's shown in the photos.

## Modifying Your ML and Capacitor

Figure 15.16 — Completed controller in case — PCB Version.

Truth be told, we spent almost a year of hamfest searching to find the variable capacitors we ended up using in the ML project. We traveled from Dayton to Orlando, with stops in between, to amass a mixed bag of almost two dozen air and vacuum variable capacitors that looked like candidates for a ML. We both feel like we kissed a lot of frogs to locate and buy these caps, and, in the end, we only used two from the lot. Some of the others have found good homes with members of our builders group, but others remain unemployed. Keep your eyes open at the flea markets as some of them will eventually end up there.

Try as you might, you may end up purchasing a capacitor whose capacitance range is not suitable for use in a ML. However, not all is lost. Some modification to the ML or capacitor can often be made to bring it into spec. Some guidelines can help you assess the avenues to explore further.

### Modifying the ML

If the upper maximum frequency is below 14.35 MHz, the ML loop inductance is too great for the capacitor for the current state of the project. There are two possible remedies to solve the inductance issue. First, you can reduce the ML inductance by decreasing its diameter slightly. Usually this only requires removing a couple of inches from the circumference of the ML.

### Modifying the Capacitor

A second way to save the capacitor is to reduce its capacitance. You can do this by removing one or two plates from the capacitor. Only remove plates in the event that the fully-meshed capacitor yields readings are well below 7 MHz so you have enough capacitance range after the modification. Note that the residual capacitance of some capacitor designs limits the lower range adjustment, so altering the loop inductance might be a better solution.

Removing plates from the capacitor is easy, just time consuming. Carefully disassemble the capacitor and remove the rotor. Good transmitting capacitors often have bolts fitted with spacers that hold the rotors and stators in place. Undo the bolts and slip off one or two stator plates and their spacers. Now do the same for the rotor plates. If the unit is a dual stator type, probably only one side needs to be modified. Alternatively, the rotor plates are often press-fit onto the rotor shaft. It will likely take some gentle persuasion using a small (pin) hammer to undo the rotor plates. If you're lucky, both the rotor and stator plates are mounted on shafts with spacers holding them in place.

Once you have removed the required plates, clean and carefully reassemble the capacitor, making sure the plate spacing is uniform. Rotate the capacitor through its range and make sure the spacing doesn't change with rotation.

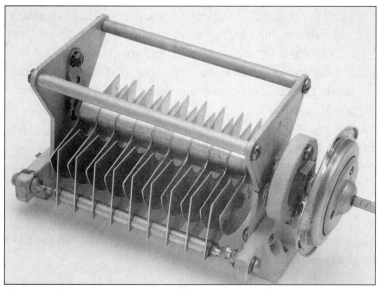

Figure 15.17 — Unmodified capacitor.

Now repeat the capacitor testing to assess the impact of the change you just made. If the cap is still out of range, repeat the surgery process as many times as necessary.

If the total range of capacitance is insufficient and the plate spacing is greater than required, the spacing may be reduced to increase the fully meshed capacitance. For instance, one of our capacitors had a 4 mm spacing, as shown in **Figure 15.17**. The unmodified range was 14.4 pF to 82 pF, which was fine on the low end, but not quite enough to reach 7 MHz resonance. By reducing the spacing to 3 mm, it was still OK for the expected voltages.

We also added another stator blade. The spacing was changed by replacing the spacers with nuts and washers. The threaded rods on the stator allowed the stator spacing to be adjusted to match the rotor spacing. The final capacitance range was now 15.3 pF to 99.7 pF, sufficient on both ends to cover both 40 meter and 20 meter bands. The pieces for the modification are shown in **Figure 15.18,** and the final modified version is shown in **Figure 15.19**.

This plate-removal surgery technique only works for capacitors that use individual spacers between the

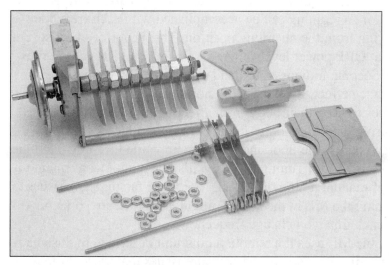

Figure 15.18 — Disassembled capacitor

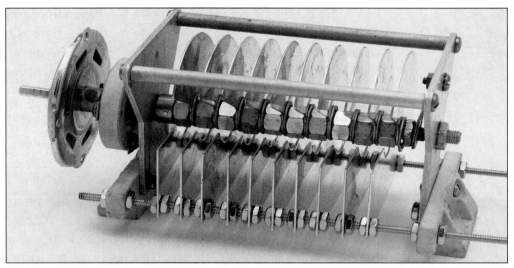

Figure 15.19 — Modified capacitor.

plates. The idea is to replace the spacers with ones of smaller width. How much to change the spacing can be determined by referring to one of the online capacitor design calculators, such as **www.66pacific.com/calculators/capacitor-calculator.aspx**. Replacement spacers can be washers, threaded nuts, different spacers or some combination thereof. Just make sure all of the new spacers are identical, so the plate gaps are uniform.

Sometimes a combination of the previous techniques may be required. This process is by trial and error. Expect to spend a few hours getting it right. (It's still faster and cheaper than driving to Orlando!)

If your capacitor has the plates pressed in, the plates can still be removed by tapping on them near the rotor carefully until they come out. Start with the rotor plates first and retest. This may be enough to get in range. If not, remove the stator plates as well. In either case, keep all of the plates and spacers in case you need to go back to the previous state.

## Software

Many commercial MLs require the operator to be within arm's length of the loop's tuning capacitor so retuning can be accomplished with each frequency change without moving from the operating position. This is not good for several reasons. First, even at QRP power levels, there is a significant RF voltage present at the capacitor. Second, moving your hand can de-tune it because of your body's stray capacitance effect. It's not uncommon for users to couple "something" to the capacitor to keep their body from affecting tuning. (We've even seen a stick used during tuning! Another member of our builders' group was doing a military crawl across the floor and reaching up while face-down to tune the cap. We dubbed this Stealth Tuning.) Finally, it's just plain inconvenient to require some form of manual tuning of the ML. For those facing HOA issues and who end up placing the ML in the attic, it might get a little tiresome running up to the attic each time you change frequency.

The Double-Double ML uses the remote tuning unit described in this chapter. In this section, we discuss some of the software issues that arose during construction and how we overcame them.

## Design Goals

There were a number of design goals we had for the ML and most of those affect the way the software was developed.

**1) Remote Tuning.** We wanted to be able to tune the ML remotely using a stepper motor to vary the capacitor setting. We knew that, because the granularity of most inexpensive stepper motors was not good enough to effectively tune the ML, a micro-stepper motor controller would be required. Because the stepper motor needs to be isolated from the capacitor, we also knew that some form of insulated mechanical coupling between the stepper motor and capacitor would be required. Finally, we wanted the remote unit to work reliably at a distance of at least 50 feet between the control unit and the ML. Our design meets these requirements.

**2) Ease of Use.** Most of us have 'watering holes" on various bands where we tend to gravitate each time we fire the rig up. That's especially true for QRP operators or those who check into a net on a regular basis. As much as possible, we wanted the operator to be able to select a preset frequency and have the stepper motor "fast-tune" to that frequency. In truth, this can never be fully automatic because the ML is sensitive to environment conditions that are subject to change (such as rain). Still, we wanted to auto-tune to a point where the SWR would be less than 1.25 under most conditions. The software should then allow the user to quickly fine-tune the ML from that point.

**3) Accurate and Easy Tuning.** Another issue is the feedback mechanism used for tuning. Many suggest you tune for "maximum noise" as heard on the receiver...a tad on the fuzzy side of things. Our tuning mechanism is shown in Figure 15.4, which is a TFT color display of the desired operating frequency and the SWR at that frequency. Under most circumstances, fine tuning allows you to obtain an SWR of 1.1:1 or less.

**4) Reasonable Cost.** As you probably know, there are a number of commercial MLs available that support remote tuning. However, at the time of this writing, their cost is between $450 and $3600. Depending upon the type of materials and options you choose to use in your Double-Double ML (for example, power handling capabilities, capacitor type and voltage rating, 1/2-inch versus 3/4-inch copper tubing, PVC versus 3-D printed support parts, and so on), the cost for a 100 W version of the DD-ML should be around $150 or less, even if you buy everything brand new.

While there are other factors that entered into the final design, these were among the most important.

## User MagLoop.h Header File Edits

There are a number of hardware/software elements to the ML system that can vary between users. Many of these either don't matter, or Al and I have no control or influence over them (for example, soil type or proximity to metal objects). However, there are elements where we can help improve the utility of the ML. These are discussed in this section.

### Band Edges

Because we do not know the frequency allocations that may apply to your location, the software needs to have the operating band edges for 40, 30, and 20 meters defined according to the regulations in your country. Around line #19 in the MagLoop.h header file you will see:

```
#define LOWEND40M       7000000L    // Define these frequencies for
                                    // your licensing authority
#define HIGHEND40M      7300000L    // The 'L' helps document that
                                    // these are long data types
#define LOWEND30M       10100000L
#define HIGHEND30M      10150000L
#define LOWEND20M       14000000L
#define HIGHEND20M      14350000L
```

If your band edges are different than those shown above, you need to edit the affected values in the MagLoop.h header file. These frequencies are used to estimate stepper motor settings during the calibration process, so these values must be changed before you attempt to calibrate the remote control unit.

As the software presently stands, you can also use these symbolic constants to more-properly define your own particular operating restrictions. For example, if you are licensed by the US and hold a General class license, you could set LOWEND40M to 7025000L because the lowest 25 kHz of the 40-meter band is reserved for higher license class holders. Perhaps down the road we could find the time to give some indication when you are tuning the ML outside of your legal operating frequencies. Even more cool would be to take into account the signal width for the mode in use and indicate that your current frequency selection is starting to impinge on that "signal edge" frequency. As it presently stands, such enhancements fall under the dreaded "this is left as an exercise for the reader."

### Stepper Motor, Controller, and Encoder Differences

We expect many of you to use components that have been living in your junk box since the Y2K problem years ago. Because of that expected variance, some elements of the software must act in concert with your choices. For example, around line #26 of the MagLoop.h header file you will find:

```
#define FASTMOVESPEED       900
#define NORMALMOVESPEED     100
#define MAXBUMPCOUNT        2       // Detent pulses to
                                    // get "real" bump
```

The first two symbolic constants above determine how fast the stepper motor moves from one position to the next. Our experience is that the faster speed is fine for resetting the motor (resetting the stepper motor to a zero position), but may not work well when tuning. Therefore, we use two different speeds at various execution points in the code. It may well be that your particular combination of stepper motor and/or stepper controller behaves differently. We suggest that you try the values above to see how they work for you. If you find that your stepper motor/controller combination is difficult to tune, try experimenting with these constants.

All rotary encoders are not the same. The granularity of your encoder (the number of detents per 360° of rotation), physical construction (mechanical versus optical), and other factors can influence the way the software interacts with your encoders. To minimize the impact of such variations, we suggest you use the same type of encoder for both menu and frequency use. Some encoders, such as the KY-040, come with a small circuit board attached with pull-up resistors in place. If your encoder doesn't perform as expected, you may need to remove the pull-up resistors.

Encoders may send different pulse chains as they are rotated from one detent to the next. The pulse may vary from 1 to 4 pulses per detent. We have set MAXBUMPCOUNT to 2 because our encoders send a 2-pulse chain for each detent. Your encoder may be different. If your encoder seems super-sensitive or sluggish when rotated, try changing this constant.

Finally, the stepper motor chosen and its controller can also affect the way the ML tunes. Any gearing or drive differences can also affect the way the ML tunes. None of these variations are show-stoppers. We are simply pointing out where you might begin your investigation if there's some element of the ML performance with which you are unhappy. While you can always use different hardware elements, tinkering around with the software should be your first line of experimentation before replacing any hardware.

Our selection of stepper motor is, as previously mentioned, is a NEMA 17 unit with a native resolution of 1.8 degrees. That is far from fine enough, so we added two refinements: First, the drive ratio is reduced by about 2.5 to 1 using a toothed belt drive like those found in 3D printers. Second we employed a micro-stepper controller that allows up to a 32:1 reduction in the step per pulse. We found that 16:1 is adequate for our capacitor when used with the belt drive reduction. Other combinations may require a different ratio. For example, when using a vacuum variable capacitor, the ratio might only need to be 4:1. Some trials might be necessary to determine the best combination.

### Other Factors

Finally, around line #15 in the MagLoop.h header file you will find:

```
#define DEBUG                         // Comment out when not debugging
#define VERSION         21            // Software version
#define SPLASHDELAY     4000L         // Normally, 4000L
```

These constants should be familiar by now. We allow the splash screen to stay on the display for 4 seconds. You may wish to change this constant to a shorter time period. Also, you will likely want to personalize the *Splash()* function which is defined in the project's .ino file.

You may wish to spend some time looking at the MagLoop.h header file to see what's there. If you plan on using a different display, there are constants in that file that you may need to change (such as pixel resolution constants). By this stage of the book, most of the code in the header file should look familiar to you. By the way, you may have gathered that selecting the right combination of microcontroller/display/library is not trivial. We have given a number of examples which do work, so you won't have to go through all of the effort we did to find workable solutions. Please see Chapter 4 for more detail.

## Software and Menu Selection

Figure 15.4 shows that the primary menu selections are: *Freq*, *Presets*, and *Calibrate*. Each of these is discussed in this chapter, but we begin with the *Calibrate* function. It is essential to calibrate the ML with the antenna located in its expected operating environment (attic, balcony, and so on). Because of the auto-tune feature and the fact that we don't know what kind of components you are using, we need to calibrate the stepper motor position with respect to the frequency ranges. Therefore, one of the first things that you need to do is find the stepper motor counts for each band edge. Because we support three bands, we store the six band-edge counts to the ML.

Store? Store where? In the EEPROM of course!

### STM32 EEPROM

Unlike the Arduino family or boards, the STM32 does not have any EEPROM directly available on its boards. Instead, the STM32 library supports an EEPROM emulation using flash memory. While it is good that we can treat some flash memory as through it were EEPROM, there are some things that are different when compared to the Arduino EEPROM stuff you may be used to.

Because there is no "real" EEPROM, we need to tell the compiler what memory segment we wish to use for the emulated EEPROM. This is done with the following function:

```
void DefineEEPROMPage()
{
  EEPROM.PageBase0 = 0x801f000;   // EEPROM base address.
                                  //   Everything indexed from
                                  //   this address
  EEPROM.PageSize  = 0x400;       // 1024 bytes of EEPROM
}
```

The *DefineEEPROMPage()* function tells the compiler the starting memory address for the EEPROM emulation (0x801F000) and that the next 1024 bytes from that address are earmarked for EEPROM use. The *DefineEEPROMPage()* function call needs to be made before any other EEPROM access is attempted.

### EEPROM Memory Organization

OK, so now we know where the EEPROM is, what are we going to put there? Well, we've already alluded to some of the data, but a memory map of the EEPROM memory space might help visualize what goes where. **Table 15.2** shows how we have organized the EEPROM memory space. Note that we are only using 100 (0 through 99) of the 1024 bytes allocated. This gives you plenty of EEPROM memory space should you want to add to the configuration data.

### The Endian Problem

Another interesting feature of the STM32 is that the "endianess" of the data can be changed to suit the vendor's needs. The Endian problem refers to the way the representation of binary values are stored and accessed in memory. For example, an *unsigned short int* on the STM32 is 16 bits. So, if the value is 1, you could use:

**Table 15.2**
**A Memory Map of the Emulated EEPROM Memory Space**

| Memory Offset | Description |
|---|---|
| 0 | The last band that was used (40, 30, or 20 meters). This is updated whenever you change bands. The *int* value is stored in *currentBand*. Therefore, bytes 0-3 hold this value since an *int* is 4 bytes on the STM32. |
| 4 | These are the stepper motor counts for each band edge. Because this is a min/max for each band, there are 6 values. These values are stored in *bandLimitPositionCounts[3][2]*. These are stored as a *long* data type, even though they could be an *int* because they use the same memory length for storage (4 bytes each). The array occupies bytes 4-27. |
| 28 | These are the six presets that you are allowed to have for each band. These values are stored in the *presetFrequencies[3][6]* array. The array occupies 72 bytes, or addresses 28 through 99. |

| 00 | 01 |
|---|---|

which is called a Little Endian representation. The *Little Endian* representation places the bit values starting with the least significant byte.

On the other hand, a *Big Endian* representation of the same value is:

| 01 | 00 |
|---|---|

which means the starting bit value is found in the most significant byte. Clearly, if we want to store configuration data in EEPROM, we need to know how the chip expects to see the data going into and coming out from memory. Here's the problem: The vendor of your board is free to organize the data using either the Little or Big Endian representation.

Fortunately, there is an easy way to abstract away from this problem. We solve the problem by using the following data structure:

```
union {
  uint16 myBytes[sizeof(uint16)];
  int val;
} myUnion;
```

Think of a *union* in C as a small chunk of memory that is capable of holding different data types. What the compiler does is examine all of the members of the *union* to determine which member requires the most memory for storage. In the case of *myUnion* defined above, a single *uint16* takes 2 bytes, and an *int* takes 4 bytes. Because we want the array to be able to hold an *int*, we need two 2-byte array elements in the *myBytes[ ]* array to be able to store an *int* data type.

Now, suppose you want to store the value of the last-used band (for example, 40 meters) in EEPROM. Is it stored as a Little Endian value:

|  |  |  | 40 |
|---|---|---|---|

or is it stored as a Big Endian value:

| 40 |  |  |  |
|---|---|---|---|

The answer depends upon how the vendor of your board wants things done. If we don't solve this issue the data are going to be confusing.

The solution is easy once you realize we don't really care about the Endian order as long as we are consistent in the way we treat it. If we store a value in EEPROM as Big Endian, not a problem as long as we retrieve it from EEPROM as a Big Endian value. The same is true for a Little Endian value. The statement:

```
int num       = 40;
myUnion.val = num;
```

places the *int* value of *num* into the *union* using the endian type currently in operation. If we later want to read the value from EEPROM, we can use:

```
for (j = 0; j < intSize; j++) {         // Write int to EEPROM as intSize-d bytes
   myUnion.myBytes[j] = EEPROM.read(j);
}
num = myUnion.val;
```

Think about it. By running the data through the *union*, we are consistently reading and writing the EEPROM data, regardless of the way it is actually represented in memory.

So, why bother with the *union* at all? The reason is because the *read()* and *write()* functions in the EEPROM library are based on 16-bit memory chunks. Therefore, the EEPROM library forces us to read and write the EEPROM *int* data in "half-an-int" chunks. (The Arduino family uses 8-bit chunks for EEPROM data.) So any time a data object is more than 16 bits, we need to use a 16-bit array to store that object so we can read/write it 16 bits at a time. Also keep in mind that each data item must start on a word (16-bit) boundary. For example, suppose you have a data object that uses 12 bytes of memory. It would seem that you could lay down the next data item starting at address 13. While that would work in the Arduino world, it will not work for the STM32 because of the word boundary restriction. Therefore, the next data items would have to start at memory address 14, not 13.

### The First Run of the Software

As mentioned above, there are a number of data items that are stored in the EEPROM memory segment. The current band is stored in EEPROM as a 16-bit value and must have the value of 40, 30, or 20. If the code reads the current band EEPROM address and does not find one of these three values, the code assumes that the EEPROM page has not been initialized. (Many manufacturers set "virgin" memory to 0xFFFF from the factory.) Because the uninitialized EEPROM value sets the high bit, the interpretation as an integer value is negative. Upon seeing any value other than 40, 30, or 20, the program automatically calls the *DoCalibrate()* function.

Before you run the software the first time, you should edit the MagLoop.h header file for the band edges mentioned earlier (for example, `#define LOWEND40M 7000000L`). Set the six band edge values to whatever is required by your license privileges. You should inspect the MagLoop.h header file completely with an eye toward constants that you may wish to change. If you plan on making code changes (such as changing the *Splash()* function), you may want to uncomment the DEBUG directive at the top of the file. This will cause the Serial object to be compiled into the code and cause some variable information to be display on the Serial monitor as the program runs. You've probably done this in other projects.

The *DoCalibrate()* function is primarily interested in storing the data that make it possible to fast tune the ML. By storing the stepper motor counts for each band edge, we can use those counts to approximate where to place the variable capacitor when doing a fast tune stepper move. If you look near the top of the ML .ino file, you will see the following:

```
long bandLimitPositionCounts[3][2];
float countPerHertz[3];

long presetFrequencies[MAXBANDS][PRESETSPERBAND] =
{
  { 7030000L,  7040000L,  7010000L,  7197000L,  7250000L,  7285000L},   // 40M
  {10106000L, 10116000L, 10120000L, 10130000L, 10140000L, 10145000L},   // 30M
  {14030000L, 14060000L, 14100000L, 14200000L, 14250000L, 14285000L}    // 20M
};
```

These array values are crucial to the proper running of the remote control unit.

The first thing that *DoCalibrate()* does is reset your stepper motor to the zero position. This is done by reading the pin on the STM32 you have set for ZEROSWITCH. The code then sets the frequency to be sent from the DDS to the LOWEND40M frequency you have defined in the header file. It then measures and records the SWR for that frequency with the stepper motor at position 1. However, because the micro-stepper motor controller can subdivided the native resolution of a NEMA 17 more (which is typically 1.8°), it might only move the capacitor by 0.05° depending upon the physical setup for your stepper motor (direct direct, geared, and so on).

The code then advances the stepper motor counter to 2 and remeasures the SWR. If the new SWR is lower than the previous SWR (and the SWR is greater than 1.05), it holds the new lower SWR value, advances the stepper motor by 1 again and remeasures the SWR. This process continues until either the SWR starts to increase or the SWR falls below 1.05. (It's a horseshoes-and-handgrenades thing.)

If the SWR starts to rise above 1.05, it breaks out of that loop and advances to the next band edge frequency (such as HIEND40M). You can change this target SWR in the MagLoop.h header file by changing TARGETMAXSWR (around line #30). If you lower the target too far, it is possible that the ML system simply can't tune to such a low value. If the calibration process finishes without displaying all of the band edges, as seen in **Figure 15.20**, you should set the target SWR value a little higher and rerun the Calibrate option. If the calibration runs to completion, the *Count* values you see in Figure 15.20 are written to EEPROM.

Note that each *Count* value is slightly higher than the previous value. Because we know the relationship between the stepper motor count and frequency is reasonably linear, we do not have to reset the stepper motor to a count of 0 for

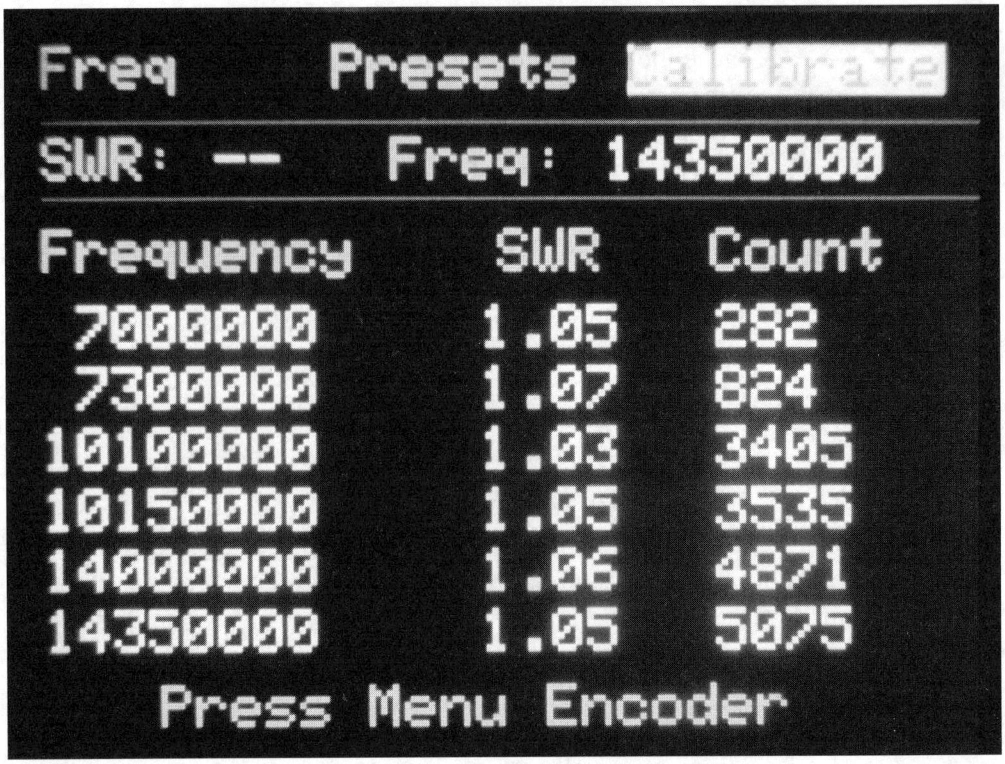

**Figure 15.20 — A completed run of the Calibrate menu option.**

each band edge. Instead, we set the count for the next band edge to be equal to 0.95 of the last band edge count. So, if LOWEND40M ends up with a minimum SWR count at 200, the count to find the HIEND40M value starts at 180 rather than 0. Doing this approximation saves time during the calibration process. The process is fairly slow and, depending upon your stepper motor and its controller, could take as long as 10 minutes to run. Most calibration runs are less than 3 minutes.

When the calibration process ends, the resulting *bandLimitPositionCounts[ ][ ]* array is written to EEPROM and appears as:

```
long bandLimitPositionCounts[3][2] = {
  { 282L,  824L},                  // Sample values from test loop
  {3405L, 3535L},
  {4871L, 5075L}
};
```

(There is no reason to expect your values to be even close to those presented here because of mechanical, electrical, and environmental differences between our ML and yours. Note the "L" data type designator is not actually part of the data, but rather serves to document their data type.)

The *bandLimitPositionCounts[ ][ ]* values can be used to boot-scoot the capacitor to a new frequency fairly quickly. For example, suppose you want to move to the frequency 7.1 MHz. Given the data above and knowing there's a linear relationship, all we need at the intercept and the slope coefficient. The slope coefficient is calculated:

```
slope = (824 - 282) / (7300000 - 7000000)
slope = .000181
```

The intercept for the low end of 40 meter was calculated to be a count of 282. Because we want to move 100,000 Hz from the band edge, we find:

```
count = intercept + slopeCoefficient * frequencyOffset
      =    282    +    .000181       *    100000
      =    282    +      18
count = 300
```

The code then can call the *stepper.moveTo(300)* function and then the *run()* function and the stepper motor moves the capacitor to that position. The user can then fine-tune the stepper motor setting if needed. Rather than calculated the slope each time, we use following array:

```
float countPerHertz[] = {
    .000180,            // (824 - 282) / (7300000 - 7000000)
    .002600,            // (3535 - 3405) / (10100000 - 10150000)
    .000582             // (5075 - 4871) / (14350000 - 14000000)
}
```

Because we don't have to calculate these constants at run time, having them stored in an array speeds things up just a smidgen.

The final thing to be stored in EEPROM are the preset frequencies you may want to use. The present frequencies should reflect your commonly used frequencies. We have set these to:

```
long presetFrequencies[MAXBANDS][PRESETSPERBAND] =
{
  { 7030000L,  7040000L,  7010000L,  7197000L,  7250000L,  7285000L},   // 40M
  {10106000L, 10116000L, 10120000L, 10130000L, 10140000L, 10145000L},   // 30M
  {14030000L, 14060000L, 14100000L, 14200000L, 14250000L, 14285000L}    // 20M
};
```

We have set the first two elements for each band to be the commonly-used CW QRP frequencies. You may use frequencies for nets that you check into often or other regularly scheduled contact frequencies. If you don't have any preferences, you might pick some in the middle of the CW or SSB segments of the bands. When you do a *Freq* menu option, you select the band to be used first. Once that selection is made, the code uses the first element of the *presetFrequencies[ ]* array for the selected band as the starting point for the auto-tune. The code performs the linear transform based on *presetFrequencies[bandSelected][0]* element's value.

The *presetFrequencies[ ][ ]* array appears in the project's .ino file around line #53. The Preset menu option that is visible in Figure 15.4 allows you to changes the presets "in the field" if the need arises. However, it is much easier and faster to just edit them in the IDE and recompile the code. You can change the values shown above to whatever you would like them to be. You should use the Arduino IDE to make any changes to the header or source code files. Once you are done making changes, recompile and upload the code to your ML remote unit.

### Freq Menu Option

Of the three menu options shown in Figure 15.4, we've discussed the Calibrate option in considerable detail. We have also stated that the Preset option allows you to edit the preset frequencies at run time, although editing with the IDE's editor is probably faster. The remaining menu option is the *Freq* menu. (Before using the *Freq* menu option, make sure you have the remote unit set to the Tune setting, not Operate.)

The first thing you must do is select the band you wish to use from the submenu shown in **Figure 15.21**. After you have selected the band, the display changes to show a page that is similar to **Figure 15.22**. The block cursor (located below the "0" digit character in Figure 15.22) indicates that the frequency increment is currently set to 1000 Hz. Therefore, each bump of the Frequency encoder changes the frequency by plus 1000 Hz (clockwise) or minus 1000 Hz (counterclockwise). For you to tune from the frequency shown in the figure to the upper end of the SSB segment of the 40-meter band would take approximately a bazillion spins of the Frequency encoder. Not good.

At this point in the code, rotating the Menu encoder to the left one detent places the increment block cursor under the "3" digit character. (The message under the current frequency tells the function of the encoder and buttons.) The new position means the increment increases to 10,000 Hz per detent. In similar fashion, moving the increment cursor over the "3" digit character raises the increment to 10,000 Hz. Depending upon the magnitude of the change you want to make, you can adjust the increment fairly easily to so that the frequency

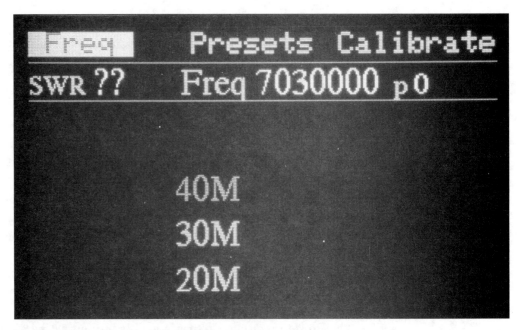

Figure 15.21 — The Band Selection submenu.

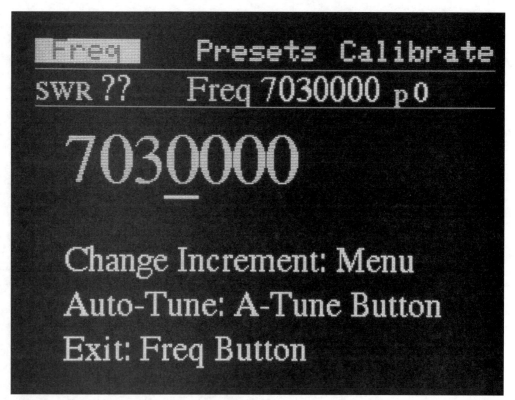

Figure 15.22 — Setting the new frequency.

change takes less than a bazillion turns. That's a good thing.

Pressing the Auto-Tune switch sets the new frequency and the auto-tune process is initiated. When the lowest SWR is found, the display shows the newly found SWR low. Figure 15.4 shows the optional frequency scan of the SWR, which graphically displays the bandwidth at the selected frequency.

There are no other menu options at this time, although there's plenty of memory for you to add options if you wish.

## Results

That's all there is to it. We have successfully modified several used capacitors to bring them into a range that was required for our ML project. Our final ML and Controller are shown in **Figure 15.23**.

We have exercised the ML pretty hard and gave it a trial-by-fire during the 2019 Field Day. We also recorded over 10,000 WSPR reverse beacon reports on the ML and compared those reports with the reports on an end-fed half-wave (EFHW) antenna which was at a height of about 40 feet. We tried as best we could to make the tests comparable, although we know that even a few minutes between antenna changes violates the *ceteris paribus* (other things being equal) assumption(s). The ML was about 4 feet above the ground at its lowest point. **Figure 15.24** shows a summary of the results.

As you would expect, there is a large cluster of observations at 2500 km or less, obviously representing contacts throughout the US. A second cluster starts around 5000 km and a third cluster around 13,000 km. The second and third clusters represent reports from Europe and Asia, respectively. The sparse

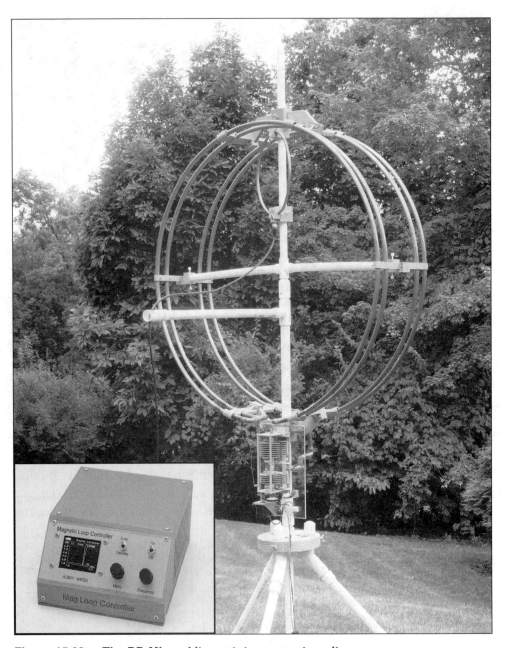

**Figure 15.23 — The DD-ML and its prototype remote unit.**

gaps in between clusters are the two oceans, with a few island reports sprinkled into the results. Relatively fewer reports using a larger single-loop with a 6-foot diameter are also present in Figure 15.24, but the sample size, while statistically significant, is still relatively small compared to the sample size for the EFHW and the ML.

A Least-Squares regression line was fitted to all three antenna data sets. The large-diameter ML and the DD-ML estimated parameters for the regression equations ($\alpha > 0.975$) are virtually the same for the two loops. The EFHW does outperform our ML by about 2 dB. To the human ear, the 2 dB difference between the EFHW and the DD-ML is probably not noticeable.

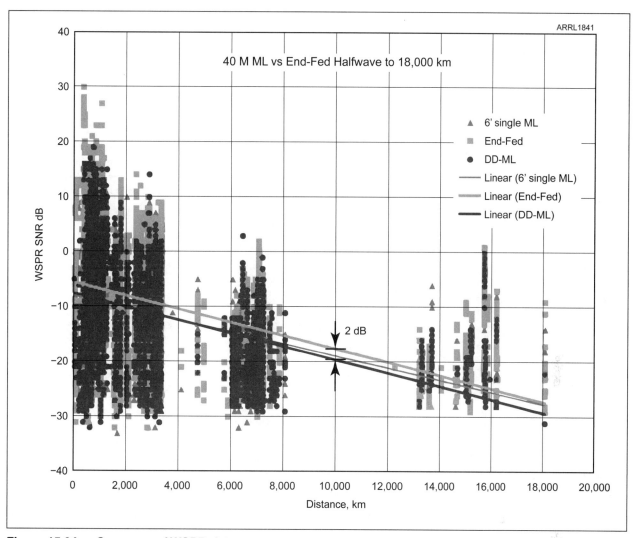

Figure 15.24 — Summary of WSPR data reports.

## Conclusion

So, is it worth your time, effort, and money to build your own ML? Well, it better be or the last two chapters were an unnecessary denuding of the nation's forests. In reality, however, the answer should be: "It depends." If you live where there's enough room for a half-wave dipole up 70 feet in the air, you're going to have some level of advantage over the ML. If you're attempting to load your condo's rain gutter for your antenna, we bet the ML will do a better job for you. If you want to take the ML in your van to your favorite camp site, taking the ML will be easier than taking the gutters. Those of you facing strict HOA limitations, the ML might be the difference between getting on the air or not.

We think you'll find our DD-ML is a worthy performer with a lot more features at a fraction of the cost of a commercial ML.

# CHAPTER 16

# Finishing Your Projects

We all like to be proud of the projects we build, but sometimes getting the final finished product to look great is just too much work or too expensive. Still, after doing the hard work of constructing the circuit boards and wiring it all up, the final step should be to put the dressing on the project inside and out.

It turns out that, with ordinary materials and not too much expense, we can create one-of-a-kind projects that look really good — almost professional! We have learned the hard way how to do the finishing steps and we share some of those tricks in this final chapter.

There are two aspects to getting a good final product, one inside and the other outside. First, we tackle the inside portion. What we really are addressing here is neat and tidy construction. Caring construction means wiring that is neatly arranged and doesn't look like a damaged spider's web, with thoughtful placement of parts, neat soldering, finishes to the cabling, cleaned PCBs, and connections to components. We address each of these topics in this chapter. **Figure 16.1** shows the final product for our DSP Post Processor (DPP) in a custom case.

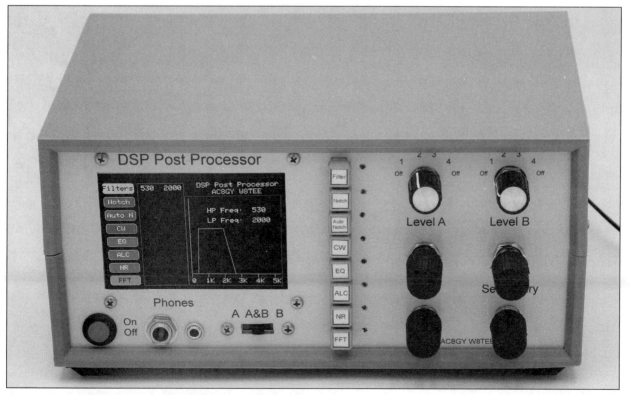

Figure 16.1 — DSP Post Processor in a custom case.

## Circuit Layout

Probably the most important aspect of getting your project right is component layout. By layout we mean the arrangement of components in the project. Layout should start with main circuit board. Component layout is also an important task from a performance point of view, too. Generally, we make a perf board or proto board version of each project first, even if we intend to do a PC board later. While it takes a bit of time to wire up by hand, making changes is easier and, unless you make your own PCBs, there isn't the delay to have a PCB fabricated.

When doing a board layout, there are several things to keep in mind:
1) Use good RF practices, which means you should
    a) keep signal paths short
    b) separate inputs and outputs
    c) create a ground plane, if feasible
2) Separate analog and digital circuits to prevent cross-talk and noise.
3) Place the PCB edge connectors for front panel items on one side and rear panel stuff on the other. We encourage the use of multi-pin connectors and ribbon cable for ease of construction and neatness. Connectors also allow the PCB to be removed and replaced if necessary, without unsoldering a bunch of wires.
4) Use sockets for major components on the PCB, such as the microcontroller and other modules. While others speak of the "unreliability" of sockets, that has not been our experience.
5) It really helps to label your connectors and sockets on the perf board.

Figure 16.2 — Socket and connector labels.

This prevents lots of wiring mistakes and saves time, as we can attest to from experience. **Figure 16.2** shows an example of our labels. We use old address labels cut to fit next to the connector or socket.

**Figure 16.3** shows a finished perf-board construction layout for our Programmable Power Supply. Use larger diameter wire for power and ground connections. For low current connections, we tend to use #30 solid wire-wrap wire because pin spacing is close and larger wires are harder to manage.

Once you are happy with a proto-board or perf-board layout, perhaps a PCB is in order. There are several good and inexpensive PCB design programs available. We use the free version of *DipTrace* for both PCB design and schematic capture. Finished designs can be fabricated by sending the Gerber files to any of the many online vendors. Sample quantity boards from overseas vendors cost around $5 but may take two to three weeks to get. In the end, it is well worth the wait. We started a builder's club and those members are happy to share the cost of the PCBs.

**Figure 16.4** shows the PCB layout for one of our projects that was designed in *DipTrace*. Al is the expert on *Diptrace* and Jack has only "dabbled" with it, but agrees that a little effort with the software is more than worth the effort, especially if you're going to share the PCB.

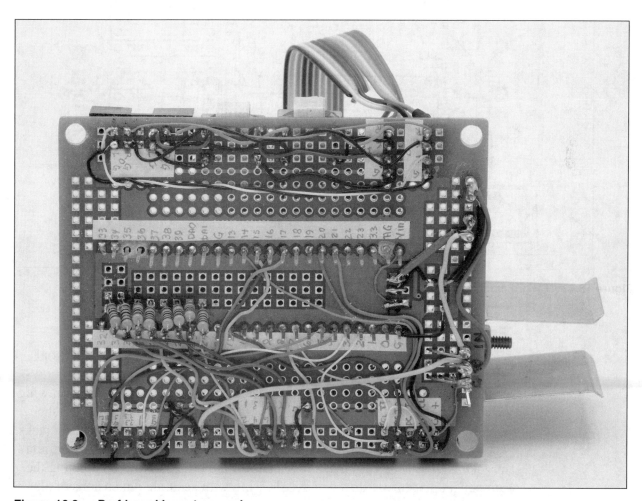

**Figure 16.3 — Perf-board layout example.**

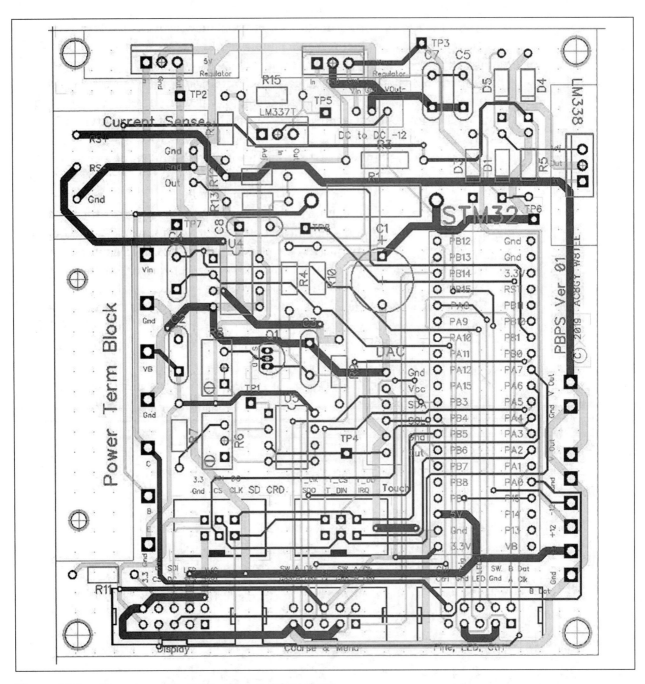

Figure 16.4 — Programmable Power Supply PCB layout.

## Front Panel Layout

Once the main circuit components are complete, the next task is the front panel. The best way to begin this process is with a couple of sketches. First, create a rough drawing of the case outline on a piece of paper. Lay that drawing on a flat surface and then do a rough placement of project's components, such as the display, encoders, switches, and connectors such as key and banana jacks. The goal here is to make sure there is enough room for everything. A common mistake is to select a case that's too small for the project. After completing the component placement, do a detailed drawing with accurate dimensions of all parts, holes, etc.

Place the knobs you intend to use on the drawing and determine if there is enough room for your fingers to use the controls comfortably. Also, some controls are used a lot more than other controls and that should be factored into their placement. I prefer to have power supply connections on the back of the unit, but would like to shoot the person who puts key, mic, and headphone jacks on the backplane. It's even more stupid with rigs that use the key insertion to deactivate something else (for example, the mic). Such an arrangement forces you to move the rig anytime you want to change modes. That's just plain stupid RDD (Really Dumb Design).

To some extent, if you're making this project just for yourself, perhaps the fact that you are left or right handed may influence how you think the controls should be placed. Also, think about *how* the controls are used. For example, for the CW Messenger project, it makes more sense to place the push button message switches on the top of the case rather than on the front panel. The reason is because many users will "slap" the switches during use and you don't want the case to be "walking" away from you as you use it.

When you are satisfied with the layout, transfer it to your front panel. We usually put masking tape over the whole front panel and then mark the positions of holes and write the diameter and other dimensions on the masked panel. A typical hand drawing on the front panel is shown in **Figure 16.5**. If you look closely at Figure 16.5, you can see near the bottom how we redrew four banana plug holes from their initial position. Don't be afraid to change your mind.

A better way to do this is to use a drawing program such as Open Office *Draw* to sketch up your layout. Then print the drawing, cut it out, and tape to the front panel. An added bonus of this approach is that you now have a template for creating an overlay for the front panel, including text and other graphics. **Figure 16.6** shows a drawing/sketch of the front panel for our Mag Loop Controller done in OpenOffice *Draw*.

Once you have a hand drawn or graphic faceplate done, use a center punch to mark the holes for drilling if you are using a metal or blank front panel. Drill

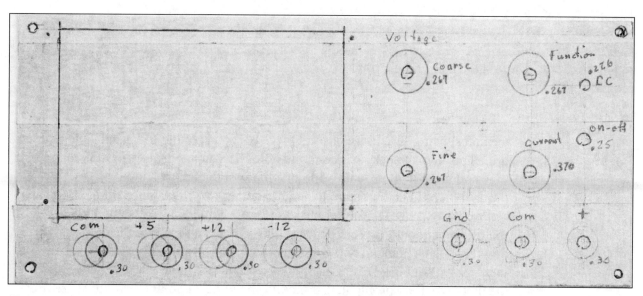

Figure 16.5 — Front panel sketch for Programmable Power Supply.

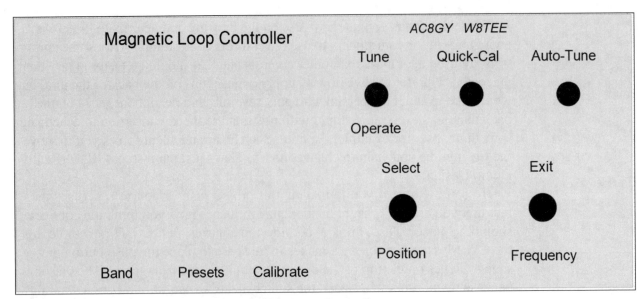

**Figure 16.6 — Front panel *Draw* sketch for Mag Loop Controller.**

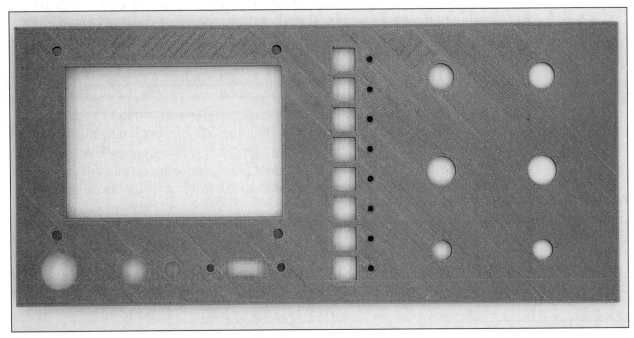

**Figure 16.7 — 3D printed front panel for the DPP.**

the holes, being sure to use a solid piece of wood behind the panel when you drill. This helps to ensure nice round holes. If you are drilling plastic, take your time and use a light pressure. Too much pressure and you run the risk of having the drill bit bind and, potentially, crack the plastic ... been there, done that.

We have been 3D printing our cases for a while now, so we don't have to drill and cut out the front panel. Al just creates a model of the layout in a program such as *Fusion 360* and print the final version. The plastic filament is inexpensive, and changes or error correction is easy and fast. If you don't like the first layout, no problem, just modify it and make another. No more fretting because you messed up your only front panel. **Figure 16.7** shows the final 3D printed front panel for the DPP.

## Metal versus Plastic Cases and Panels

Metal cases are very nice, no doubt, but nice ones are expensive. Cheap thin aluminum cases don't hold up as well as a more robust steel case. Our own custom 3D printed cases usually have a wall thickness of 3 mm. The result is a sturdy, attractive case that is exactly what the project needs.

The only downside is the lack of shielding. There are two ways around the lack of shielding. The first is to use aluminum or copper foil tape to cover the inside. We did this on our Antenna Tuner project. The second way is to get some conductive paint and lightly spray the inside with an even coat. The paint is expensive, but makes a very nice job. In either case, if there are "gaps" between pieces of the case, you need to make sure the pieces are electrically tied together with grounding wire.

Finally, do the rear panel in the same way as the front panel.

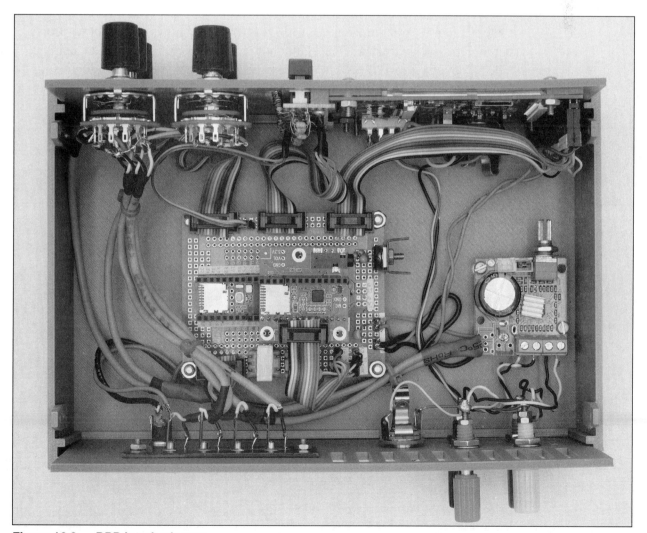

**Figure 16.8** — DPP interior layout.

## Interior Layout

Once the project components are finished and the front panel is designed, it is time to place those components on the bottom of the case and do the layout. You already created a circuit board with connectors for front and back panels. Now the task is to place everything to make the wiring paths as short as possible.

We start by covering the bottom of the case with masking tape. Then place each component on the base and find the best arrangement that keeps the wire lengths short, provide air paths for components that heat up, and allow access to adjustments as needed. Don't forget that USB connectors on µC boards should be positioned so it is easy to do software upgrades. Now mark the position of each attachment hole on the case. Try to avoid awkward screw hole locations that are hard to reach. Leave enough room for attachment to front and back panel components.

When you are satisfied with the layout, drill the holes and do a test to make sure the components fit. You can now mount the components or wait until later in the process. **Figure 16.8** shows the actual interior layout for the DPP.

## Wiring

We have all seen wiring jobs that look like an explosion in a wire factory. Other examples are almost an art form. **Figure 16.9** shows a homebrew project by David Richards, AA7EE. Note how his circuits are uncrowded, using a "clean" layout. Most of his projects use the Manhattan style of building using PCB material for the case, but all of his work is absolutely beautiful. Our projects strive to achieve Dave's level of neatness, but seem to fall a bit short in one aspect or other, but they do work.

Figure 16.9 — Beautiful layout by David Richards, AA7EE.

The best way to hook up things from a performance point of view is to route the wires from point to point, while leaving enough slack to tuck the wires neatly around components. "Neatly tucking" stranded wire is quite similar to herding cats. Our experience is that solid copper wire almost always yields a neater layout. When bundles of wires are going to the same location, they can be grouped together with cable ties and even secured to supports or held in place with cable clamps. Cable ties are cheap and go a long way toward reducing the clutter. Some pre-planning is required to pull off attractive cable layout successfully.

What about good RF practices? Wires close to each other are capacitively, and possibly inductively, coupled. Inputs and outputs need to be separated and there is always the possibility of cross-coupling. Digital signals need to be kept away from low level signal cables as well. Signal wires from the front to the back should either be carried with shielded cable or twisted-wire pairs. Having said this, it is possible to keep RF paths short and still look good. Making things look neat has the added benefit of making troubleshooting easier when necessary.

Another good practice is to use multiple colors of wire so that connections can be traced easily. Wherever possible, we use black wire colors for ground and red wires for positive supply connections. We try to keep the supply and ground connections at the bottom of the case and routed around the edges. Position and length of power supply cables is less critical than other connections.

**Figure 16.10** is an example of routing wires for both utility and good ap-

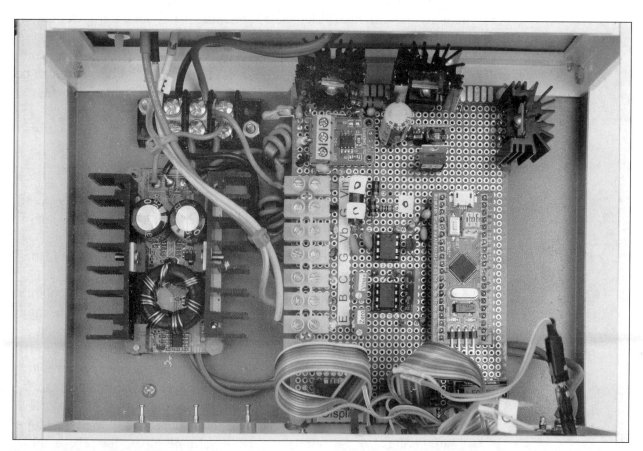

**Figure 16.10 — Project wiring.**

pearance. Note the use of ribbon cable, routing of power wiring and colored wires.

## Making Connections

Many of our projects have lots of inter-connections of components. For both utility and good appearance, we use multi-pin connectors and ribbon cable. We also make liberal use of heat-shrink tubing to provide isolation and strain relief at the connection points. This prevents breakage as you move components around and do the assembly.

We also often use hot glue to secure wires and even to coat vulnerable connections to with it to prevent flexure. A hot air gun makes it easy to "undo" hot glue additions.

## Front Panel Finishing

After you have spent all of that time building your project, you want the first impression to be the best you can. This means dressing up the front panel with annotations and other text. There are a number of ways of doing this:
• Create artwork for silk screened panels. This is probably the most professional and most expensive way to go. If you are making a large quantity, it is feasible, but not for one-of-a-kind projects.
• Use a label maker to create stick-on labels. This is better than nothing but is certainly not professional looking.
• Use transfer lettering. The end result can be very good, but getting it right is very tedious and difficult to change if something is not quite right. Our experience is this is almost always zero fun.
• Use clear overhead projection film and print your artwork on a laser printer. Then glue the film to the front panel. We have done this and the result can be quite good. However, we have had some issues with the film lifting over time. We use spray-on adhesive for the attachment. The trick is to get the glue on evenly. Do this only in a well-ventilated space. The fumes can be pretty nasty. Finally, it is best to coat the artwork with a lacquer-based fixative for protection.

### Graphic Faceplate

Our favorite process, however, is to print our artwork on good quality photo papers with an inkjet printer. The artwork is prepared using a drawing program, perhaps using the same template used for front panel layout. Text, graphic primitives, color blocks and text can all be employed. The big advantage is the flexibility of design.

The process is simple. First do the design and print off some tests on plain paper. Use the test pieces to overlay on the front panel to ensure that all of the holes and cutouts line up. Then print a version on the heavy photo paper, but don't print the outline of the holes or cutouts. These holes are actually cut out later, after the paper is glued to the front panel. Now coat the photo paper with the fixative mentioned previously. This waterproofs the artwork and helps to keep off fingerprints.

Next, carefully align the paper to the panel and glue in place. We attach one or two central points using a couple of dabs of gel-type super glue. These gel

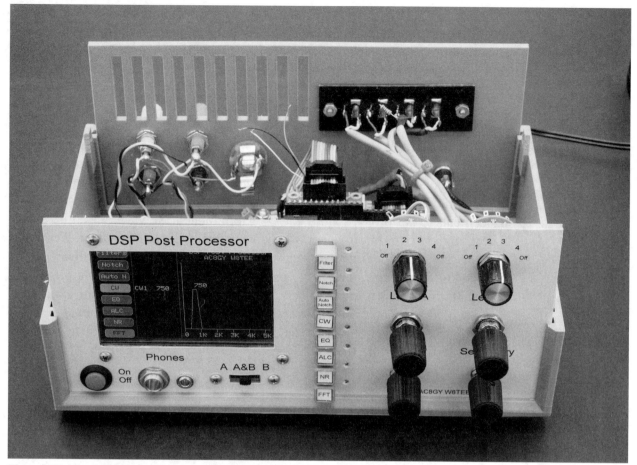

Figure 16.11 — Example of front panel finishing.

glues give a few seconds of working time to make sure the positioning is correct. Once the first glue spots are secure, glue the rest one side at a time. It is not necessary to glue it all, just use dabs at key locations, such as edges, around large cutouts and a few in between.

After the glue is dry, place the new sandwich upside down on a cutting mat or piece of wood and using a very sharp small utility (i.e. X-acto) knife, cut out the edges, using the panel as a guide. Go slow and try to make straight first cuts that go all the way through the paper. Second cuts sometimes veer off in other directions. Then cut out the holes and other openings making sure to keep the knife straight. If you concentrate on clean cuts, you can do away with the need for bezels around the displays. (Most of our display projects use this technique.)

Finally, turn the assembly over and clean up the holes for mounting pots, encoders, etc. Try to avoid handling the front surface to prevent marring.

Wow! You now have a very professional-looking front panel that really shows off your new project. **Figure 16.11** shows an example of this technique.

## Final Touches

There are a few remaining things to do to complete the project. We like to add feet to the case if it does not already have them. Rubber or soft plastic stick-on feet are easy, or screw-on feet could be used for more security.

Also, look at the knobs you selected — do they complement the rest of the look? If not add some nice-looking metal or other fancier knobs of the same size. Hamfests always seem to have some nice knobs available. Look for attractive knobs even if you don't need them now so they're available for the next project!

## Conclusion

This is one of those good-news-bad-news things. We've come to the end of the book. Our work stops here, but yours just begins. We really hope you've read this book from cover-to-cover, even those chapters you weren't initially interested in (Chapter 2 and 3?). We hope you have picked up some ideas for enhancements to our projects along the way, too. Indeed, we view this entire book as a springboard for improvements/enhancements by you, the readers. We know there are engineers out there who can make improvements to the hardware and we also know there are some really bright software people who can improve the code that drives many of the projects. Our hope is that you share your improvements with the rest of us via the website mentioned earlier in the book.

So, why are you sitting there looking at this sentence? Get out there, start building, and share with the rest of us!

## APPENDIX A

# Products and Component Sources

## Products

Most of the projects in this book directly augment our ham radio hobby. Many of us also enjoy kit building and most QRP products. We are listing several of those here because of our experience with the products plus some of the book projects augment the products shown here.

**CR Kits (www.crkits.com)** — Their HT-1A (**Figure A.1**) is a 40 and 20-meter dual band CW transceiver kit that comes with all of the SMD parts already mounted on the board. Jack was interested in this transceiver because it was a dual band QRP rig from the same company that produced the CRK 10-A, which had a very good, low noise, receiver that we modified for our VFO in the first *Projects* book. The HT-1A features 5 W output, RIT and XIT, AGC and S-meter, built-in keyer, full break-in, side tone selection, and other features.

**HF Signals (www.hfsignals.com)** — This company sells the μBITX, (**Figure A.2**) a 5-band, 10 W, CW/SSB transceiver, as a "semi-kit." That is, all SMD parts, toroids, electrolytics, and other components are already soldered in place. You only place the off-board components (switches, connectors, etc.). Al and Jack each own at least two of these transceivers. It is an amazing kit for the price.

Figure A.1 — The CR Kits HT-1A dual-band CW transceiver.

Figure A.2 — The HF Signals µBITX transceiver.

**MyAntennas (myantennas.com)** — Al and Jack both use the MyAntennas 8010 antenna for their home stations. It's a multiband end-fed (EFHW) antenna and built like a tank. The antenna and matching transformer are shown in **Figure A.3. Figure A.4** shows an actual SWR scan of the antenna installed at Al's house, and it is probably pretty typical of a home installation as neither of us has done anything special to the antenna. We're using them just as they come out of the box.

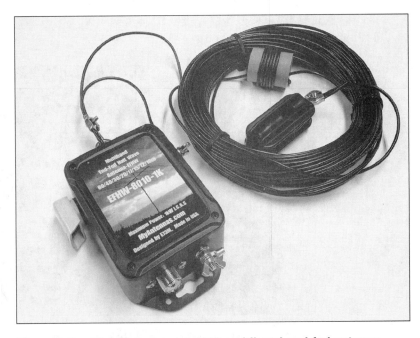

Figure A.3 — The MyAntennas 8010 multiband end-fed antenna.

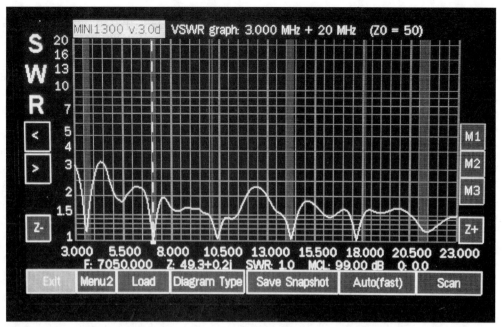

Figure A.4 — Scan of the MyAntennas 8010 multiband end-fed antenna showing low SWR on the 80, 40, 30, 20, 17, and 15 meter bands.

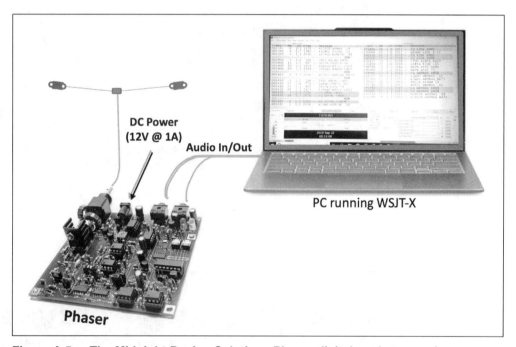

Figure A.5 — The Midnight Design Solutions Phaser digital mode transceiver.

**Midnight Design Solutions (midnightdesignsolutions.com/phaser/)** — The Phaser Digital Mode Transceiver (**Figure A.5**) is a single-board, 4-W SSB radio specifically designed for using digital modes with computers running *WSJT-X*, *DigiPan*, *fldigi* and most other digital mode applications. The Phaser can transmit and receive on the 80, 60, 40, 30, 20 or 17-meter amateur bands. Each of these monoband transceivers is programmed to operate first at the popular FT8 frequencies. A push button entry shifts the operating frequency to a programmable alternate frequency, which is initially programmed for the JS8 "watering hole." The alternate frequency can be reprogrammed for the user's mode of choice (for example, FT4, PSK31, JT65A, WSPR, Feld-Hell, Olivia, or SSTV). The transmitter features an adjustment-free phasing single-sideband (SSB) design and provides unwanted sideband suppression in excess of 30 dB across its operating range. The use of SSB also eliminates the issue of out-of-phase signal cancellation at the Phaser's direct-conversion receiver. As such, it allows Phaser users to communicate with each other — a clear advantage over the use of simpler double-sideband (DSB) transceivers.

**QRP Guys (qrpguys.com)** — The company sells a wide variety of kits from transceivers to bare PC boards, including the JackAl board and several of Al and Jack's PCBs. Please note that this is a small company and they stack orders up for once-a-week shipping, so delivery may be up to a week longer than expected if you order late on the pickup day. We wouldn't let them sell our boards if they weren't reliable.

Figure A.6 — The QRP Labs QCX+ CW transceiver.

Figure A.7 — The SOTABEAMS DSP filter.

Figure A.8 — The WA3RNC 40 Meter CW transceiver.

**QRP Labs (qrp-labs.com)** — The company sells the QSX 5 W, single-band, CW transceiver kit for most HF bands. As this was written in mid-2020, the company announced that it would soon offer its QSX kit, which will be an SSB transceiver, with a multiband version down the road. Their products are first rate and Jack owns two QCX transceivers and will own a QSX when it becomes available. In mid-2020, QRP Labs announced the QCX+ (**Figure A.6**), which includes a larger PCB for easier assembly, an extra "experimenters" PCB, and case for $80.

**SOTABEAMS (www.sotabeams.co.uk)** — The company sells a variety of products, many which are involved with Summits On The Air (SOTA) activities. While this book has a DSP post processor project, SOTABEAMS has a DSP filter (**Figure A.7**) that is about the size of a deck of cards, which makes it more suitable for backpacking and it works very well. The filter is easy to use and very effective.

**WA3RNC (www.wa3rnc.com)** — The WA3RNC 40 meter CW QRP transceiver (**Figure A.8**) is an incredibly small, but very effective rig. It is a semi-kit with over 100 SMD parts already soldered in place. The builder adds a couple of dozen other components and you're done. It features adjustable output from 0 to 5 W, 0.15 microvolt sensitivity, factory assembled digital dial, pre-wound toroids, sharp IF crystal filter and many other features. This is a very "backpackable" rig.

## Components

- **All Electronics (www.allelectronics.com)** — Component supplier.
- **Arrow (www.arrow.com)** — Complete supplier of components.
- **Debco Electronics (www.debcoelectronics.com)** — Components. A local store we buy from often.
- **Digi-Key Electronics (www.digikey.com)** — Component supplier. No minimum order.
- **Jameco Electronics (www.jameco.com)** — Component supplier.
- **Kits and Parts (kitsandparts.com)** — Components, especially toroids and hard-to-find mica caps. Diz, the owner, also sells quality QRP kits.

- **Marlin P. Jones & Associates (www.mpja.com)** — Component supplier. Monthly email specials are interesting, and they are a good source for all components including power supplies.
- **Mouser Electronics (www.mouser.com)** — Component supplier.
- **Newark (www.newark.com)** — Component supplier.
- **Surplus Electronics Sales (www.surplus-electronics-sales.com)** — Supplies new components with very good prices and fast delivery.
- **Tayda Electronics (www.taydaelectronics.com)** — Components. Most items ship fairly quickly from their US warehouse.

### eBay

There are probably thousands of component suppliers on eBay, and probably 98% of them are reputable vendors. However, some components, like power transistors, DDS modules (such as the AD9851 and Si5351), and similar items sometimes are "floor sweepings." Chinese vendors have learned that US buyers prefer to buy from a US company, so some Chinese vendors now have drop-ship facilities in the US that show a US address. Pay close attention to delivery dates as eBay expects their vendors to ship on time.

# Index

Note: The letters "ff" after a page number indicate coverage of the indexed topic on succeeding pages.

Page numbers starting with I refer to the introductory section, About This Book, prior to Chapter 1. Page numbers starting with A refer to Appendix A.

*#define*: .................................................2-25, 6-8, 9-28
*#endif*: .................................................................9-28
*#ifndef*: ...............................................................9-28
*#include*: ............................... 2-21, 4-7, 4-15, 9-28
% (C operator): ..................................................9-25
*****/ sequence: ..................................................2-10
*/ pair: ..................................................................2-10
.cpp file (C++): ........................... 2-25, 3-15, 9-27
.h (header) file: .......................................... 4-7, 9-27
.ino file: ...................................... 2-25, 3-15, 9-27
/* and */: ..............................................................2-28
/* pair: .................................................................2-10
/***** sequence: ..................................................2-10
// pair: ..................................................................2-29
:: ................................................................3-5, 8-29
µBit-X transceiver kit: ..........................................A-1
µC: .....................................................................1-1ff
16x2 LCD display: .......................................4-1, 4-3
2.4-inch TFT LCD display: ....................................4-1
2.8-inch TFT LCD display: ....................................4-1
20x4 LCD display: ................................................4-1
2N3904 transistor: ..............................................10-7
3 dB cutoff frequency: ......................................11-12
3.3 V regulator: ..................................................10-5
3.5-inch TFT LCD display: ....................................4-1
3D printed case: .......................................11-8, 12-9
3D printing: .........................................................16-6
4N25 optocoupler: ..............................................10-7
5-inch TFT LCD display: ......................................4-1
6-pin IC: ..............................................................10-7
7-segment LED display: ........................................4-1

## A

AC mains: .............................................................8-1
Accuracy: ..............................................................6-7
AD8397 log amp: ................................................13-4
AD9833 DDS: ............................................13-4, 13-8
AD9850 DDS: ..........................................15-8, 15-13
AD9851 DDS: ............................................13-4, 13-8
   Library: ........................................................13-4
Adafruit
   GFX graphic library: ......................................8-26
   ILI9341 library: .............................................7-11

ADC (*see* Analog-to-digital converter)
Air path: .............................................................16-8
Air variable capacitor (AVC)
   Dual stator: ...................................................15-6
   Modification: ...............................................15-22
   Removing plates: .........................................15-23
   Single stator, single rotor: .............................15-6
   Spacers: ......................................................15-23
   Split stator: ...................................................14-3
Algorithm: ...................................................2-1, 2-5
   Squaring a number: .....................................2-15
Analog style meter emulation: ..........................13-23
Analog-to-digital converter (ADC): ........... 11-6, 12-3
Angle brackets (<,>): ........................................2-22
Ant: ....................................................................10-8
Antenna analyzer (W8TEE): .............................15-18
Antenna Tuner with Graphical SWR Analyzer (ATA): 9-1ff
   Calibration: ...................................................9-21
   Main menu: ...................................................9-14
   Oper (Operate) menu: ..................................9-20
   Software: ......................................................9-22
Arduino
   Clones: ...........................................................1-8
   I2C and SPI interfaces: ..................................4-8
   Mega 2560 Pro Mini: .....................................1-1
   Nano: .............................................................1-1
   Program anatomy: .........................................2-8
Arduino Integrated Development Environment
   (IDE): ...........................................1-8ff, 4-7, I-1
   .ino source code file: ...................................2-25
   Arduino board selection: ................................1-9
   Compile icon: ...............................................1-11
   Executable program: ...................................1-11
   Incremental compile: ...................................2-26
   Installation path: ............................................1-9
   Linker: ................................................1-11, 2-19
   Multiple code files: ......................................2-25
   Multiple project files: ..................................2-25
   Port selection: .............................................1-10
   Program stats window: ................................1-12
   Serial Monitor: .............................................1-12
   Source code window: ..................................1-11
   Tab: ..............................................................7-15
   Upload icon: ................................................1-12

*Arduino Projects for Amateur Radio*: ........................... I-1
ARRL Field Day: ................................................... 14-11
ASCII Table program: ............................................. 2-29
Assembler: ............................................................ 1-16
Atmel: ..................................................................... 1-5
*attachInterrupt()*: ..................................... 4-22, 10-19
ATX style power supply: ........................................ 5-1
    Color codes: ................................................... 5-3
Audio amp: ........................................................ 7-10
Audio Library: ..................................................... 11-1
Audio microphone signal conditioner: ................ 12-1
Auto notch: ........................................................ 11-13
Automatic gain control (AGC): ............................ 8-32
Automatic level control (ALC): ........... 11-16, 11-19, 12-13
    Compressor: ................................................ 12-10
    Threshold: ...................................... 11-16, 12-10
Automatic notch filter: ....................................... 11-13
Axis labels: .......................................................... 9-24

## B

Back panel: ........................................................ 16-8
*backlight()*: ....................................................... 4-15
Backpack: ........................................... 2-14, 3-9, 3-19
Band edges: ...................................................... 15-25
Band-pass (BP) filter: ........................................ 11-18
*bandLimitPositionCounts[][]* array: ..................... 15-32
Base class: ......................................................... 3-14
Battery: ...................................................... 8-31, 10-3
BAV21 diode: ....................................................... 6-5
BC (Bad Code): .................................................... 2-6
*begin()*: ...................................................... 3-9, 4-14
*Beginning C for Arduino*: ................................... 2-30
Bezel: ......................................................... 6-9, 8-24
Bias: .................................................................... 13-8
Big Endian: ....................................................... 15-29
Bill of Materials (BOM): ...................................... 7-13
Binary instructions: ........................................... 1-16
Binary search: ................................................... 7-17
Binary tree: ............................................... 7-18, 7-19
Biquad filter: .................................... 11-17, 12-13
Biquadratic cascaded filter: ............. 11-17, 12-13
Bit: ....................................................................... 1-1
Black box: ......................................................... 2-14
Blink program: .................................................. 1-12
Blocking: ........................................................... 15-7
Blue Pill (BP): ..................................... 1-1, 7-2, 15-7, I-3
    I2C and SPI interfaces: ............................... 4-10
Bluetooth: .................................................... 1-7, I-1
BNC connector: ................................................. 6-3
Board Manager: ................................................ 1-22
Bode Plotter: .................................................... 13-9
Boost converter: .............................................. 8-14
Boost regulator: ................................................. 8-5
Boost/buck converter (B/B): .............................. 5-2
Bootloader: ........................................................ 1-8
    Old: ................................................................ 1-8
Bourns encoder: .............................................. 10-6
Bubble Sort: ...................................................... 2-6
Bucket: ............................................................... 3-4
Bucket analogy: ............................................... 2-18

Building class: ................................................. 3-13
Buzzer: .............................................................. 10-2
Bypassing: ....................................................... 13-10
Byte: ................................................................... 1-2

## C

C programming language: .............................. 2-1
    .cpp file (C++): .......................................... 2-25
    Keywords: .................................................... 2-2
    Standard Library: ....................................... 2-2
    Symbol Table: ........................................... 2-16
    Symbol Table with Memory Address Filled In: ...... 2-17
C++ programming language: ......................... 3-1ff
    Class: .......................................................... 3-2
    *cls*: ............................................................. 3-2
Cable ties: ....................................................... 8-24
*CalculateWatts()*: ............................................. 6-8
Calibration ...................................................... 8-24
Called: .............................................................. 3-11
Caller: ............................................................... 3-11
Camel notation: ...................................... 4-17, I-6
Cascading *if* statement: ................................ 2-27
Case material: ................................................ 16-7
Case statement block: ................................... 2-27
Cast: ..................................................... 2-18, 10-15
Cat5 cable: ..................................................... 15-15
Center frequency: .......................................... 11-14
Ceramic capacitor: ......................................... 11-7
CH340 device driver: ............................ 1-8, 1-10
Child class: ..................................................... 3-14
Circuit layout: .................................................. 16-2
Clark, Roger: ................................................... 1-17
Class attributes: ............................................... 3-3
    Assigning: ................................................. 3-19
    Retrieving: ................................................ 3-19
Class constructor: ............................................ 3-4
Class D amplifier: ............................................ 11-6
Class member: ................................................. I-7
Class methods: ........................................ 3-3, 3-7
Class properties: .............................................. 3-3
Clock speed: ..................................................... 1-5
*cls*: ...................................................................... 3-3
Clutter: .............................................................. 2-29
Code quality: ..................................................... 2-6
    BC (Bad Code): ........................................... 2-6
    RDC (Really Dumb Code): ......................... 2-6
    SDC (Sorta Dumb Code): ........................... 2-6
    WC (Wow Code): ......................................... 2-7
Code walkthrough: ......................................... 4-12
Coding conventions: .................................... 2-22ff
Coding styles: ................................................ 2-23
Cohesion: ....................................................... 9-26
Cohesive function: ........................................ 2-26
Comma-separated values (CSV): ................. 8-10
Comment
    /* and */: .................................................... 2-29
    // pair: ........................................................ 2-29
    Multi-line: ................................................... 2-29
    Single line: ................................................. 2-29
*commit()*: ....................................................... 10-15

Compile icon: ........................................................ 1-11
Compiler: ................................................... 1-3, 3-21
Component sources: ............................................... A-5
*Compressor()*: ....................................................... 12-13
Computer program: ................................................... 2-1
Computer programmer: .............................................. 2-1
*Config* menu: ............................................................. 7-7
Connectors: ............................................................ 16-2
    DB15: ........................................................... 15-15
*const*: ...................................................................... 2-25
*contestExchanges[]*: ............................................. 10-12
Continental drift: .................................................... 7-21
Continuously variable filter: ................................... 11-3
Conventions used in this book: ................................ I-6
Converter: ................................................................ 5-2
Cookie cutter: ....................................................... 3-14
Copper foil: .............................................................. 9-9
Copy: ..................................................................... 3-19
*CopyCat* submenu: ................................................. 7-7
Core libraries: ....................................................... 1-17
*cos()*: .................................................................... 13-23
Coupling: .............................................................. 2-26
Coupling (function): ............................................... 9-25
CR Kits HT-1A: ....................................................... A-1
Cross-coupling: ..................................................... 16-9
Cross-talk: ..................................................... 9-9, 16-2
Cutoff
    Filter response: ........................................... 11-12
    Filters (HP and LP): ..................................... 11-12
    Frequencies: ................................................. 11-3
CW: ......................................................................... 7-1
    Farnsworth Method: ....................................... 7-4
    Flashcard approach: ..................................... 7-7
    Koch Method: ............................................... 7-4
    PARIS: ........................................................... 7-4
    Rhythm: ........................................................ 7-3
    Word: ............................................................ 7-4
CW filter: ............................................................. 11-14
    Response plot: ............................................ 11-14
    Width: ............................................................ 11-7
CW Messenger (CWM): ....................................... 10-1ff
    Edit Message option: ................................. 10-14
    Main menu: .................................................. 10-10
    Schematic diagram: ..................................... 10-3
    Software: ....................................................... 10-8
CWM.h: ................................................................... 10-8

# D

DAC (*see* Digital-to-analog converter)
Data caching: ...................................................... 10-21
Data declaration: ........................................ 2-18, 2-19
Data definition: ..................................................... 2-19
DB15 connector: .................................................. 15-15
DD-ML (*see* Double-Double Magnetic Loop)
Dead-bug style construction: ................................ 15-9
*DEBUG*: ............................................................... 4-23
*DEBUG* flag: .......................................................... 6-8
Debugging: ................................................ 3-9, 3-18, 7-16
Decade: ................................................................ 12-13
Declaration: ......................................................... 3-17

Declaration of an object: ........................................ 3-4
*DefineEEPROMPage()*: ................................ 7-9, 15-28
*delay()*: ............................................. 1-14, 4-16, 6-8
Desoldering: ........................................................... 6-6
Destructive: ......................................................... 3-20
Device firmware update (DFU): .......................... 1-18
Digital recorder: ................................................... 12-14
Digital signal processing (DSP): .............. 11-1, 11-16
Digital-to-analog converter (DAC): ..... 8-6, 11-6, 12-3
*digitalWrite()*: ..................................................... 1-14
*DipTrace*: ............................................................ 16-3
Direct digital synthesis (DDS): .............................. 9-4
*displaykbdout()*: ................................................. 13-22
Displays: ............................................................. 4-1ff
    16x2 LCD: ..................................................... 4-1
    2.4-inch TFT LCD: ....................................... 4-1
    2.8-inch TFT LCD: ....................................... 4-1
    20x4 LCD: ..................................................... 4-1
    3.5-inch TFT LCD: ....................................... 4-1
    5-inch TFT LCD: .......................................... 4-1
    7-segment LED: ........................................... 4-1
    Choosing: ..................................................... 4-2ff
    Interfacing: .................................................... 4-4
    Nokia 5110: .................................................. 4-2
    OLED: ............................................................ 4-1
Distortion: ............................................................. 13-11
*DoCalibrate()*: ..................................................... 15-30
Dominant hand: ................................................... 10-6
Donald Duck: ....................................................... 11-15
Dot effect: ............................................................. 9-24
Dot operator: ................................................ 3-6, 3-19
Double quote marks (″): ..................................... 2-22
Double-Double Magnetic Loop (DD-ML): 14-1ff, 14-7, 15-1ff
    Auto-tune: ..................................................... 15-8
    Back panel: ................................................ 15-22
    Base panel: ................................................ 15-21
    Calibrate function: ...................................... 15-28
    Construction details: ..................................... 14-8
    Controller: ................................... 14-18, 15-1ff, 15-2
    Controller block diagram: ............................ 15-2
    Controller schematic: .................................. 15-4
    Controller software: ................................... 15-24
    Controller testing: .................................... 15-15ff
    Copper pipe: ................................................. 14-8
    Cost of remote tuning controller: .............. 15-25
    Drive belt: ................................................... 15-27
    Faraday loop (FL): .............................. 14-9, 15-18
    Faraday loop (FL) position: ....................... 15-18
    Freq menu: ................................................ 15-33
    Frequency encoder: ................................... 15-34
    Front panel: ................................................ 15-21
    FT8 signal reports: ..................................... 14-12
    Limit switches: ................................... 15-7, 15-17
    Loop inductance: ........................................ 15-3
    MagLoop.h header file: ............................. 15-25
    Main board testing: .................................... 15-16
    Main control unit: ........................................ 15-7
    Main loop (ML): .......................................... 15-3
    NEMA 17 stepper motor: ..................... 15-2, 15-6
    On-air testing: ............................................ 14-11
    Operate mode: ............................................ 15-9

Power supply voltages: .......................... 15-14, 15-16
PVC pipe: ............................................................. 14-9
Remote cables: .................................................. 15-15
Remote tuning: .................................................. 15-25
Software: ............................................................ 15-15
Splash screen: ................................................... 15-16
Stepper motor testing: ....................................... 15-17
SWR plot: ........................................................... 15-19
SWR readout: ...................................................... 15-2
Tripod: ................................................................. 14-9
Tubing bender: .................................................... 14-9
Tuning capacitor: ................................................. 15-3
Tuning unit (controller): ....................................... 15-2
WSPR testing: ................................................... 14-13
Double-sideband suppressed-carrier
   modulation: ................................ 13-1, 13-5, 13-14
*DrawGraphGrid()*: ................................................ 9-24
*drawKeypad()*: .................................................... 13-21
Drift-free: ............................................................. 13-2
DS3231 real time clock: .................................... 10-22
DSP Audio Mic-Processor (DMP): ..................... 12-1ff
   Automatic level control (ALC) — compressor: ... 12-10
   Back panel wiring: ............................................ 12-7
   Block diagram: .................................................. 12-3
   Comp Setting screen: .................................... 12-11
   Construction: .................................................... 12-6
   Encoder interrupt functions: ........................... 12-14
   EQ Setting screen: ........................................... 12-9
   EQ skirt slopes: .............................................. 12-10
   Equalizer (EQ): ..................................... 12-9, 12-13
   Extensions: ..................................................... 12-14
   Layout: .............................................................. 12-6
   Level encoder: .................................................. 12-9
   Menu functions: .............................................. 12-14
   Operation: ......................................................... 12-9
   Schematic diagram: ......................................... 12-4
   Setup mode: ..................................................... 12-9
   Signal flow: ..................................................... 12-13
   Software: ......................................................... 12-12
DSP filter: ............................................................. A-5
DSP Post-Processor (DPP): .............................. 11-1ff
   Automatic level control (ALC): ....................... 11-16
   Automatic notch filter: .................................... 11-13
   Block diagram: .................................................. 11-3
   Channel A: ....................................................... 11-6
   Channel B: ....................................................... 11-6
   Cutoff filters (HP and LP): .............................. 11-12
   CW filter response plot: ................................. 11-14
   CW Filters (CW) menu: .................................. 11-14
   EQ Setting screen: ......................................... 11-15
   Equalizer (EQ): ............................................... 11-15
   Fast Fourier Transform (FFT) display: ............ 11-16
   Features: .......................................................... 11-1
   Hold button: .................................................... 11-13
   Noise reduction (NR): ..................................... 11-16
   Notch filter: ..................................................... 11-12
   Operation: ....................................................... 11-11
   Schematic diagram: ......................................... 11-4
   Signal flow: ..................................................... 11-18
   Software: ......................................................... 11-16
   User interface (UI): ........................................... 11-2

*dtostrf()*: .............................................................. 6-8
Dummy load: ........................................................ 6-1ff
Duplicate definition error: .................................... 3-11
DuPont wire cables: ................................................ I-4
Dynamic microphone: ......................................... 12-5

# E

eBay: ..................................................................... A-6
EEPROM memory: .... 1-4, 7-8, 10-13, 12-9, 12-11, 15-28
   Emulation: .......................................... 10-15, 15-28
   Space: ............................................................ 15-28
   Storing data: ..................................................... 7-9
EFHW antenna: .................................................... A-2
Electrically Eraseable Programmable Random Access
   Memory (see EEPROM memory)
Electret microphone: ......................................... 12-5
Empire State building: ........................................ 2-28
EN (Reset) pin: ................................................... 1-22
Encapsulated: ...................................................... 3-6
Encapsulation: ......................................... 3-10, 3-14
Encoder: ............................................. 4-19, 7-5, 7-8
   Bourns: ............................................................ 10-6
   Granularity: ..................................................... 15-26
   Interrupt: .......................................................... 4-20
   Interrupt Service Routine (ISR): ...................... 4-22
   Keypad: .......................................................... 13-18
   KY-040: ................................................. 4-19, 15-7
   MAXBUMPCOUNT: ...................................... 15-26
   Optical: ............................................................ 10-6
   Pin assignments with various µCs: ................. 4-18
   Polling: ............................................................ 4-20
   Rotary: ............................................................. 10-6
   Threaded shaft: ................................................ 15-8
*ENCODER1PINA*: ............................................... 4-22
Encoding option: ................................................... 7-8
End-fed half-wave (EFHW) antenna: .... 14-12, 15-35, A-2
Endian problem: ................................................ 15-28
Epiphany: ............................................................ 3-10
Equalizer (EQ): ...................................... 11-15, 12-9, 12-13
   8-band: ........................................................... 11-19
   Skirt slopes: ................................................... 11-15
ESP32: .......................................... 1-1, 1-19, 13-4
   Board Manager: .............................................. 1-22
   Board variations: ............................................. 10-8
   EEPROM: ...................................................... 10-13
   EEPROM processing: ................................... 10-15
   EN (Reset) pin: ............................................... 1-22
   General Purpose Input Output (GPIO) Pins (table): 1-20
   I2C and SPI interfaces: ................................... 4-10
   SPI with TFT display: ...................................... 4-17
ESP32 Node MCU: ............................................. 13-5
ESP32-WROOM-32: ........................................... 10-3
   240 MHz clock speed: .................................... 10-8
   Flash memory: ................................................ 10-8
   SRAM memory: ............................................... 10-8
Executable program: .......................................... 1-11
Exit option: ........................................................... 7-7
Exogenous factors: ............................................. 15-8
Expansion board: ............................................... 1-19
*extern*: ..................................................... 2-19, 3-17

## F

Faceplate: ...................................................6-9, 12-9, 16-5
Fading: ......................................................................11-16
Faraday loop (FL): ...............................................14-4, 15-6
Farnsworth Method: .........................................................7-4
    Delay: ...........................................................................7-4
    Spacing: ........................................................................7-4
Fast Fourier Transform (FFT): ....11-6, 11-16, 12-3, 12-12
    Real-time display: .......................................................11-3
FAT16 format: .................................................................8-10
FAT32 format: .................................................................8-10
Field Day .......................................................................14-11
Filter
    3 dB cutoff frequency: ..............................................11-12
    Automatic notch: .......................................................11-13
    Band-pass (BP): ........................................................11-18
    Biquadratic cascaded filter: ........................11-17, 12-13
    Center frequency: .....................................................11-14
    Continuously variable filter: ........................................11-3
    Cutoff (HP and LP): ..................................................11-12
    Cutoff frequencies: .....................................................11-3
    Cutoff response: .......................................................11-12
    CW: ............................................................................11-14
    CW filter response plot: ............................................11-14
    CW width: ....................................................................11-7
    Flat response: ...........................................................11-12
    High-pass (HP): ............................................11-3, 11-18
    Low-pass (LP): .........................................................11-18
    Notch: ........................................................................11-12
    Roll-off: ..........................................................11-17, 12-13
    Spectrum plot: ..........................................................11-18
Filtering (power supply): ................................................11-8
Finished product: ............................................................16-1
Five Program Steps: .........................................................2-3
    Initialization Step: ........................................................2-3
    Input Step: ....................................................................2-4
    Output Step: .................................................................2-7
    Processing Step: .........................................................2-5
    Termination Step: .........................................................2-7
Flash memory: ......................................................1-3, 10-8
Flashcard (learning CW): ................................................7-7
Flat filter response: .......................................................11-12
Flat forehead mistake: ...........................................1-21, 2-6
FM modulation: .............................................................13-22
*fmDelay*: .......................................................................13-22
Font (Courier): .................................................................I-7
Front panel: ..........................................................11-8, 16-4
    Finishing: ..................................................................16-10
    Labels: ......................................................................16-10
    Layout: ..............................................................12-9, 16-5
    Masking tape: .............................................................8-19
    Template: .........................................................8-19, 16-5
FT8 signal reports: .......................................................14-12
Function: ........................................................................2-8ff
    Body: ............................................................................2-9
    Header: .......................................................................2-23
    Name: ......................................................I-7 2-9, 2-12, 2-23
    Scope: ........................................................................3-17
    Signature: ..................................................2-9, 2-15, 3-10
    Type specifier: ...................................................2-8, 3-11

## G

Gammon, Nick: ..............................................................4-14
Gardner, Glenn, AA8C: .................................................14-6
General class license: ....................................................7-1
General purpose input/output (GPIO): .......................1-19
    Pins for ESP32 (table): ............................................1-20
GIGO (Garbage In, Garbage Out): ................................2-5
Global scope: ................................................................3-15
Global variable: .............................................................9-28
Goal speed (CW): ............................................................7-4
Good RF practices: ...........................................16-2, 16-9
Graph.cpp: .....................................................................8-26
Graphical design tool: .................................................11-17

## H

Harmonic: ........................................................................13-2
    Distortion: ...................................................................13-2
Header file (.h): ...........................................2-22, 2-25, 9.27
Heat shrink tubing: ..............................................15-14, 16-10
Heat sink: .............................................6-2, 11-7, 15-12
Herding cats: ...................................................................16-9
Hexadecimal: ..................................................................1-15
HF Signals: ......................................................................A-1
HIEND40M: ....................................................................15-31
High-impedance load: .....................................................13-8
High-pass (HP) filter: ............................................11-3, 11-18
High-Q antenna: ..............................................................15-1
HiLetgo ESP-WROOM-32: ............................................1-19
Homeowner association (HOA) restrictions: ..............14-1
Hot glue: ...............................................................8-17, 15-13
Human hearing: ............................................................12-13
Hysteresis: .....................................................................8-30
    Curves: .......................................................................8-30

## I

I2C interface: ...................................................................4-4
    Arduino Nano: .............................................................4-8
    Blue Pill (BP): ............................................................4-10
    Data lines: .................................................................4-17
    ESP32: .......................................................................4-10
    Pros and cons: ............................................................4-6
    Simple LCD program: ..............................................4-12
    STM32F103: ..............................................................4-10
    Teensy 3.6: .................................................................4-8
I2C Scanner: ..................................................................4-14
IDC connector: .........................................8-17, 11-7, 12-6, 15-12
IDE (*see* Arduino Integrated Development Environment)
ILI9341 display driver: ..........................................7-11, 8-5, 10-1
    Adafruit library: .........................................................7-11
ILI9431 display driver: ...................................................4-4, 10-5
ILI9488 controller: ..................4-4, 8-5, 11-6, 13-5, 13-20
Impedance: ......................................................................9-3
Include file: .....................................................................2-24
Increment: ....................................................................15-34
Incremental compile: .........................................2-26, 7-15
Indexing: ......................................................................10-18
Inductor, T157-2 core: .....................................................9-9
Inheritance: ...........................................................3-10, 3-13
Initialization: ..................................................................4-14

Initialization Step: 2-3
Input Step: 2-4
Instantiation: 3-4, 3-5
Instruction pointer: 3-21
*int*: 3-11
*int* data type: 2-10
Inter-Integrated Circuit bus (*see* I2C interface)
Interconnections: 11-7
Interior layout: 16-7
Interrupt: 2-28, 4-20, 10-5, 10-19, 15-7
    Menu example: 4-20
    vs polling: 2-28, 10-19
Interrupt Service Routine (ISR): 4-21, 10-7, 10-19
    Encoder: 4-22
    Guidelines: 4-21
*interruptServiceRoutine()*: 4-22
IRAM memory: 10-7
IRAM_ATTR: 10-7
*is a*: 3-14
*is a* relationship: 3-14
Isolation (power supply): 13-10
ISR (*see* Interrupt Service Routine): 4-21

## J

J176 FET: 8-6
JackAl board: 8-32
Jalowiczor, Jakub: 14-6
JMP: 1-16
Junk box: 5-5, I-5

## K

Keypad: 13-18, 13-20, 13-21
Keywords: 2-2
Kissed frogs: 15-22
Knobs: 16-5, 16-11
Koch Method: 7-4
Koch, Ludwig: 7-4
KY-040 encoder: 4-19, 15-7

## L

Labels: 16-2
LCD display: 4-1
led pin: 1-12
Level shifter: 1-6
Libraries: 4-7
Library: 2-20ff
    Table of Contents (ToC): 2-10
Library functions: 2-11
Limit switch: 15-7
Linear regulator: 13-10
Linker: 1-11, 2-19, 3-17, 3-21
*LiquidCrystal_I2C library*: 4-14
Little Endian: 15-29
LM338 regulator: 8-5
LM386 audio amp: 7-10
LM7805 regulator: 10-3
Logic level shifter: 1-6
Long data type: 1-2
*loop()*: 1-14, 2-8, 4-15, 4-16

Low-pass (LP) filter: 11-18
LOWEND40M: 15-31
Luggable antenna: 14-1
lvalue: 2-17, 3-17

## M

Mag Loop Design Calculator (IW5EDI): 14-5
MagLoop.h header file: 15-25
Magnetic loop (ML): 14-1ff
    Coupler: 14-5
    Double-Double Magnetic Loop (DD-ML): 14-7
    Drive belt: 14-5
    Efficiency: 14-5
    Faraday loop (FL): 14-4
    High-Q: 15-1
    Loop: 14-2
    Manual tuning: 15-1
    NEMA 17 stepper motor: 14-4
    Pipe diameter: 14-8
    Q factor: 14-2
    Reduction drive: 14-4
    Remote control unit: 14-4
    RF coupling loop: 14-4
    Rloss: 14-7
    Stealth tuning: 15-24
    Tuning capacitor: 14-2
Magnetic loop (ML) also *see* Double-Double Magnetic Loop (DD-ML)
Maidenhead locator: 10-2, 10-13
Main controller unit: 14-19
Manhattan style construction: 15-9, 16-8
Mannini, Simone, IW5EDI: 14-5
Manual tuning: 15-1
Masking tape: 8-19, 16-5
MAXBUMPCOUNT: 15-26
MAXMESSAGELENGTH: 10-12
MAXMESSAGES: 10-12
MC1496 modulator: 13-5, 13-9
Mega-munch: 1-2, 7-16
Member method: 3-12
Memory contents: 2-18
Memory location: 2-18
Menu encoder: 8-8
Menu example
    Interrupt: 4-20
    TFT display with SPI interface: 4-18
Menu.cpp: 8-28
Menuing library: 8-27
Menuing system: 7-5
    Exit option: 7-7
    Submenu: 7-7
*message[]*: 4-16
Method call: 3-20
Method name: I-7
Method signature: 3-12
Methods: 3-3
    Purpose of: 3-3
Micro-stepper controller: 15-27
Microcontroller (µC): 1-1ff
    Arduino clones: 1-8

Arduino Mega 2560 Pro Mini: .................................... 1-1
Arduino Nano: ................................................................ 1-1
ESP32: ................................................................. 1-1, 1-19
HiLetgo ESP-WROOM-32: ................................... 1-19
Nano Pro Mini: ........................................................... 1-7
STM32F103: ..................................................... 1-1, 1-17
Table of resources: .................................................. 1-3
Teensy 4.0 (T4): ............................................. 1-1, 1-16
Microcontroller Resources (table): .................................. 1-3
Microcontroller Voltage and Current Information
  (table): ........................................................................... 1-5
Microphone
  Bias: ........................................................................... 12-5
  Dynamic: .................................................................. 12-5
  Electret: .................................................................... 12-5
MicroSD card: ...................................................................... 8-10
Midnight Design Solutions: ................................................ A-3
Mini Dummy Load (MDL): ..................................... 4-1, 6-1ff
  Accuracy: .................................................................... 6-7
  Bezel: .......................................................................... 6-9
  Calibration: ................................................................ 6-6
  Construction: ............................................................ 6-3
  Faceplate: .................................................................. 6-9
  Power Measurements table: .................................. 6-7
  Safe Power Rating table: ....................................... 6-2
  Schematic diagram: ................................................ 6-5
  Software: ................................................................... 6-8
  Source code: .......................................................... 6-10
Mixer: ....................................................................................... 12-13
MJ2955 transistor: ..................................................... 8-5, 8-6
Mnemonics: ............................................................................ 1-16
Modulation
  AM: ............................................................................ 13-14
  Sine wave: ............................................... 13-14, 13-22
Modulator: ..................................................... 13-5, 13-9, 13-22
Modulo operator: .................................................................. 9-25
Morse code: ............................................................................ 7-1
Morse Code Tutor (MCT): ................................................. 7-1ff
  Bill of Materials (BOM): ....................................... 7-13
  Feature set: ............................................................... 7-5
  Menuing system: ..................................................... 7-5
  PC board: ................................................................... 7-1
  Schematic diagram: .............................................. 7-14
  Software: ................................................................. 7-15
  Source code: .......................................................... 7-20
  TFT Display to BP table: ...................................... 7-11
Multiband end-fed antenna .............................................. 14-12
Multiline comment: ................................................ 2-10, 2-29
Multimeter: ............................................................................. 13-1
Multiple code files: .............................................................. 2-25
Multiple project files: .......................................................... 2-25
MyAntennas EFHW antenna: ......................... 14-12, A-2

# N

Name collision: ...................................................................... 2-9
Naming conventions ............................................................ I-6
  Methods: .................................................................. 4-17
Nano: .......................................................................................... 6-3
  Safe handling: .......................................................... 6-6
Nano Pro Mini: ........................................................................ 1-7

NEMA 17 stepper motor: ........................ 14-4, 15-2, 15-6
  Controller: ................................................... 15-7, 15-20
  Hums: ....................................................................... 15-17
  Micro-stepper controller: .................................... 15-27
  Testing: ................................................................... 15-17
  Zero position: ........................................................ 15-17
Nibble: ....................................................................................... 1-1
Noise: ....................................................................................... 16-2
Noise reduction (NR): ............................... 11-16, 11-19
Nokia 5110: ............................................................................ 4-2
Notch filter: ......................................................................... 11-12
  Broad response: ................................................... 11-12
  Q setting: ............................................................... 11-12
  Sharp response: ................................................... 11-12
Nouns: ...................................................................................... 3-4

# O

Object Oriented Programming (OOP): ............. 3-1ff, I-3
Object Oriented Programming Trilogy
  Encapsulation: ...................................................... 3-10
  Inheritance: ............................................................ 3-10
  Polymorphism: ...................................................... 3-10
Octave: .................................................................................. 12-13
OLED display: .................................................. 4-1, 6-2, 6-3
Open Source: ....................................................................... 2-21
Opening brace ({): ............................................................. 2-24
Optical encoder: ................................................................ 10-6
Optocoupler: ....................................................................... 10-7
Oscilloscope: ....................................................................... 13-1
Output Step: ......................................................................... 2-7
Overheat components: ................................................... 15-13

# P

Paddle key: ........................................................................ 10-22
Parameter list: ...................................................................... 2-9
Parameterized macro: ...................................................... 3-11
Parent class: ....................................................................... 3-14
PARIS: ..................................................................................... 7-4
Patch: .................................................................................... 1-15
PC board (PCB): .................................................................. I-5
  Design software: ................................................... 16-3
Peak-to-peak voltage: ....................................................... 6-6
Perf board construction: ................................................. 16-3
Phaser Digital Mode Transceiver: ................................ A-3
Photo paper: ........................................................ 8-19, 16-10
*pinMode()*: ......................................................................... 1-12
Plus screwdriver (Phillips screwdriver): ..................... 2-1
Point-to-point wiring: ...................................................... 15-12
Polling: .............................................. 2-28, 4-20, 10-5, 10-21
Polymorphism: ...................................................... 3-10, 3-12
Power supply project: ..................................................... 5-1ff
Preprocessor directive: ................... 2-21, 4-7, 4-14, 9-28
Preset voltages: .................................................................. 8-9
*presetFrequencies[][]* array: .................................... 15-33
*presetFrequencies[MAXBANDS]*
  *[PRESETSPERBAND]*: ................................... 15-31
Presets: ................................................................................... 8-2
*print()*: ............................................................................... 4-16
Private: ................................................................................ 8-28

*process()* method: 10-20
Processing Step: 2-5
Processor bits: 1-1
Products
    Antennas: A-1
    Transceivers: A-1
PROGMEM directive: 1-4
Program
    Comments: 2-28
    Flow: 3-20
    Output: 4-1
    Statement: 2-16
    Stats window: 1-12
Program steps
    Initialization: 4-14
    Output: 4-15
    Process: 4-15
Programmable Bench Power Supply (PBPS): 8-1ff
    Applications: 8-4
    Cal menu: 8-13
    Calibration: 8-24
    Construction: 8-13ff
    Feature set: 8-2
    Config menu: 8-12
    Menu encoder: 8-8
    PC board: 8-34
    Pre-defined functions: 8-8
    Preset voltages: 8-9
    SD menu: 8-10
    Software: 8-26
    Testing: 8-24
    Time menu: 8-12
    User 2, User 3: 8-12
Project errors (hiccups): I-6
Projects Power Supply (PPS): 5-1ff
    Construction: 5-6
    Cost comparison table: 5-5
    Multi-pin connector: 5-3
    Performance: 5-5
    Schematic diagram: 5-3
    Specifications: 5-2
Prototype board: 15-10
Pseudo-random number generator: 7-6
Public: 8-28
Pulses per revolution (PPR): 10-6
Push button switch: 11-7
    SPST: 10-13
Push-pull output stage: 13-8
PVC pipe: 14-9
PWM signal: 1-19

# Q

Q constant: 11-12
Q factor: 14-2
Q setting (notch filter): 11-12
QCX transceiver kit: A-4
QRP: 6-6, 7-1
QRPGuys: I-4, 15-18, A-4
QRP Labs: A-4
QWERTY keyboard: 10-1

# R

RDC (Really Dumb Code): 2-6, 2-10, 2-11, 2-27
*ReadWordsPerMinute()*: 10-15
Real time clock (RTC): 10-22
Rectified RF: 6-5
Registers: 1-1
*ResetStepperToZero()*: 15-16
Resistive touch TFT screen: 10-5
Reuse code: 8-29
RF amplifier stage: 13-9
RF power resistor: 6-1
    Safe handling: 6-5
RG-174: 15-15
Rhythm of CW: 7-3
Ribbon cable: 12-6, 15-12, 16-10
Richards, David, AA7EE: 16-8
Right-Left Rule: 2-23
Ritchie, Dennis: 2-2
Rloss: 14-7
Roll-off: 11-17
Rotary encoder: 9-4, 10-6
*rotate()*: 10-19
*RotateISR()*: 4-22
*run()*: 15-33
rvalue: 2-17, 3-17

# S

Scaffolding: 4-23
Schematic capture software: 16-3
Schematic diagram
    Checking connections: 15-14
Schmidt Trigger: 8-30
Scope: 3-14
Scope level: 8-30
Scope qualifier: 3-5
Scope resolution operator: 8-29
Scroll management code: 10-10
SD card: 8-10
    Reader: 8-8
SDC (Sorta Dumb Code): 2-6, 2-12
SDO pin: 10-5
Secondhand store: 8-14
Semantic error: 2-16
Semicolon: 2-16
*sendfrequency(freqFM)*: 13-22
Sentinel: 7-17, 10-17
Sentinel bit: 10-18
Serial Monitor: 1-12
Serial object: 1-12, 3-7
Serial Peripheral Interface bus (*see* SPI interface)
*Serial.begin(9600)*: 1-15
*Serial.print()*: 1-12
*setCursor()*: 4-16
*setup()*: 1-12, 2-4, 2-8, 4-14
Shielded cable: 11-8, 12-7
Shielding: 16-7
Si5353 DDS: 15-8
Sidewalk: 3-20
Sign bit: 2-10

Signal Generator (SG): .................................... 13-1ff
    AM modulation: ...................................... 13-14
    Analog style meter: ....................... 13-18, 13-23
    Attenuator block: ..................................... 13-4
    Block diagram: ........................................ 13-4
    Channel A outputs: ................................ 13-12
    Channel B: ............................................. 13-8
    Channel B outputs: ................................ 13-13
    Coarse adjustment: ............................... 13-18
    Double-sideband suppressed-carrier
        modulation: ..................................... 13-14
    Encoder: .............................................. 13-18
    Feature Set table: .................................. 13-3
    FM modulation: ..................................... 13-22
    MHz button: .......................................... 13-18
    Modulation outputs: ............................... 13-14
    Multiplier button: ................................... 13-18
    Schematic diagram: ............................... 13-6
    Sine wave modulation: ................ 13-14, 13-22
    Software: ............................................. 13-20
    Specifications: ....................................... 13-4
    Square wave: ....................................... 13-14
    Sweep function: ................................... 13-18
    Sweep mode: ....................................... 13-23
    Sweep parameters: ............................... 13-18
    Touch screen calibration: ..................... 13-23
    Triangle wave: ..................................... 13-14
    User interface (UI): .............................. 13-18
Signal generator (SG): ..................................... 13-1
Signal-to-noise (SNR) reports: ..................... 14-13
*sin()*: ............................................................ 13-23
Sine wave: ............................... 9-4, 13-2, 13-14
    Modulation: ......................................... 13-22
    Output: ................................................. 8-10
Single-line comment: ..................................... 2-29
sizeof operator: ............................................. 3-11
Slab: ............................................................. 3-20
SMD (*see* Surface mount device)
Sockets: ................................. 11-7, 12-6, 15-10, 16-2
    20-pin: ................................................. 15-11
    Point of failure: ................................... 15-10
Software: ....................................................... 2-1ff
    *DipTrace*: ............................................. 16-3
    Double-Double Magnetic Loop (DD-ML): ........... 15-15
    Double-Double Magnetic Loop (DD-ML)
        controller: ..................................... 15-24
    DSP Audio Mic-Processor (DMP): ....... 12-12
    Mini Dummy Load (MDL): ...................... 6-8
    Morse Code Tutor (MCT): ..................... 7-20
    PCB design: ......................................... 16-3
    Schematic capture: ............................... 16-3
    Signal Generator (SG): ....................... 13-20
Software engineer: ......................................... 2-1
SoftwareControlledHam Radio website: .......... I-2
SOTA (*see* Summits On The Air): .................. 10-2
SOTABEAMS: .................................................. A-5
Source code window: .................................... 1-11
Spacers .......................................................... 15-23
*spaces[]*: ....................................................... 4-15
Spectral content: ............................................ 11-3
Spectrum plot: ............................................... 11-18

SPI display: ..................................................... 10-1
SPI interface: .............................. 4-5, 7-11, 15-7
    Arduino Nano: ........................................ 4-8
    Blue Pill (BP): ...................................... 4-10
    ESP32: ................................................. 4-10
    Menu example with TFT display: ........ 4-18
    Pros and cons: ....................................... 4-6
    STM32F103: ......................................... 4-10
    Teensy 3.6: ............................................ 4-8
    TFT display: ......................................... 4-17
Splash screen: ................................................ 8-8
    Double-Double Magnetic Loop (DD-ML): ........... 15-16
*Splash()*: ...................................................... 15-27
Square wave: ............................................... 13-14
SRAM memory: ..................................... 1-4, 10-8
Statement block scope: ................................. 3-18
Static Random Access Memory (see SRAM memory)
Step attenuator: ............................................ 13-8
Stepper motor (*see* NEMA 17 stepper motor)
*stepper.moveTo(300)*: ................................... 15-33
Stick-on feet: ................................................ 16-11
STM32F103: ............ 1-1, 1-17, 7-2, 9-4, 15-7, 15-10, I-1
    Device firmware update (DFU): ............. 1-18
    EEPROM: .............................................. 7-8
    EEPROM memory: ............................... 15-28
    I2C and SPI interfaces: ......................... 4-10
    Pins: .................................................... 15-12
    SPI with TFT display: .......................... 4-17
Stroustrup, Bjarne: .......................................... 3-9
Subclass: ....................................................... 3-14
Submenu: ........................................................ 7-7
Summits On The Air (SOTA): ................. 10-2, A-5
Surface mount device (SMD): ......... 13-9, 15-12, I-5
Swiss Army knife: .......................................... 9-26
*Switch/case* statement block: ....................... 2-27
Switches: ...................................................... 10-6
    Normally open (NO): ........................... 15-7
    Push button: ........................................ 11-7
    Push button switch (SPST): ................ 10-13
SWR: ..................................................... 9-1, 15-31
SWR impedance bridge: .................................. 9-3
Symbol table: ................................................ 2-16
Symbolic constant: ......................... 1-21, 2-25, 6-8
Syntax rule: ................................................... 2-16

## T

T_IRS pin: ...................................................... 10-5
Tables
    A C Symbol Table: ............................... 2-16
    A C Symbol Table with Memory Address Filled In: 2-17
    BOM for MCT: ...................................... 7-13
    Connections Between TFT Display and ESP32: .. 10-5
    General Purpose Input Output (GPIO) Pins
        for ESP32: ..................................... 1-20
    Microcontroller Voltage and Current Information: ... 1-5
    Mini Dummy Load Power Measurements: ............. 6-7
    Pin Assignments (TFT Display and Encoder): ...... 4-18
    Power Supply Cost Comparison: ........... 5-5
    Resources by Microcontroller: ............... 1-3
    Safe Power Rating for DL: .................... 6-2

Signal Generator Feature Set: ............................. 13-3
Teensy 4.0 Pin Assignments: ................. 11-6, 12-6
TFT Display to BP: ............................................ 7-11
Tapped toroid inductor: ........................................... 9-2
Teensy
    SPI with TFT display: ......................................... 4-17
Teensy 3.6: .............................................................. I-1
    I2C and SPI interfaces: ........................................ 4-9
Teensy 4.0 (T4): ....................... 1-1, 1-16, 11-3, 12-1, I-1
    Audio Library: ..................................................... 11-1
    Pin assignments: ..................................... 11-6, 12-6
Teensy Audio Adapter (TAA): ......................... 11-3, 12-3
Template: ............................................................. 16-5
Termination Step: ................................................... 2-7
TFT display: ............................................... 4-1, 15-7
    ESP32 SPI interface: ......................................... 4-17
    Menu example with SPI interface: ..................... 4-18
    Pin assignments with various µCs: ..................... 4-18
    SPI interface: ..................................................... 4-17
    STM32F103 SPI interface: ................................ 4-17
    Teensy SPI interface: ........................................ 4-17
TFT_eSPI library: ................................................ 10-8
Third-order intermodulation distortion (IMD) testing: . 13-1
Thompson, Ken: ..................................................... 2-2
Toroid: ................................................................... 9-9
Touch screen: ............................................. 4-2, 11-2
    320x240 pixel: ................................................... 12-3
    Calibration: ........................................................ 13-23
*touch_calibrate()*: .............................................. 13-23
Trial fit: ............................................................... 15-12
Triangle wave: .................................................... 13-14
Trim pot: .............................................................. 7-10
Tripod: ................................................................. 14-9
Tubing bender (Harbor Freight): ............................ 14-9
Tuning capacitor: .................................................... 9-2
Two-tone signal: .................................................... 13-1
Type-checking: ..................................................... 7-16

## U

Union: ................................................................ 15-29
Upload icon: ......................................................... 1-12
USB connector: ..................................................... 16-8
USB extension cable: .......................................... 15-21
User interface (UI): .............................................. 11-2
    DSP Audio Mic-Processor (DMP): ..................... 12-1
User_Setup.h: ...................................................... 10-8

## V

Vacuum variable capacitor (VVC): ................. 14-3, 15-6
Validate: ................................................................ 2-5
Variable: .............................................................. 2-15
    Declaration: ....................................................... 2-15
    Definition: .......................................................... 2-15
    Global: ............................................................... 9-28
    Name: ................................................................ 2-23
Variable gain amplifier (VGA): .............................. 8-32
Verbs: ................................................................... 3-4
Volatile keyword: ................................................ 10-21
Voltage divider: ............................................ 6-5, 8-33
Voltage range: ....................................................... 1-5
Voltage regulator: .................................................. 1-5
Voltage-time pairs: .............................................. 8-10
Volume control: .................................................... 12-2

## W

WA3RNC 40 Meter QRP transceiver kit: ................. A-5
Waveform: ........................................................... 13-2
    Square: ............................................................ 13-14
    Triangle: ........................................................... 13-14
WC (Wow Code): ................................................... 2-7
White smoke: ..................................................... 15-17
Wi-Fi: ............................................................ I-1, 1-7
Wire: .................................................................... 16-9
    #24 AWG or larger: ................................... 11-7, 12-6
    #24 AWG solid: ............................................... 15-13
    #30 AWG solid: ...................... 11-7, 12-6, 15-12, 15-13
    Cat5 cable: ...................................................... 15-15
    Colors: ............................................................... 16-9
    DuPont cables: ................................................... I-4
    RG-174: ........................................................... 15-15
    Ribbon cable: ........................... 12-6, 15-12, 16-10
    Routing: ............................................................ 16-9
    Shielded cable: ........................................ 11-8, 12-7
Word: .................................................................... 7-4
Words per minute (WPM): ............................ 7-4, 10-6
*writeByte()*: ........................................................ 10-15
WSPR signal reports: .......................................... 14-13
    Signal-to-noise ratio (SNR): .............................. 14-13

## Y

Yellow marker: .................................................... 15-14

## Z

Zener diode: ........................................................ 8-31

# Notes

# Notes

# Notes

# Notes

# Notes

# Notes

# Notes

# Notes

# Notes

# Notes